Numerical Methods in Chemical Engineering Using Python® and Simulink®

Numerical methods are vital to the practice of chemical engineering, allowing for the solution of real-world problems. Written in a concise and practical format, this textbook introduces readers to the numerical methods required in the discipline of chemical engineering and enables them to validate their solutions using both Python and Simulink.

- Introduces numerical methods, followed by the solution of linear and non-linear algebraic equations.
- Deals with the numerical integration of a definite function and solves initial and boundary value ordinary differential equations with different orders.
- Weaves in examples of various numerical methods and validates solutions to each with Python and Simulink graphical programming.
- Features appendices on how to use Python and Simulink.

Aimed at advanced undergraduate and graduate chemical engineering students, as well as practicing chemical engineers, this textbook offers a guide to the use of two of the most widely used programs in the discipline.

The textbook features numerous video lectures of applications and a solutions manual for qualifying instructors.

Numerical Methods in Chemical Engineering Using Python® and Simulink®

Nayef Ghasem

CRC Press is an imprint of the
Taylor & Francis Group, an **informa** business

MATLAB® is a trademark of The MathWorks, Inc. and is used with permission. The MathWorks does not warrant the accuracy of the text or exercises in this book. This book's use or discussion of MATLAB® software or related products does not constitute endorsement or sponsorship by The MathWorks of a particular pedagogical approach or particular use of the MATLAB® software.

First edition published 2024
by CRC Press
6000 Broken Sound Parkway NW, Suite 300, Boca Raton, FL 33487-2742

and by CRC Press
4 Park Square, Milton Park, Abingdon, Oxon, OX14 4RN

CRC Press is an imprint of Taylor & Francis Group, LLC

ISBN: 978-1-032-41946-6 (hbk)
ISBN: 978-1-032-41951-0 (pbk)
ISBN: 978-1-003-36054-4 (ebk)
ISBN: 978-1-032-52408-5 (eBook+)

DOI: 10.1201/9781003360544

Typeset in Times
by codeMantra

Access the Instructor and Student Resources/Support Material:
http://www.routledge.com/9781032419466

Contents

Acknowledgment

The author would like to thank Allison Shatkin, acquiring editor for this book, and Hannah Warfel, editorial assistant, for their help and cooperation. The author would also like to thank the Matlab engineers for their help and kind cooperation. The comments and suggestions of the reviewers were highly appreciated.

Author

Nayef Ghasem is a professor of chemical engineering at the United Arab Emirates University. He teaches mainly process modeling and simulation, natural gas processing, and reactor design in chemical engineering, as well as other graduate and undergraduate courses in chemical engineering. He has published primarily in modeling and simulation, bifurcation theory, polymer reaction engineering, advanced control, and CO_2 absorption in gas–liquid membrane contactors. He has also authored *Principles of Chemical Engineering Processes* (CRC Press, 2012), *Modeling and Simulation of Process Systems* (CRC Press, 2019), and *Computer Methods in Chemical Engineering, 2nd Edition* (CRC Press, 2021). He is a senior member of American Institute of Chemical Engineers (AIChE).

1 Introduction

Numerical methods attempt to find approximate solutions to problems rather than exact solutions. Numerical methods are procedures used to estimate mathematical procedures. We need an approximation because either we cannot solve the system analytically or the analytical method is difficult. This chapter explains the sources of mathematical equations that are derived from chemical engineering subjects. This chapter offers an advanced introduction to numerical method.

LEARNING OBJECTIVES

1. Realize what a numerical method is.
2. Know when to use numerical methods.
3. Find roots of equations.
4. Identify systems of linear and nonlinear algebraic equations.
5. Detect ordinary differential equations.
6. Identify boundary value problems.

1.1 BACKGROUND

The numerical method is a mathematical implement intended to solve numerical problems. The numerical method is an approximate computational method for solving a mathematical problem that is often complex or has no analytic solution. Numerical methods use computers to solve problems by incremental, repetitive, and iteration methods, which may be dull or unsolvable by manual calculations. This chapter intends to give scientists and engineers an overview of numerical methods of interest.

We can distinguish several distinct stages in the numerical method problem-solving process. While formulating a mathematical model of a physical system, engineers must consider that they expect to solve a problem on a computer. It will therefore provide specific objectives, appropriate input data, adequate checks, and the type and quantity of output. Once the problem has been formulated, numerical methods and the initial error analysis must be developed to solve the problem. A numerical method used to solve a problem is called an algorithm. An algorithm is a complete set of procedures to solve a mathematical problem. The programmer should convert the proposed algorithm into a step-by-step instruction for the computer. The numerical analysts consider all sources of error that might affect the results and how much precision is required to determine the appropriate step size or the number of repetitions required.

DOI: 10.1201/9781003360544-1

1.1.1 What Is the Numerical Method

Defining analytical and numerical methods helps understand the difference between them. Analytical means we apply a set of logical steps to solve our problem that is proven to find the exact answer to the problem. For example, if we had $x + 1 = 0$, we could easily find what x is equal to one by adding minus one to both sides; this works for many derivatives, integrals, and everything else you learned at school so far. However, what if finding that exact answer took, say, 20 hours or even after all that work, you find out it is not even possible to find an answer when you know there should be one? Well, that is where numerical methods come in.

Numerical methods are mathematical ways to solve specific problems. They allow us to approximately solve math problems, specifically those we cannot solve analytically. In contrast, analytical solutions apply a set of logical steps to solve a problem that are proven to find the exact answer to the problem. It studies numerical methods that try to find approximate solutions to problems rather than exact solutions. Numerical analysis is a branch of mathematics that solves ongoing problems using numerical approximation. The numerical method used to solve a problem is called an algorithm. An algorithm is a procedure that solves a mathematical problem or selects or constructs an appropriate set of rules that correctly fall within the scope of numerical analysis. After deciding on an algorithm or group of algorithms to solve a numerical problem, the analyst must consider all sources of error that may affect the results. Given the amount of precision required, calculate the magnitude of the estimated error and select the appropriate step size or number of repetitions required. Adequate checks for accuracy and corrective measures operate in cases of non-convergence. The process of digital computing involves building a mathematical model and an appropriate numerical system. Finally, implement and check the solution.

1.1.2 Why Are Numerical Methods Necessary

Numerical methods essentially plug in inputs to a problem, check how close the output is to a solution, change our input, and repeat this process until we get as close as we desire to our approximated solution. How long our computer takes to plug in as many numbers as it needs versus accuracy? How close are we to the solution to the problem? So to make this clear, let us go back to our $x + 1 = 0$ problem and begin with, say, a guess of 1/2; well, we plug in 1/2 for x and find that 1.5 does not equal 0, so we would try a new guess and so on until our left side gets close to our right side. This may sound like more work in this simple example, but using computers and working with more complicated problems using numerical methods are a rescuer.

The use of numerical methods is more efficient and faster than the use of analytical methods. Many engineering problems are too complex to be solved manually, so numerical methods are needed. Examples include distillation column design (linear algebra), differential equations, and statistics. Many tools are available to reduce the complexity of solving complex differential equations. Very few things have an exact analytic solution, and more have approximate solutions. Many things have no solution but are only numerical. Any modern problem in engineering has to be solved to a large extent numerically or by doing a lot of costly experiments [1].

1.1.3 When to Use Numerical Methods

Numerical methods should be used if the problem is multidimensional (e.g., three-dimensional flow in complex mixing or extrusion elements, temperature fields, rheology) and if the geometry of the flow area is very complex. They need a high degree of mathematical drafting and programming. The main advantage of the numerical method is that a numerical solution to problems can be obtained since there is no analytic solution. We generally prefer the analytical method because it is faster and because the solution is accurate. However, sometimes we recourse to a numerical method due to time or hardware limitations.

1.2 TYPES OF NUMERICAL METHODS

Numerical analysts and mathematicians use a variety of tools to develop numerical methods for solving mathematical problems. The most crucial idea mentioned earlier, which includes all mathematical problems, is to change a particular problem into one that can be easily solved. Other ideas vary about the type of mathematical problem that was solved. There are two numerical methods.

1.2.1 Direct Method

This method solves a problem in a limited number of steps and gives an accurate answer if performed with precise arithmetic operations, such as the Gaussian elimination method. In practice, maximum accuracy is used, and the result is close to the solution. Without rounding errors, the direct method provides an accurate solution.

In conclusion, direct numerical methods are ways of solving problems in a finite number of steps. The method solves any non-singular matrix. When solving a system of linear equations using these direct methods, we take a set number of steps, such as converting a system of linear equations to an augmented matrix, then trying to convert it to an upper-triangular matrix, and then solving. After a set number of steps, we arrive at a final solution.

1.2.2 Indirect Numerical Methods (Iterative Method)

Iterative numerical methods are different from direct numerical methods in that they do not necessarily stop after a certain number of steps; they will continue to iterate, which means to do again or repeat with a better guess, until they get enough error. Some examples of iterative numerical methods are the Bisection method, Newton's method, and Jacoby iteration. Additionally, when using iterative solvers, begin with an initial guess.

In actual computation, direct methods are much more computationally heavy, as everything in the matrix is solved in one computational step, which requires a whole matrix that is actively used continually. The problem is broken down into a set of small steps in an iterative manner, which is why iterative methods are usually preferred for larger arrays.

In summary, for the indirect numerical methods, one cannot stop after a certain number of steps and say that it is good enough because what we will have is an unfinished problem and no answer, and will need to complete the set of steps to achieve any answer that makes any sense. By contrast, in the direct numerical methods, before we begin, we can find out how many steps we need to get to an answer, and one must complete that number of steps to achieve that answer. Consider solving the following matrix,

$$\begin{bmatrix} 1 & 2 & 3 \\ 4 & 5 & 6 \\ 7 & 8 & 9 \end{bmatrix}$$

We need to turn these three elements (4,7,8) to zero and then use backward substitution, and then we will have to solve a system of linear equations; otherwise, it is not invertible and has no unique solution. However, with iterative numerical methods, setting the number of steps is not required; stop whenever needed. The longer the computation period, the more accurate the solution becomes, which offsets computation time versus accuracy. In conclusion, direct numerical methods use a set number of steps to achieve a final solution; however, iterative numerical methods use an initial guess to approximate a final solution.

1.3 LINEAR AND NONLINEAR SYSTEMS OF ALGEBRAIC EQUATIONS

A linear equation can take the form

$$Ax + By + C = 0 \tag{1.1}$$

Any equation not written in this form is considered a nonlinear one. The substitution method we used for linear systems is the same as the one we will use for nonlinear systems. We solve one equation for one variable and then substitute the result into the second equation to find another variable, and so on. However, there is a variance in the possible outcomes. Table 1.1 shows examples of linear and nonlinear equations.

Understanding the difference between linear and nonlinear equations is extremely important. Table 1.2 shows linear and nonlinear equations.

1.3.1 LINEAR ALGEBRAIC EQUATION

A system of linear equations is just a collection of independent equations that cannot derive from one another, which means we cannot say we have two equations if both of our equations are $y + 3 = 0$ and also $y = -3$; they are the same thing, just three is subtracted from both sides. Linear equations must be related and share the same set of variables. An equation is linear if each term is simply a constant or a constant multiplied by a variable, similar to the equation of a line $y = ax + b$, which is a constant time a variable plus b, which is simply a constant. If we know that our equation

TABLE 1.1
Examples of Linear and Nonlinear Equations

Linear Equations	Nonlinear Equations
$y=8x-9$	$y=x^2-7$
$y+3x=1$	$\sqrt{y}+x=6$
$y+x=7$	$y^2-x=9$

TABLE 1.2
Differences between Linear and Nonlinear Equations

Linear Equations	Nonlinear Equations
The equation has a maximum order of one degree.	The equation has the maximum degrees of two or more.
A straight line is formed on the graph.	The equation forms a curve on the graph.
The general form of linear equation is $y=mx+c$ where x any y are the variables, m is the slope of the line, and c is a constant value.	The general form of nonlinear equations is $ax^2+by^2=c$ where x and y are the variables and a, b, and c are the constant values.

is linear, you may ask what it would mean to solve a system of linear equations. Accordingly, solving a system of linear equations essentially means finding and thus assigning a value to each of your unknown variables because there is sometimes more than one solution to a system of linear equations. The total set of solutions that can solve our system of linear equations is called the solution set. In general, a linear algebraic equation (LAE) is one in which the highest power of the variables is always 1. The standard form of a LAE is one variable of the form

$$ax+b=0 \tag{1.2}$$

where a and b are constants and x is a variable, a is a coefficient of the variable x, and b is a constant. A system of linear equations is a collection of more than one equation. The linear equation is an algebraic equation where each term has an exponent of 1, and when this equation is plotted, it always results in a straight line; that is why it is called a 'linear equation'. There are linear equations in one variable and linear equations in two variables. Let us learn how to define linear and nonlinear equations with the help of the following examples. To solve the system of linear equations, the minimum number of unknowns should be equal to or less than the number of equations we have. If the number of unknowns is more than the number of independent equations, then we cannot solve the system of equations.

Example 1.1 A Linear Equation in One Variable

Linear equations in one variable are expressed as $ax+b=0$, where a and b are integers, and x is a variable with only one solution. For example, $2x+3=8$ is a linear equation with one variable; calculate the value of x.

Solution

The equation is organized as follows by putting the variable x on one side

$$2x = 8 - 3$$

Divide both sides of two

$$x = \frac{8-3}{2} = 2.5$$

Example 1.2 Linear Equation of One Variable

Solve the linear equation in one variable:

$$3x + 6 = 18$$

Solution

To solve the given equation, we put the numbers on the right side of the equation and keep the variable on the left side. This means:

$$3x = 18 - 6$$

Then, when we find the value of x, we get

$$3x = 12$$

Finally, the value of x:

$$x = \frac{12}{3} = 4$$

Example 1.3 Sum of Two Numbers

The sum of the two numbers is 44. Find the numbers by forming a linear equation if one number is ten times more than the other.

Solution

Suppose the number is x, so the other is $x + 10$.

We know that the sum of both numbers is 44.

Therefore, the linear equation can be outlined as follows:

$$x + (x + 10) = 44$$

The result is

$$2x = 44 - 10$$

Simplifying the right-hand side,

$$2x = 34$$

so the value of x is 17.

This means that one of the numbers is 17, and the other is $17+10=27$.

1.3.2 Nonlinear Algebraic Equations

Nonlinear algebraic equations involve multiple variables; an equation includes non-linear terms such as polynomials of degree 2 or more, exponential, logarithm, or other terms. All other types of equations where $f(x)=0$, when $f(x)$ is not a linear function of x, are called nonlinear. A typical way to learn about a nonlinear equation is to note that x is 'not alone' as in ax, but shares a product with itself, as in

$$x^3 + 2x^2 - 9 = 0 \tag{1.3}$$

We can see that x^3 and $2x^2$ are nonlinear terms

Example 1.4 System of linear Equation

Solve the following system of linear equations

$$3x+6y=12$$

$$5x-8y=2$$

Solution

From the first equation,

$$3x=12-6y \rightarrow\rightarrow x=4-2y$$

Substitute x in the second equation

$$5(4-2y)-8y=2$$

Simplify

$$20-10y-8y=2$$

Rearrange

$$18y=18$$

Divide both sides by 18 to get: $y = 1$

Substitute $y = 1$ in the first equation: $3x + 6y = 12$

$$3x + 6(1) = 12$$

Hence: $x = 2$

Example 1.5 System of Linear and Nonlinear Equation

Solve the system of equations by substitution

$$3x - y = -2$$

$$2x^2 - y = 0$$

Solution

Rearrange y as a function of x

$$y = 3x + 2$$

Substitute y in the second equation

$$2x^2 - 3x - 2 = 0$$

The value of x

$$x = \frac{-b \mp \sqrt{b^2 - 4ac}}{2a} = \frac{3 \mp \sqrt{3^2 - (4)(2)(-2)}}{(2)(2)} = \frac{3 \mp \sqrt{25}}{4}$$

$$x = \left(-\frac{1}{2}, 2\right), y = \left(\frac{1}{2}, 8\right)$$

Example 1.6 Application of Nonlinear Algebraic equations

The hydrolysis of acetic anhydride forms acetic acid took place in a continuously stirred tank reactor ($V = 1250$ L). The species concentration of feed to the reactor is 2.5 mol/L of acetic anhydride and 50 mol/L of water. The feed flow rate is 15 L/s (Figure 1.1). The reaction

$$(CH_3CO)_2O + H_2O \leftrightarrow 2CH_3COOH$$

The reaction rate is first in acetic anhydride and first in the water. The reaction rate constant is 0.075 L/(mol s). Derive the nonlinear equations.

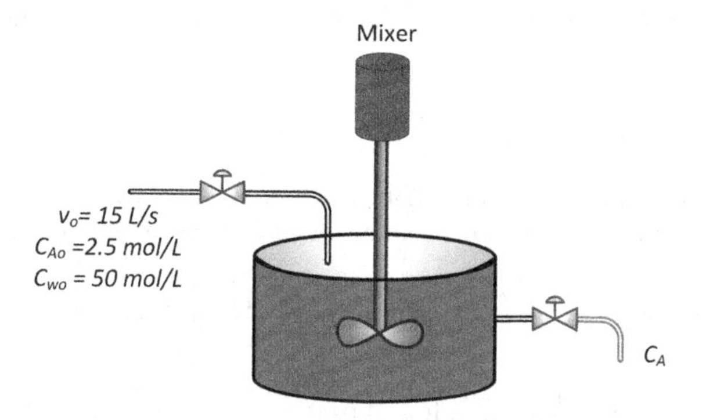

FIGURE 1.1 Schematic of the continuous stirred tank reactor (CSTR).

Solution

Putting the reaction in symbolic form

$$(CH_3CO)_2\,O + H_2O \leftrightarrow 2CH_3COOH$$

$$A + W = 2B$$

The rate of reaction is first order in *A* and *W*, hence

$$r_A = -kC_AC_W$$

Component mole balance on *A*

$$F_{Ao} - F_A + R_AV = 0$$

Substitute the rate of reaction

$$v_oC_{Ao} - v_oC_A - kC_AC_WV = 0$$

Component mole balance on *W*

$$F_{Ao} - F_A + R_AV = 0$$

Substitute the rate of reaction

$$v_oC_{Wo} - v_oC_W - kC_AC_WV = 0$$

The given data are

$$v_0 = 15\ \frac{\text{L}}{\text{s}},\ k = 0.0025\ \frac{\text{L}}{(\text{mol s})},\ V = 1250\ \text{L},\ C_{Ao} = 2.5\ \frac{\text{mol}}{\text{L}},\ \ C_{wo} = 50\ \frac{\text{mol}}{\text{L}}$$

The following two nonlinear algebraic equations are to be solved simultaneously

$$v_o C_{Ao} - v_o C_A - k C_A C_W V = 0$$

$$v_o C_{Wo} - v_o C_W - k C_A C_W V = 0$$

After the substitution of the known data, there are two unknown variables (C_A, C_W) and two equations. The unknown can be found by the substitution method.

$$\left(15\frac{\text{L}}{\text{s}}\right)\left(2.5\frac{\text{mol}}{\text{L}}\right)-\left(15\frac{\text{L}}{\text{s}}\right)C_A-\left(0.0075\frac{\text{L}}{\text{mol s}}\right)C_A C_W\,(1250\text{ L})=0$$

$$\left(15\frac{\text{L}}{\text{s}}\right)\left(50\frac{\text{mol}}{\text{L}}\right)-\left(15\frac{\text{L}}{\text{s}}\right)C_W-\left(0.0075\frac{\text{L}}{\text{mol s}}\right)C_A C_W\,(1250\text{ L})=0$$

Divide both sides by 15

$$\left(2.5\frac{\text{mol}}{\text{L}}\right)-C_A-(29.6875)C_A C_W=0$$

$$\left(50\frac{\text{mol}}{\text{L}}\right)-C_W-(29.6875)\,C_A C_W=0$$

Subtract the first equation from the second one

$$C_A - C_W + 47.5 = 0$$

$$C_W = C_A + 47.5$$

Substitute C_w in the first equation

$$\left(2.5\frac{\text{mol}}{\text{L}}\right)-C_A-(0.625)C_A(C_A+47.5)=0$$

Simplify

$$2.5 - C_A - 0.625C_A^2 - 29.6875C_A = 0$$

Simplify further

$$2.5 - 0.625C_A^2 - 30.6875C_A = 0$$

Solving the quadratic equation, the concentration of A and W are:

$$C_A = 0.081\ \frac{\text{mol}}{\text{L}}$$

$$C_w = 47.6\ \frac{\text{mol}}{\text{L}}$$

Example 1.7 Using Numerical Integration

The elementary gas phase reaction occurs in a plug flow reactor (PFR) at constant temperature and pressure.

$$A + B \rightarrow 2C$$

The total inlet concentration is 0.5 mol/L. The feed is equal molar in A and B. The inlet molar rate of A is 5.0 mol/s. The reaction rate constant is $k = 0.1$ L/mol/s. Find the volume of the PFR to achieve 90% conversion of A.

Solution

This is a reactor design example where the design equation of the PFR for the gas phase is to be used.

$$V = F_{Ao} \int_0^{0.9} \frac{dx}{-r_A}$$

The rate law is elementary, and hence

$$-r_A = kC_A C_B$$

Putting C_A and C_B as a function of x using the stoichiometric table

$$C_A = \frac{F_A}{v}$$

where F_A is the molar flow rate of A, v is the volumetric flow rate. For gas phase,

$$v = v_0 (1 + \epsilon x)\left(\frac{P_0}{P}\right)\left(\frac{T}{T_o}\right)$$

where $\epsilon = y_{A0}\delta$, $\delta = 2 - 1 - 1 = 0$, under isothermal and isobaric reaction conditions,

$$C_A = \frac{\frac{F_A}{v} = F_{A0}(1 - x)}{v_0 (1 + 0)} = C_{Ao}(1 - x)$$

The rate law is as follows,

$$-r_A = kC_{A0}^2 (1 - x)^2$$

Combining the rate law with the design equation,

$$V = \frac{F_{A0}}{kC_{A0}^2} \int_0^{0.9} \frac{dx}{(1 - x)^2}$$

Analytical integration,

$$\int_0^{x_1} \frac{dx}{(1-x)^2} = \frac{x_1}{1-x_1}$$

Substitution of known values.

$$V = \frac{F_{A0}}{kC_{A0}^2} \int_0^{0.9} \frac{dx}{(1-x)^2} = \frac{F_{A0}}{kC_{A0}^2} \frac{x_1}{1-x_1} = \frac{5}{0.082\,(0.5*0.5)^2} \frac{0.9}{1-0.9} = 8780.45 \text{ L}$$

The integral can also be found using numerical integration techniques in later chapters.

Example 1. 8 Liquid Reaction in Continuous Stirred Tank Reactor

The liquid phase reaction

$$A + B \rightarrow C$$

Occurs in $1\,m^3$ continuous stirred tank reactor (CSTR) reactor. The rate law of this reaction is $r_A = -kC_A$, if $k = s^{-1}$, $C_{Ao} = 0.02$ mol/L in an inert. The inlet volumetric flow rate is 10.0 L/s.

What is the conversion of A under steady-state conditions at 350°K and 5 atm? The ideal gas law R=0.0821 L·atm/mol/K.

Solution

The question is a reactor design problem, where the single pass conversion of a liquid phase reaction in a CSTR reactor is required (required knowledge in reactor design).

$$V_{\text{CSTR}} = \frac{F_{Ao}X}{-r_A}$$

The rate of reaction is first order, the liquid concentration of A in terms of x

$$C_A = C_{Ao}(1-X)$$

Combining

$$V_{\text{CSTR}} = \frac{(F_{Ao}X)}{kC_{Ao}(1-X)}$$

Rearranging to solve for X

$$\frac{(V_{\text{CSTR}} \times kC_{A0})}{F_{Ao}} = \frac{V_{\text{CSTR}} \times kC_{A0}}{v_o C_{Ao}} = \frac{V_{\text{CSTR}} \times k}{v_o} = \frac{X}{1-X}$$

Substitution of known values

$$\frac{1\,\text{m}^3\left(1000\,\dfrac{\text{liter}}{\text{m}^3}\right)\times 0.005\,\text{s}^{-1}}{10\,\dfrac{\text{liter}}{\text{s}}} = \frac{X}{1-X}$$

Simplifying

$$\frac{1\,\text{m}^3\left(1000\,\dfrac{\text{liter}}{\text{m}^3}\right)\times 0.005\,\text{s}^{-1}}{10\,\dfrac{\text{liter}}{\text{s}}} = 0.5 = \frac{X}{1-X}$$

Solving for X

$$X = \frac{0.5}{1.5} = \frac{1}{3}$$

1.4 ORDINARY DIFFERENTIAL EQUATIONS

The ordinary differential equation (ODE) is helpful for students in engineering and physics because they will need the concepts to follow derivations and homework problems in their major courses in mass transfer, heat transfer, and many more. The ODE is an equation that involves some ordinary derivatives of a function. Differential equations require a solid understanding of previous calculus and algebraic manipulation concepts. Differential equations are not easy, and they acquire knowledge in familiar and unfamiliar contexts. ODE equations are used in many models to determine how the state of that model changes (concerning time or another variable). Thus, explanatory equations are essential for many scientific fields because they arise whenever a relationship is given to changing a model/system. Differential equations involve derivatives of one or more functions.

$$\frac{dy}{dx} = 6x + 1 \tag{1.4}$$

$$\frac{dy}{dx} + \frac{3}{x}y = x^2 \tag{1.5}$$

$$\frac{d^3y}{dx^3} = 64\,x \tag{1.6}$$

$$\frac{d^2y}{dx^2} - 2\frac{dy}{dx} - 3y = 0 \tag{1.7}$$

1.4.1 Order of Differential Equation

The order of a differential equation is the highest order of derivatives present in the equation.

First order:

$$\frac{dy}{dx} + 2y = \sin x \tag{1.8}$$

$$\left(\frac{dy}{dx}\right)^2 = 4x \tag{1.9}$$

Second order:

$$\frac{d^2y}{dx^2} + 4xy = e^x \tag{1.10}$$

Third order:

$$\frac{d^3y}{dx^3} - 6\frac{d^2y}{dx^2} + 9\frac{dy}{dx} = 0 \tag{1.11}$$

Example 1.9 Generation of First ODE

Consider the following simple series of reactions taking place in a batch reactor

$$A \xrightarrow{k_1} B$$

$$B \xrightarrow{k_2} C$$

Assuming that the reaction rate is first order concerning A and B, derive the concentration change of components A, B, and C with time.

$$r_1 = k_1 C_A$$

$$r_2 = k_2 C_B$$

Solution

The complete set of differential equations for a well-mixed batch reactor are as follows:

$$\frac{dC_A}{dt} = -r_1 = -k_1 C_A$$

$$\frac{dC_B}{dt} = r_1 - r_2 = k_1 C_A - k_2 C_B$$

$$\frac{dC_c}{dt} = r_3 = k_2 C_B$$

1.4.2 Analytical Solution of First-Order Differential Equation

Often the goal is to solve the ODE equation, that is, to determine a function or many functions that fulfill the equation. Consider the following first-order ODE:

$$\frac{dy}{dx} = 5\ y \tag{1.12}$$

The exact solution can be found as follows:

Rearrange the equation as follows

$$\frac{dy}{y} = 5dx \tag{1.13}$$

Integrating

$$\ln(y) = 5x \tag{1.14}$$

The general solution is

$$y = e^{5x} \tag{1.15}$$

The general solution of

$$\frac{dy}{dx} = 2x \tag{1.16}$$

is

$$y = x^2 + c \tag{1.17}$$

The particular solution of

$$\frac{dy}{dx} = 2x,\ \ y(1) = 4 \tag{1.18}$$

is

$$y = x^2 + c \tag{1.19}$$

Using the initial conditions to find c, when $x = 1$, $y = 4$ (initial conditions)

$$4 = (1)^2 + c \tag{1.20}$$

Solving for c, $c = 3$. Accordingly, the general solution

$$y = x^2 + 3 \tag{1.21}$$

Example 1.10 Derive the Differential Equations for a Mixing Tank

A liquid level system where liquid enters a cylindrical tank at the rate of F_f and leaves at a rate F_o. The mathematical model for the system is a differential equation that shows the relationship between the height (h) of the liquid and the input flow rate. The surface area of the bottom of the cylindrical tank is A (Figure 1.2). Develop the differential equation.

Solution

Applying the unsteady state material balance around the tank,

$$\frac{dm}{dt} = \dot{m}_f - \dot{m}_o$$

In terms of flow rate, assuming constant density,

$$\frac{d(\rho V)}{dt} = \rho F_f - \rho F_o$$

Simplify and replace V with Ah. The following ODE is generated:

$$A\frac{dh}{dt} = F_f - F_o$$

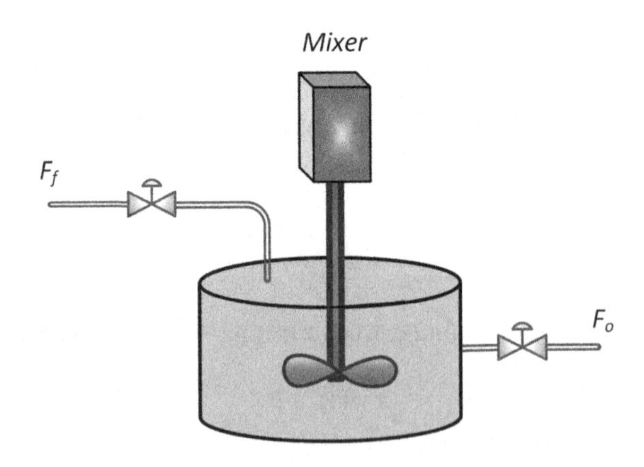

FIGURE 1.2 Schematic diagram of stirred tank.

Example 1.11 The Concentration of Salt in a Tank

Purified water flows into a tank at a rate of 10 L/min. The contents of the tank are well mixed; they flow out at a rate of 10 L/min. Add salt to the tank at the rate of 0.1 kg/min. Initially, the tank contains 10 kg of salt in 100 L of water. The volume of the tank is 100 L. How much salt is in the tank after 30 minutes?

Solution

Here, the setup is very similar to the previous example. The only difference from the previous one is adding 0.1 kg/min salt to the tank. As a result, we can modify the differential equation to consider this. Let the concentration of salt in the tank, C_s

Total material balance,

$$\frac{d(VC_s)}{dt} = 0 + 0.1 - C_s F_f$$

Simplify

$$(100\text{ liter})\frac{dC_s}{dt} = 0.1\frac{\text{kg}}{\text{min}} - C_s\frac{\text{kg}}{\text{liter}} \times 10\frac{\text{liter}}{\text{min}}$$

Rearrange

$$\frac{dC_s}{dt} = \frac{0.1}{100} - \frac{C_s\,(10)}{100}$$

The generated ODE is

$$\frac{dC_s}{dt} = 0.001 - 0.1C_s$$

The initial condition is at time = 0, and the concentration of salt is

$$C_s(0) = \frac{10\text{ kg salt}}{100\text{ liter water}} = 0.1\frac{\text{kg}}{\text{liter}}$$

Integrating the generated to ODE

$$C_s(t) = 0.01 + 0.09e^{-0.1t}$$

The concentration of salt in the tank after 30 min

$$C_s(t) = 0.01 + 0.09e^{-0.1(30)}$$

$$C_s(t) = 0.01448\frac{\text{kg}}{\text{liter}}$$

The amount of salt

$$m_s = 100\text{ liter} \times C_s(30) = 0.01448\frac{\text{kg}}{\text{liter}} \times 100\text{ liter} = 1.448\text{ kg}$$

Example 1.12 Mixing Tank with Variable Volume

Pure water flows into a tank at 12 L/min. The tank's contents are thoroughly mixed and flow out at 10.0 L/min. Initially, the tank contains 10 kg of salt in 100 L of water. Develop the model ODE equation and then calculate the amount of salt in the tank after 30 min.

Solution

The inlet flow rate is greater than the outflow rate. As a result, size is not constant. Using the initial conditions and flow rates, we can say that the volume V of liquid in the tank is

$$V = V_0 + v_o\ t = 100 + (12 - 10)t = 100 + 2t$$

Total component balance (salt)

$$\frac{d(C_s V)}{dt} = F_f(0) - F_o C_s$$

Since the volume varied with time

$$V\frac{dC_s}{dt} + C_s\frac{dV}{dt} = -F_o C_s$$

Overall mass balance at constant density

$$\frac{dV}{dt} = F_f - F_o$$

Replace dV/dt with $F_f - F_o$

$$V\frac{dC_s}{dt} + C_s(F_f - F_o) = -F_o C_s$$

Rearrange

$$(V_o + v_o t)\frac{dC_s}{dt} = -F_o C_s - C_s F_f + F_o C_s$$

Simplify further

$$(V_o + v_o t)\frac{dC_s}{dt} = -C_s F_f$$

Rearrange

$$\frac{dC_s}{C_s} = -F_f\left(\frac{dt}{(V_o + v_o t)}\right)$$

Using separable integration

$$\ln(C_s) = -\left(\frac{F_f}{v_o}\right)\ln\left(\frac{V_o}{v_o} + t\right) + C_1$$

Substitute known values to find C_1

$$\ln(0.1) = -\left(\frac{12}{2}\right)\ln\left(\frac{100}{2}+0\right)+C_1$$

$$C_1 = 21.169$$

The concentration of salt at any time

$$\ln(C_s) = -\left(\frac{F_f}{v_o}\right)\ln\left(\frac{V_o}{v_o}+t\right)+21.169$$

After 30 min, the salt concentration

$$\ln(C_s) = -\left(\frac{12}{2}\right)\ln\left(\frac{100}{2}+30\right)+21.169$$

The concentration after 30 min

$$C_s = 0.00596\frac{\text{kg}}{\text{liter}}$$

The amount of salt in 30 min

$$C_s = 0.00596\frac{\text{kg}}{\text{liter}}\times 100 \text{ liter} = 0.596 \text{ kg}$$

1.5 BOUNDARY VALUE PROBLEMS

A boundary value problem (BVP) is a system of ODEs with solutions and derivative values defined at more than one point. The solution and derivatives are defined at only two points (the boundaries), a two-point BVP. A differential equation with given conditions allows us to find the specific function that satisfies a given differential equation, rather than a family of functions. These types of problems are called initial value problems (IVP). If the provided conditions are at more than one point and the differential equation is of order two or higher, it is called a BVP. This section aims to give a brief look at the idea of BVPs and to give enough information to allow us to do some basic partial differential equations in the next chapter. With BVPs, we will have a differential equation and define the function and derivatives at different points, which we will call boundary values. The following systems represent BVPs frequently found in chemical engineering (Figure 1.3).

An essential problem in chemical engineering is the prediction of diffusion and reaction in a porous catalyst pellet. The goal is to predict the overall reaction rate of the catalyst pellet. The conservation of mass in a spherical coordinate provides

$$D\left[\frac{1}{r^2}\frac{d}{dr}\left(r^2\frac{dc}{dr}\right)\right] = R \tag{1.22}$$

with the following boundary condition

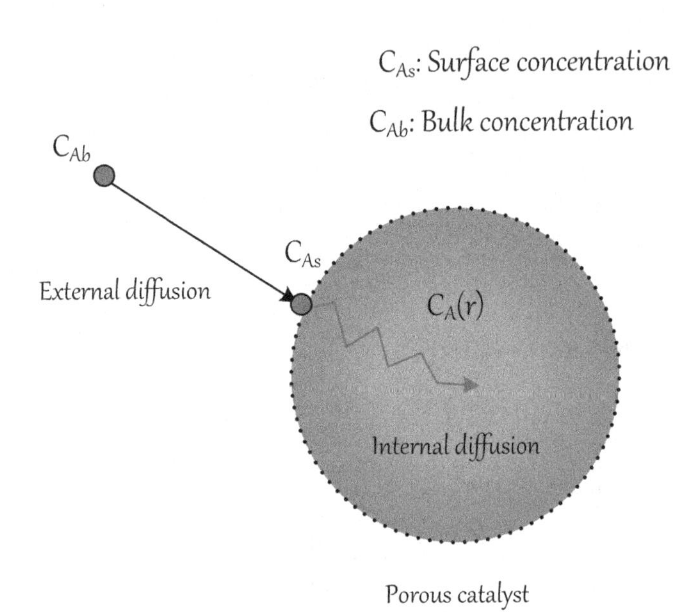

FIGURE 1.3 Reaction in spherical porous catalyst.

$$\text{at } r = 0, \frac{dc}{dr} = 0 \quad \text{(Axial symmetry about the origin)}$$

$$\text{at } r = r_p, \quad c = c_{As} \quad \text{(Concentration at the catalyst surface)}$$

where D is the diffusivity, r is the radial coordinate, c is the concentration of a given chemical species, and R is the reaction rate function. Consider a cylindrical porous solid catalyst particle of radius R and length L, as shown in Figure 1.4.

Assuming the catalyst has a sizeable length-to-diameter ratio ($L >> 2R$), the diffusion of substance A is mainly in the r-direction. Substance A undergoes a first-order decomposition reaction

$$A \rightarrow B, \; r_A = -kC_A$$

C_A is the local molar concentration of A in the catalyst, and k is the first-order rate constant. Considering molecular diffusion and neglecting bulk diffusion. Assuming isothermal operation, the A concentration at the catalyst's surface is constant and equal to C_{AS}. The following differential equation governs the steady-state equation mass balance on cylindrical coordinates considering the given assumptions

$$\frac{1}{r}\frac{d}{dr}\left(rD\frac{dC_A}{dr}\right) - kC_A = 0 \qquad (1.23)$$

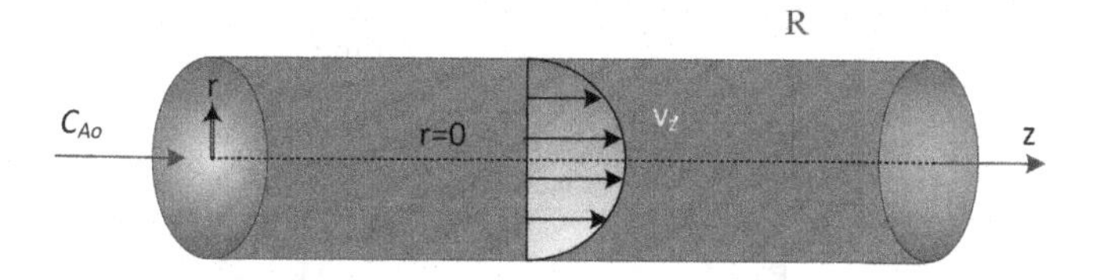

FIGURE 1.4 Porous cylindrical porous solid catalyst.

The relevant boundary conditions are

$$\text{at } r = 0, \ \frac{dC_A}{dr} = 0 \tag{1.24}$$

Heat transfer is an essential subject in engineering. Fins are frequently used in heating houses, cars, and industry.

Figure 1.5 shows a fin used in heat transfer to increase the surface area available for heat transfer between metal walls and poorly conductive fluids such as gases. A simple and practical application of thermal conductivity is a calculation-efficient cooling fin. Assuming $L >> B$, the fin will not lose heat from the end of the edges; the heat flows to the surface by the temperature profile in the fin. In which h is the constant convective heat transfer coefficient as is the surrounding fluid temperature T_a, then the governing differential equation is

$$k\frac{d^2T}{dz^2} = \frac{hA}{V}(T - T_a) = \frac{h\,(2W\ L)}{2BWL}(T - T_a) = \frac{h}{B}(T - T_a) \tag{1.25}$$

where k is the thermal conductivity of the fin, and the boundary conditions are

$$\begin{aligned} &\text{at } z = 0, \quad T(0) = T_w \\ &\text{at } z = L, \ \frac{dT}{dz}(L) = 0 \end{aligned} \tag{1.26}$$

Example 1.13 Boundary Value Problems

Find a solution to the BVP

$$\frac{d^2y}{dx^2} - y = 0; \quad y(0) = 0,\ y(1) = 1$$

Consider a general solution to the differential equation

$$y(x) = c_1e^x + c_2e^{-x}$$

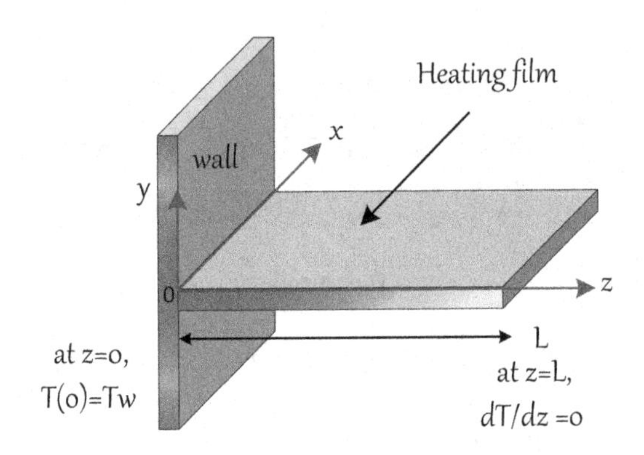

FIGURE 1.5 Schematic of a heating fin [1].

Solution

The characteristic equation is

$$m^2 - 1 = 0, \quad m_{\mp} = \mp 1$$

The general solution is

$$y(x) = c_1 e^{x} + c_2 e^{x}$$

Using the boundary condition 1,

$$y(0) = 0 = c_1 e^{0} + c_2 e^{-0} = c_1 + c_2$$

Simplifying,

$$0 = c_1 + c_2$$

Using boundary condition,

$$y(1) = 1 = c_1 e^{1} + c_2 e^{-1}$$

Simplify

$$1 = c_1 e + \frac{c_2}{e}$$

Substitute $c_1 = -c_2$

$$1 = -c_2 e + \frac{c_2}{e}$$

Rearrange

$$1 = c_2\left(\frac{1}{e} - e\right) = c_2\left(\frac{1-e^2}{e}\right)$$

$$c_2 = \left(\frac{e}{1-e^2}\right)$$

Then

$$c_1 = -c_2 = \left(\frac{-e}{1-e^2}\right)$$

The final solution

$$y(x) = \left(\frac{-e}{1-e^2}\right)e^x + \left(\frac{e}{1-e^2}\right)e^{-x}$$

Example 1.14 Boundary Value Problem

Consider the following second-order BVP

$$\frac{d^2y}{dx^2} + 4y = 0, \quad y(0) = 1, \frac{dy}{dx}\left(\frac{\pi}{2}\right) = 2$$

If the general solution of the differential equation is

$$y(x) = c_1 \cos(2x) + c_2 \sin(2x)$$

Using the boundary conditions, find the expression for c_1 and c_2

Solution

The characteristic polynomial

$$p(m) = m^2 + 4 = 0 \rightarrow m_{\mp} = \mp 2i$$

The general solution

$$y(x) = c_1 \cos(2x) + c_2 \sin(2x)$$

Substitution of boundary condition 1

$$y(0) = 1 = c_1 \cos(0) + c_2 \sin(0)$$

Hence,

$$1 = c_1 + 0 => c_1 = 1$$

To substitute boundary condition 2, first, differentiate the proposed solution equation

$$y'^{(x)} = -2c_1 \sin(2x) + 2c_2 \cos(2x)$$

Substitution of boundary condition 2

$$y'\left(\frac{\pi}{2}\right) = 2 = -2c_1 \sin\left(2 \times \frac{\pi}{2}\right) + 2c_2 \cos\left(2 \times \frac{\pi}{2}\right)$$

Simplify

$$y'\left(\frac{\pi}{2}\right) = 2 = -2c_1 \sin(\pi) + 2c_2 \cos(\pi) = 0 - 2c_2$$

Accordingly,

$$c_2 = -1$$

Substitution of c_1 and c_2

$$y(x) = (1)\cos(2x) + (-1)\sin(2x) = \cos(2x) - \sin(2x)$$

Example 1.15 Unsteady State Mass Transfer in a Tubular Liquid Flow

Develop the unsteady state BVP model equation for the diffusion of a laminar flow liquid solute A into water. Starting with the following equation of change,

$$\frac{\partial C_A}{\partial t} + v_z . \nabla C_A = D\nabla^2 C_A + r$$

Solution

The schematic diagram is shown in Figure 1.6

Consider the following assumptions:

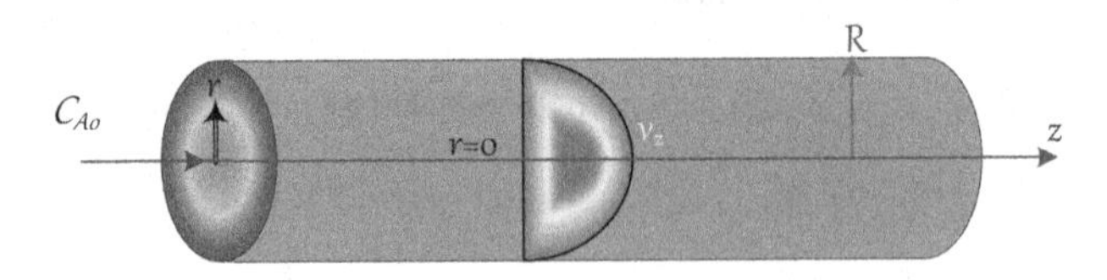

FIGURE 1.6 Schematic of the unsteady state mass transfer in a tubular liquid flow.

1. Steady-state operation.
2. No reaction ($r = 0$).
3. Neglect the diffusion in the radial direction.
4. Neglect velocity in the r-direction.
5. Laminar flow ($N_{R_e} < 2100$), the parabolic velocity profile.

$$v_z = 2v_o\left(1-\left(\frac{r}{R}\right)^2\right)$$

Starting with the equation change

$$\frac{\partial C_A}{\partial t} + v_z.\nabla C_A = D\nabla^2 C_A + r_A$$

Expand the equation

$$\frac{\partial C_A}{\partial t} + 2v_o\left(1-\left(\frac{r}{R}\right)^2\right)\frac{\partial C_A}{\partial z} + v_r\frac{\partial C_A}{\partial r} = D\left(\frac{\partial^2 C_A}{\partial z^2} + \frac{1}{r}\frac{\partial}{\partial z}\left(r\frac{\partial C_A}{\partial r}\right)\right) + r_A$$

Initial conditions

$$t = 0,\ \ C_A = 0$$

Boundary conditions

$$z = 0 \quad C_A = C_{Ao}$$

Assuming a very long tube,

$$z = L, \quad \frac{\partial C_A}{\partial z} = 0$$

Symmetric condition at the center

$$r = 0, \ \frac{\partial C_A}{\partial r} = 0$$

Non-reactive wall

$$r = R, \ -\frac{\partial C_A}{\partial r} = 0$$

The equation can be simplified based on the given assumption

$$0 + 2v_o\left(1-\left(\frac{r}{R}\right)^2\right)\frac{\partial C_A}{\partial z} + 0 = D\left(\frac{\partial^2 C_A}{\partial z^2} + 0\right)$$

The equation is simplified to

$$2v_o\left(1-\frac{r^2}{R^2}\right)\frac{\partial C_A}{\partial z}-D\frac{\partial^2 C_A}{\partial z^2}=0$$

Example 1.16 Flow Through the Circular Tube Using the Navier-Stokes Equation

The Navier-Stokes equation can be used in problems where the pseudo-momentum equilibrium method was initially used. For this problem, assume the pressure decreases linearly with length. Also, suppose that density and viscosity are constant. The Navier-Stokes equation can be written in the *z*-direction as

$$\rho\left(\frac{\partial v_z}{\partial t}+v_r\frac{\partial v_z}{\partial r}+\frac{v_\theta}{r}\frac{\partial v_z}{\partial \theta}+v_z\frac{\partial v_z}{\partial z}\right)=-\frac{\partial P}{\partial z}+\mu\left(\frac{1}{r}\frac{\partial}{\partial r}\left(r\frac{\partial v_z}{\partial r}\right)+\frac{1}{r^2}\frac{\partial^2 v_z}{\partial \theta^2}+\frac{\partial^2 v_z}{\partial z^2}\right)+\rho g_z$$

It is better to use cylindrical coordinates here and velocity in the *z*-direction only. The continuity equation is equivalent to the following, assuming constant ρ and μ. The ρg is not necessary because gravity does not play a role. The schematic diagram is shown in Figure 1.7.

Solution

Based on the steady-state assumptions, the Navier-Stokes equation in the *z*-direction is simplified to the following:

$$\frac{\partial P}{\partial z}=\mu\left[\frac{1}{r}\frac{\partial}{\partial r}\left(r\frac{\partial v_z}{\partial r}\right)\right]$$

Since the pressure gradient is linear

$$\frac{P_L-P_o}{\mu L}r=\frac{d}{dr}\left(r\frac{dv_z}{dr}\right)$$

Perform the first integration

$$\frac{dv_z}{dr}=-\left(\frac{P_L-P_o}{2\mu L}\right)\left(\frac{r^2}{r}\right)+\frac{c_1}{r}$$

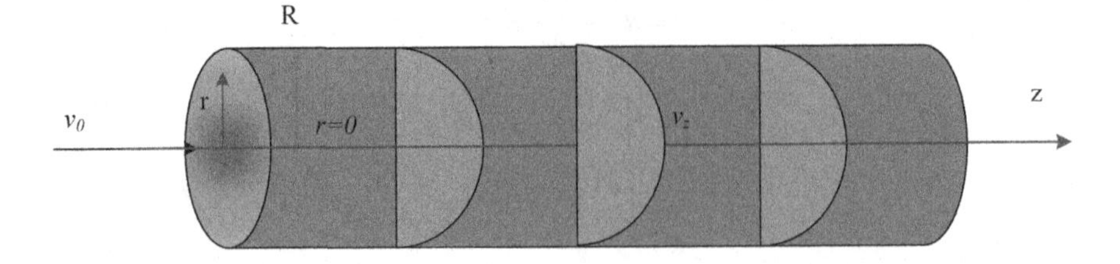

FIGURE 1.7 Schematic diagram of water flow in a pipe.

Integrating the above equation,

$$v_z = -\left(\frac{P_L - P_o}{4\mu L}\right)r^2 - c_1 lnr + c_2$$

Boundary conditions 1

at $r = 0, \frac{dv_z}{dr} = 0$

Applying boundary condition 1 in the following equation,

$$\frac{dv_z}{dr} = -\left(\frac{P_L - P_o}{2\mu L}\right)\left(\frac{r^2}{r}\right) + \frac{c_1}{r}$$

Substitute the boundary condition

$$0 = -\left(\frac{P_L - P_o}{2\mu L}\right)\left(\frac{0}{r}\right) + \frac{c_1}{r}$$

Accordingly,

$$c_1 = 0$$

at $r = R_1, v_z = 0$ since it is at a solid–liquid boundary

$$\frac{P_L - P_o}{4\mu L}R^2 = c_2$$

Substitution of c_1 and c_2

$$v_z = -\left(\frac{P_L - P_o}{4\mu L}\right)r^2 + \frac{P_L - P_o}{4\mu L}R^2$$

Rearrange

$$v_z = \left(\frac{P_L - P_o}{4\mu L}\right)\left(R^2 - r^2\right)$$

Reposition again

$$v_z = \left(\frac{P_L - P_o}{4\mu L}\right)R^2\left(\left(\frac{r}{R}\right)^2 - 1\right)$$

or

$$v_z = \frac{(P_o - P_L)R^2}{4\mu L}\left(1 - \left(\frac{r}{R}\right)^2\right)$$

1.6 SUMMARY

Numerical methods is a discipline in mathematics concerned with the development of effective methods for obtaining numerical solutions to mathematical problems that are difficult to solve by analytical methods. This chapter deals with possible equations that may appear in many topics of chemistry or other engineering subjects that need to be solved by numerical methods. Linear and nonlinear algebraic equations may appear in material and energy balances, mass transfer, heat transfer, and ordinary and partial differential equality in modeling and simulation.

1.7 PROBLEMS

1. Solve the following LAE

$$x + 7 = 12$$

   Find the value of x
   Answer: $x = 5$
2. Solve the following system of LAEs

$$2x + y = 6$$

$$x + y = 4$$

   Find the value of x, y
   Answer: ($x = 2$, $y = 2$)
3. A system of nonlinear equations consists of two or more equations with at least one equation that is not linear

$$x^2 + y^2 = 25$$

$$3x + 4y = 0$$

   Find the value of x, y
   Answer: ($x = 4$, $y = -3$)
4. Solve the system of two nonlinear equations

$$x^2 + y^2 = 45$$

$$x^2 - y^2 = -27$$

   Find the value of x, y
   Answer: ($x = 3$, $y = 6$)
5. Pure water flows into a tank at a rate of 10 L/min. The contents of the tank are preserved and mixed well. The contents flow out at a rate of 10 L/min. Initially, the tank contained 10 kg of salt in 100 L of water. How much salt will there be in the tank after 30 minutes?
   Answer: 0.5

6. Pure water flows into a tank at a rate of 8 L/min. The contents of the tank are preserved and mixed well, and the contents flow out at a rate of 10 L/min. Initially, the tank contained 10 kg of salt in 100 L of water. How much salt will there be in the tank after 30 minutes?
   Answer: 0.0248
7. Water with salt dissolved at a specified concentration of 0.1 kg/L, flowing into a tank at a rate of 10 L/min. The contents of the tank are well mixed, and the contents flow out at a rate of 10 L/min. Initially, the tank contained 10 kg of salt in 100 L of water. How much salt will there be in the tank after 30 minutes?
   Answer: 0.0248
8. Water with salt dissolved at a specified concentration of 0.1 kg/L, flowing into a tank at a rate of 10 L/min. The contents of the tank are well mixed, and the contents flow out at 12 L/min. Initially, the tank contained 10 kg of salt in 100 L of water. How much salt will there be in the tank after 30 minutes?
   Answer: 0.099
9. Write the ODE of the following two simple series reactions taking place in a batch reactor

$$A \xrightarrow{k_1} B \xrightarrow{k_2} C$$

   Write the complete set of the differential equation for a well-mixed batch reactor.
10. Solve the following BVP

$$\frac{d^2y}{dx^2} + 2y = 0, \quad y(0) = 1, \quad y(\pi) = 1$$

   Answer:

$$y = \cos\left(\sqrt{2}t\right) - \frac{\cos\left(\sqrt{2}t\right)}{\sin\left(\sqrt{2}t\right)} \sin\left(\sqrt{2}t\right)$$

REFERENCE

1. Ghasem, N., 2019. *Modeling and Simulation of Chemical Process Systems*, New York: CRC Press.

2 Numerical Solutions of Linear Systems

Numerical methods can solve systems of linear equations. Generally, they are divided into two categories: direct methods, which lead to the exact solution within a limited number of steps, such as the Gaussian elimination method, and indirect methods. Indirect methods (iterative methods) are valuable for problems involving massive private arrays, such as Jacobi and Gauss-Seidel. This chapter focuses on linear algebraic equations and how to solve them using different methods such as Cramer's rule, Gauss elimination method, and Gauss-Jordan, Jacobi, and Gauss-Seidel elimination methods.

LEARNING OBJECTIVES

1. Identify the linear algebraic equation.
2. Find the determinant of 2×2 and 3×3 matrices.
3. Use Cramer's rule to solve systems of two and three equations.
4. Use Gauss and Gauss-Jordan elimination methods.
5. Apply Jacobi and Gauss-Seidel iterative methods.
6. Make use of Simulink and Python to solve linear algebraic equations.

2.1 INTRODUCTION

A system of linear equations is a collection of linear equations that are independent of one another, and we cannot derive one equation from another. An equation is linear if each term is simply a constant or a constant multiplied by a variable. If we ever have a variable multiplied by another variable or a variable to the power of anything, then the equation is not linear. Solving a system of linear equations means finding a value for each unknown variable. There is sometimes more than one solution to a set of linear equations; the complete set of solutions is called the solution set.

The linear system of equations $ax = b$ can be solved using various numerical methods. A numerical method is a mathematical tool designed to solve numerical problems in natural and social sciences, engineering, medicine, and business. The following sections explain the system of linear algebraic equations and the appropriate methods for numerical solutions [1].

2.1.1 Definition of a Linear System of Algebraic Equations

A linear equation is an equation that is linear about its variables, such as $3x = 7$. We can also have a linear equation with more than one variable, such as $2x + 3y = 9$.

DOI: 10.1201/9781003360544-2

A linear algebraic equation is one in which the highest power of the variables is always 1. The standard form of a linear algebraic equation is one variable of the form

$$ax + b = 0$$

where a and b are constants, x is a variable, a is a coefficient of the variable x, and b is a constant. A system of linear equations is a collection of more than one equation. The system of n linear algebraic equation is as follows:

$$a_{11}X_1 + a_{12}X_2 + \ldots + a_{1n}\ X_n = b_1$$

$$a_{21}X_1 + a_{22}X_2 + \ldots + a_{2n}\ X_n = b_2$$

$$a_{m1}X_1 + a_{n2}X_2 + \ldots + a_{mn}\ X_n = b_m$$

where m is the number of rows and n is the number of columns. Here m is the number of independent equations to solve n number of variables. In a matrix structure,

$$AX = b$$

where A is the coefficient matrix:

$$\begin{bmatrix} a_{11} & a_{12}\ldots\ldots & a_{1n} \\ a_{21} & a_{22}\ldots\ldots & a_{2n} \\ a_{mn} & a_{m2}\ldots\ldots & a_{mn} \end{bmatrix}$$

The X matrix is $\begin{bmatrix} X_1 \\ X_2 \\ X_n \end{bmatrix}$

The b matrix is $\begin{bmatrix} b_1 \\ b_2 \\ b_m \end{bmatrix}$

In general, a matrix has $n \times m$ elements. If $n = m$, there may be a unique solution, and the rank of the matrix is m. Consider the following matrix, A

The determinant, $|A|$

$$|A| = ad - bc$$

The following are two linear algebraic equations:

$$2x + 3y = 5$$

$$x - y = 4$$

A 2×2 matrix represents the equations as follows:

$$A=\begin{bmatrix} 2 & 3 \\ 1 & -1 \end{bmatrix}, X=\begin{bmatrix} x \\ y \end{bmatrix}, b=\begin{bmatrix} 5 \\ 4 \end{bmatrix}$$

The following three algebraic equations represent an example of a 3×3 matrix:

$$2x+3y+2z=7$$

$$x+y-z=5$$

$$3x+3y-2z=4$$

The coefficients exist in the following matrix:

$$A=\begin{bmatrix} 2 & 3 & 2 \\ 1 & 1 & -1 \\ 3 & 3 & -2 \end{bmatrix}$$

The X and b matrices are as follows:

$$X=\begin{bmatrix} x \\ y \\ z \end{bmatrix}, b=\begin{bmatrix} 7 \\ 5 \\ 4 \end{bmatrix}$$

2.1.2 Properties of a Matrix and Operation

The matrix properties can exist as

- Symmetric matrix, $aij = aji$
- A square matrix, $n = m$
- Diagonal matrix, $\begin{bmatrix} \ddots & 0 \\ 0 & \ddots \end{bmatrix}$ (only diagonal elements)

The possible operations:
Summation:

$$[A]_{m\times n}=[B]_{m\times n}+[C]_{m\times n}$$

Multiplication (rules):

$$[A]_{m\times n}\times[B]_{n\times l}=[C]_{m\times l}$$

Therefore, multiplication is not permissible for

$$[B]_{nl}\times[A]_{mn} \quad for \quad l\neq m$$

The following multiplications are permissible:

$$\begin{bmatrix} 5 & 1 \\ 3 & 2 \\ 0 & 3 \end{bmatrix}_{3\times2} \times \begin{bmatrix} -3 & 0 \\ 1 & 0 \end{bmatrix}_{2\times2}$$

$$\begin{bmatrix} 5 & 1 \\ 3 & 2 \\ 0 & 3 \end{bmatrix}_{3\times2} \times \begin{bmatrix} 1 & 2 & 3 \\ 4 & 5 & 6 \end{bmatrix}_{2\times3}$$

$$\begin{bmatrix} 1 & 2 & 3 \end{bmatrix}_{1\times3} \times \begin{bmatrix} 0 \\ 5 \\ 4 \end{bmatrix}_{3\times1}$$

The following multiplications are not permissible:

$$\begin{bmatrix} 5 & 1 \\ 3 & 2 \\ 0 & 3 \end{bmatrix}_{3\times2} \times \begin{bmatrix} 1 & 2 \\ 4 & 5 \\ 8 & 7 \end{bmatrix}_{3\times2}$$

$$\begin{bmatrix} 5 & 1 \\ 3 & 2 \\ 0 & 3 \end{bmatrix}_{3\times2} \times \begin{bmatrix} 0 & 1 & 3 \\ 4 & 5 & 6 \\ 7 & 8 & 9 \end{bmatrix}_{3\times3}$$

$$\begin{bmatrix} 1 & 2 & 3 \end{bmatrix}_{1\times3} \times \begin{bmatrix} 1 \\ 2 \end{bmatrix}_{2\times1}$$

Rule $C_{ij} = \sum_{1}^{k} a_{ik} \times b_{kj}$

For example, the permissible multiplications of the following two matrices:

$$\begin{bmatrix} 5 & 1 \\ 3 & 2 \\ 0 & 3 \end{bmatrix}_{3\times2} \times \begin{bmatrix} 3 & 0 \\ 1 & 5 \end{bmatrix}_{2\times2} = \begin{bmatrix} (5\times3+1\times1) & (5\times0+1\times5) \\ (3\times3+2\times1) & (3\times0+2\times5) \\ (0\times0+3\times1) & (0\times0+3\times5) \end{bmatrix}_{3\times2}$$

2.1.3 Inverse of Matrix $[A]^{-1}$

The I is the identity matrix

$$I = [A]_{m\times n} \times [A]^{-1}_{n\times m}$$

The I is the identity matrix

$$I \equiv \begin{bmatrix} 1 & 0 & 0 \\ 0 & 1 & 0 \\ 0 & 0 & 1 \end{bmatrix}$$

Consider the following matrix A:

$$A = \begin{bmatrix} a_{11} & a_{12} \\ a_{21} & a_{22} \end{bmatrix}$$

The inverse of A matrix, $[A]^{-1}$

$$[A]^{-1} = \frac{1}{|A|}\begin{bmatrix} a_{22} & -a_{12} \\ -a_{21} & a_{11} \end{bmatrix}$$

where the determinant, $|A|$

$$|A| = (a_{11\times}a_{22} - a_{12} \times a_{21})$$

where the value of $|A|$ should not be equal to zero; otherwise, the matrix A is considered singular.

2.1.4 Transpose a Matrix

Matrix transposition is obtained by interchanging its rows into columns or columns into rows. The transpose of a matrix is denoted by using the letter 'T' in the superscript of the given matrix. For example, if A is the given matrix, then the transposition of the matrix A is represented by A^T. Consider the following matrix, A:

$$A = \begin{bmatrix} a_{11} & a_{12} \cdots\cdots a_{1n} \\ a_{21} & a_{22} \cdots\cdots a_{mn} \end{bmatrix}$$

The transpose matrix $[A]^T$ is the interchange of rows with columns, as shown below:

$$[A]^T = \begin{bmatrix} a_{11} & a_{21} \cdots\cdots a_{m1} \\ a_{1n} & \cdots\cdots\cdots \quad a_{mn} \end{bmatrix}_{n\times m}$$

For example,

$$[A] = \begin{bmatrix} a_1 \\ a_2 \\ \vdots \\ a_n \end{bmatrix}_{n\times 1}$$

The transpose matrix is

$$[A]^T = \begin{bmatrix} a_1 & a_2 & \dots & a_n \end{bmatrix}_{1\times n}$$

2.2 CRAMER'S RULE

Cramer's rule is a method that uses determinants to solve linear algebraic equations systems with the same number of equations as variables. Cramer's rule generally works well for a small matrix, and Cramer's rule replaces a variable column with a regular column.

$$[A][X]=[b]$$

$$X_j = \frac{|A_j|}{|A|}$$

where A_j is determined by replacing the jth column of A by b in a case of an algebraic equation in the form of the following sections.

2.2.1 Solving Two Linear Equations Using Cramer's Rule

An example of two linear equations of two variables as follows:

$$ax + by = e$$

$$cx + dy = f$$

If we solve for x, the x column is replaced with the column of constants (e, f) on the right-hand side of the equations. Using Cramer's rule, the solution for x is given as follows in which $D \neq 0$:

$$x = \frac{D_x}{D} = \frac{\begin{bmatrix} e & b \\ f & d \end{bmatrix}}{\begin{bmatrix} a & b \\ c & d \end{bmatrix}} = \frac{ed - fb}{ad - cb}$$

If we are solving for y, replace the y column with the constant,

$$y = \frac{D_y}{D} = \frac{\begin{bmatrix} a & e \\ c & f \end{bmatrix}}{\begin{bmatrix} a & b \\ c & d \end{bmatrix}} = \frac{af - ce}{ad - cb}$$

2.2.2 Solving Three Linear Equations Using Cramer's Rule

Consider a system of three linear equations to be solved using Cramer's rule.

$$a_1x + b_1y + c_1z = d_1$$

$$a_2x + b_2y + c_2z = d_2$$

$$a_3x + b_3y + c_3z = d_3$$

The solution is

$$x = \frac{D_x}{D}, y = \frac{D_y}{D}, z = \frac{D_z}{D}, \text{ where } D \neq 0$$

Putting the equations in a matrix format,

$$D = \begin{bmatrix} a_1 & b_1 & c_1 \\ a_2 & b_2 & c_2 \\ a_3 & b_3 & c_3 \end{bmatrix}, D \neq 0$$

Then,

$$D_x = \begin{vmatrix} d_1 & b_1 & c_1 \\ d_2 & b_2 & c_2 \\ d_3 & b_3 & c_3 \end{vmatrix}, D_y = \begin{vmatrix} a_1 & d_1 & c_1 \\ a_2 & d_2 & c_2 \\ a_3 & d_3 & c_3 \end{vmatrix}, D_z = \begin{vmatrix} a_1 & b_1 & d_1 \\ a_2 & b_2 & d_2 \\ a_3 & b_3 & d_3 \end{vmatrix}$$

The equations should be independent, and the determinant should not equal zero. For example, the following two equations are not independent.

$$2x_1 + 3x_2 = 11$$

$$4x_1 + 6x_2 = 22$$

We know that by checking if the matrix is independent ($|A| \neq 0$) or not independent ($|A| = 0$).

$$|A| = \begin{bmatrix} 2 & 3 \\ 4 & 6 \end{bmatrix}_{2\times 2} = 2\times 6 - 3\times 4 = 0$$

Accordingly, the two equations are not independent. There is no unique solution, and the matrix coefficient is singular; a zero determinant means that either the system has no solution or an infinite number of solutions.

Example 2.1 Implication of Cramer's Rule for Linear Algebraic Equations

Solve the following two linear algebraic equations manually using Cramer's rule.

$$2x_1 + 3x_2 = 11$$

$$x_1 + x_2 = 4$$

Verify the manual calculation with Simulink and Python programming of Cramer's rule.

Solution

First, we must check the singularity of the equation by finding the determinant.

Put the equations into matrix format.

$$A = \begin{bmatrix} 2 & 3 \\ 1 & 1 \end{bmatrix}, b = \begin{bmatrix} 11 \\ 4 \end{bmatrix}$$

The determinant of A,

$$|A| = (2 \times 1) - (3 \times 1) = -1$$

The two algebraic equations are independent, and there is a unique solution. To find the unknown values, use Cramer's rule:

$$A_x = \begin{bmatrix} 11 & 3 \\ 4 & 1 \end{bmatrix} = 11 - 12 = -1$$

$$A_y = \begin{bmatrix} 2 & 11 \\ 1 & 4 \end{bmatrix} = 8 - 11 = -3$$

Accordingly,

$$x_1 = \frac{A_x}{|A|} = -\frac{1}{-1} = 1$$

$$x_2 = \frac{A_y}{|A|} = \frac{-3}{-1} = 3$$

Solution Using the Simulink Algebraic Constraint Block

Two ways can solve the two independent equations using Simulink's 'Algebraic Constraint' block (Figure 2.1a). Figure 2.1a is to write each equation using the available blocks in Simulink (Gain, Constant, and Sum). By contrast, in Figure 2.1b, the matrix coefficients entered the Gain in a vector structure '[2 3;1 1]' and selected from the 'Multiplication' pulldown menu 'Matrix(K*u) (u vector)'. Then double click on the Constant block and enter the matrix constants such as '[11 4]'.

Solution Using the Simulink MATLAB® Function

Figure 2.2 shows the Simulink solution using Cramer's rule in the form of a MATLAB function embedded in the Simulink MATLAB function block available in the Simulink library under user-defined functions.

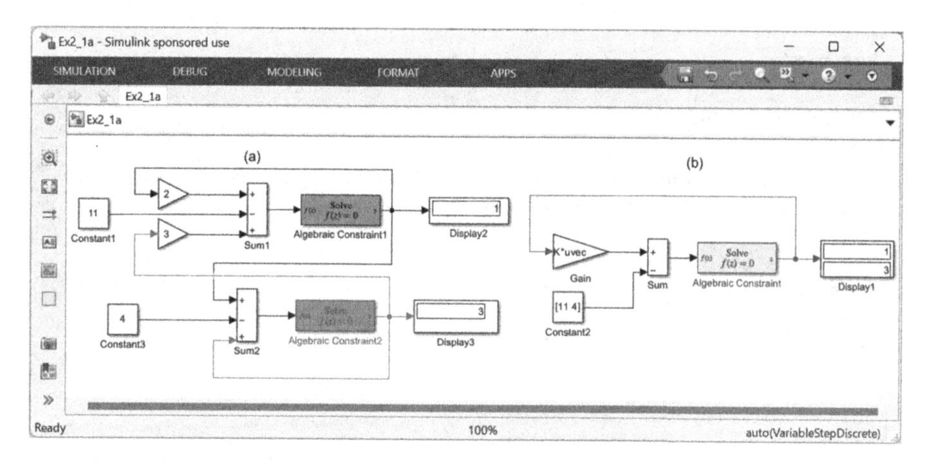

FIGURE 2.1 Simulink solution uses the Simulink algebraic constraint function block to solve the equation defined in Example 2.1.

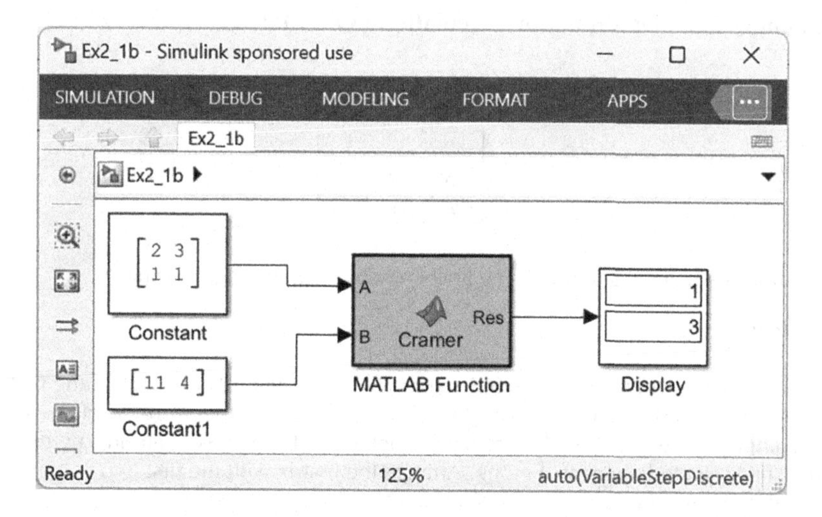

FIGURE 2.2 Simulink solution using Cramer's rule embedded in the MATLAB function to solve the equations stated in Example 2.1.

The following program is the MATLAB code describing Cramer's rule embedded into the MATLAB function.

```
function Res = Cramer(A, B)
x = ones(2,1);
a_det = det(A);
for i = 1:2
    C = A;
    C(:, i) = B;
    x(i, 1) = det(C)/a_det;
end
Res = x';
```

Python Solution

The following program is the Python code that employs Cramer's rule to solve the two algebraic equations presented in Example 2.1.

```
# Example 2.1
# Calculating the determinant
from sympy import *
x1,x2 = symbols(['x1','x2'])
system = [Eq(2*x1 + 3*x2,11),
          Eq(x1 + x2,4)]
soln = solve(system,[x1,x2])
print('\n the solution:',soln)
```

The following are the Python program executed results of Example 2.1.

```
The solution: {x1: 1, x2: 3}
```

Example 2.2 Determinant Calculation (3 × 3 Matrix)

Find the determinant of the following matrix manually.

$$A = \begin{bmatrix} 0 & 2 & 1 \\ 3 & -1 & 1 \\ 4 & 0 & 1 \end{bmatrix}$$

Confirm the manual solution using Simulink and Python programming.

Solution

To find the determinant of the 3×3 matrix, first, take the first element of the first row and multiply it by a secondary 2×2 matrix, which becomes from the elements remaining in the 3×3 matrix that do not belong to the row or column to which your first selected elements belong. Arrange the matrix with the first two columns and then follow the formula.

$$|A| = \begin{bmatrix} 0 & 2 & 1 \\ 3 & -1 & 1 \\ 4 & 0 & 1 \end{bmatrix}$$

Accordingly,

$$|A| = (0)\begin{vmatrix} -1 & 1 \\ 0 & 1 \end{vmatrix} - (2)\begin{vmatrix} 3 & 1 \\ 4 & 1 \end{vmatrix} + (1)\begin{vmatrix} 3 & -1 \\ 4 & 0 \end{vmatrix}$$

Simplify

$$|A| = +(0)[(-1)(1)-(0)(1)] - (2)[(3)(1)-(4)(1)] + (1)[(3)(0)-(4)(-1)]$$

$$|A| = 0 + 2 + 4 = 6$$

Simulink Solution

The Simulink solution of Example 2.2 is shown in Figure 2.3, followed by the embedded MATLAB code implanted in the MATLAB function.

The following program is the MATLAB code embedded in the MATLAB function.

```
function Res = Determinant(A)
x = ones(3);
a_det = det(A);

Res = a_det;
```

Python Solution

The following Python code is programmed to calculate the determinant of the 3×3 matrix of the three equations presented in Example 2.2.

```
#Example 2.2
#Calculating the determinant
#importing Numpy package
import numpy as np
```

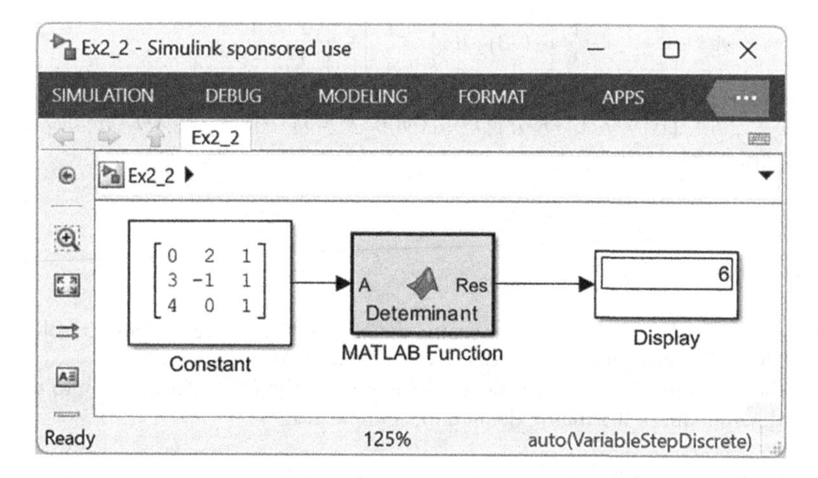

FIGURE 2.3 Simulink's solution using a MATLAB function implanted with MATLAB code to find the determinant of the matrix given in Example 2.2.

```
# creating a 3X3 Numpy matrix
n_array = np.array([[0,2,1],
                    [3,-1,1],
                    [4,0,1]])

# calculating the determinant of matrix
det = np.linalg.det(n_array)

print("\n Determinant of given 3X3 square matrix:",int(det))
```

Execution
Determinant of given 3×3 square matrix: 6

Example 2.3 Finding the Determinant of the 3×3 Matrix

Find the determinant of the following 3×3 matrix manually and using Simulink and Python programming.

$$A = \begin{bmatrix} 1 & -3 & 7 \\ 1 & 1 & 1 \\ 1 & -2 & 3 \end{bmatrix}$$

Solution

The determinant is a scalar value equal to the product of the main diagonal elements minus the product of its counter diagonal elements. The determinant of A

$$A = \begin{vmatrix} 1 & -3 & 7 \\ 1 & 1 & 1 \\ 1 & -2 & 3 \end{vmatrix}$$

Split the matrix as follows:

$$|A| = +1\begin{vmatrix} 1 & 1 \\ -2 & 3 \end{vmatrix} - (-3) + 1\begin{vmatrix} 1 & 1 \\ 1 & 3 \end{vmatrix} + 7\begin{vmatrix} 1 & 1 \\ 1 & -2 \end{vmatrix}$$

$$|A| = 1\big[\big((1)(3) - (-2)(1)\big)\big] - (-3)\big[(1)(3) - (1)(1)\big] + 7\big[(1)(-2) - (1)(1)\big]$$

Simplify

$$|A| = 1(3+2) + 3(3-1) + 7(-2-1) = 5 + 6 - 21 = -10$$

Simulink Solution

Figure 2.4 displays the value of the matrix determinant listed in Example 2.3, solved by Simulink, shown in the embedded MATLAB function.

The embedded MATLAB code is implanted into the MATLAB function to find the determinant of the matrix defined in Example 2.3.

```
function Res = Determinant(A)
  x = ones(3);
  a_det = det(A);

Res = a_det;
```

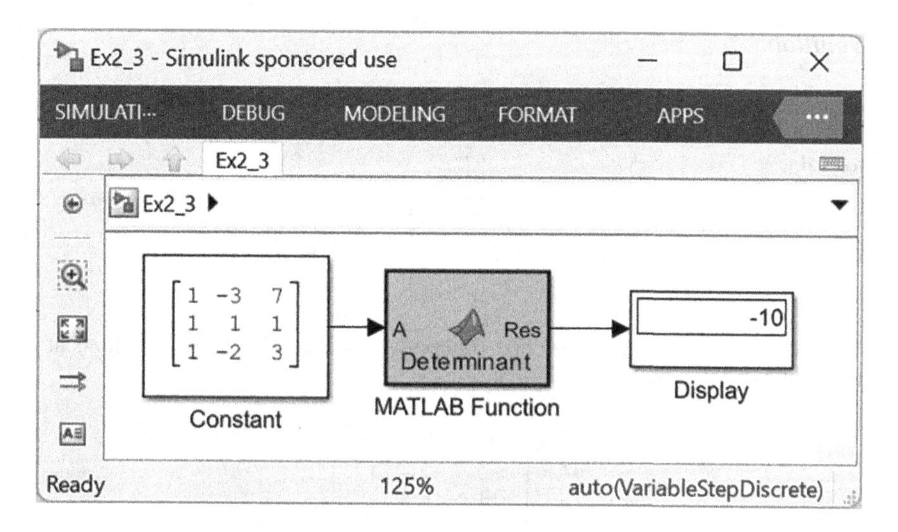

FIGURE 2.4 Simulink's solution using the MATLAB function implanted with MATLAB code to find the determinant of the matrix given in Example 2.3.

Python Solution

The Python code for the determinant of the 3×3 matrix is programmed below. It can solve any 3×3 matrix by modifying the array coefficients.

```
# Example 2.3
#importing Numpy package
import numpy as np
# creating a 3X3 Numpy matrix
n_array = np.array([[1,-3,7],
                    [1,1,1],
                    [1,-2,3]])

# calculating the determinant of matrix
det = np.linalg.det(n_array)

print("\n Determinant of given 3X3 square matrix:",int(det))
```

Execution

```
Determinant of given 3 X 3 square matrix: -10
```

Example 2.4 Cramer's Rule Implementation

Use Cramer's rule to solve the set of the following linear equation manually.

$$x + y - z = 6$$

$$3x - 2y + z = -5$$

$$x + 3y - 2z = 14$$

Confirm the manual calculation with Simulink and Python programming of Cramer's rule.

Solution

Cramer's rule is a method that uses determinants to solve systems of equations that have the same number of equations as variables. First, put the equation in matrix format.

$$D = \begin{bmatrix} 1 & 1 & -1 \\ 3 & -2 & 1 \\ 1 & 3 & -2 \end{bmatrix}, d = \begin{bmatrix} 6 \\ -5 \\ 14 \end{bmatrix}$$

Then, to solve for x, replace the first column with the d column (the constants at the right-hand side of the three equations).

$$D_x = \begin{bmatrix} 6 & 1 & -1 \\ -5 & -2 & 1 \\ 14 & 3 & -2 \end{bmatrix}$$

Replace the second column with the d column.

$$D_y = \begin{bmatrix} 1 & 6 & -1 \\ 3 & -5 & 1 \\ 1 & 14 & -2 \end{bmatrix}$$

Replace the third column with the d column.

$$D_z = \begin{bmatrix} 1 & 1 & 6 \\ 3 & -2 & -5 \\ 1 & 3 & 14 \end{bmatrix}$$

Find the determinant of the matrix D.

$$|D| = \begin{vmatrix} 1 & 1 & -1 \\ 3 & -2 & 1 \\ 1 & 3 & -2 \end{vmatrix}$$

Split the matrix.

$$|D| = +(1)\begin{vmatrix} -2 & 1 \\ 3 & -2 \end{vmatrix} - (1)\begin{vmatrix} 3 & 1 \\ 1 & -2 \end{vmatrix} + (-1)\begin{vmatrix} 3 & -2 \\ 1 & 3 \end{vmatrix}$$

$$|D| = 1 + 7 - 11 = -3$$

Augment the matrix with the first two columns and then follow the formula.

$$D_x = \begin{vmatrix} 6 & 1 & -1 \\ -5 & -2 & 1 \\ 14 & 3 & -2 \end{vmatrix}$$

Solving for D_x

$$D_x = 6[(-2)(-2)-(3)(1)]-1[(-5)(-2)-(14)(1)]-1[(-5)(3)-(14)(-2)]$$

$$D_x = 6+4-13 = -3$$

Augment the matrix with the first two columns and then follow the formula.

$$D_y = \begin{vmatrix} 1 & 6 & -1 \\ 3 & -5 & 1 \\ 1 & 14 & -2 \end{vmatrix}$$

Solving for D_y

$$D_y = 1[(-5)(-2)-(14)(1)]-6[(3)(-2)-(1)(1)]-1[(3)(14)-(1)(-5)]$$

$$= -4+42-47 = -9$$

Solving for D_z

$$D_z = \begin{vmatrix} 1 & 1 & 6 \\ 3 & -2 & -5 \\ 1 & 3 & 14 \end{vmatrix}$$

Augment the matrix with the first two columns and then follow the formula.

$$|D_z| = \begin{vmatrix} 1 & 1 & 6 \\ 3 & -2 & -5 \\ 1 & 3 & 14 \end{vmatrix}$$

Using Cramer's Rule, similarly

$$|D_z| = 6$$

Solve for x.

$$x = \frac{D_x}{D} = \frac{-3}{-3} = 1$$

Solve for y.

$$y = \frac{D_y}{D} = \frac{-9}{-3} = 3$$

Solve for z.

$$z = \frac{D_z}{D} = \frac{6}{-3} = -2$$

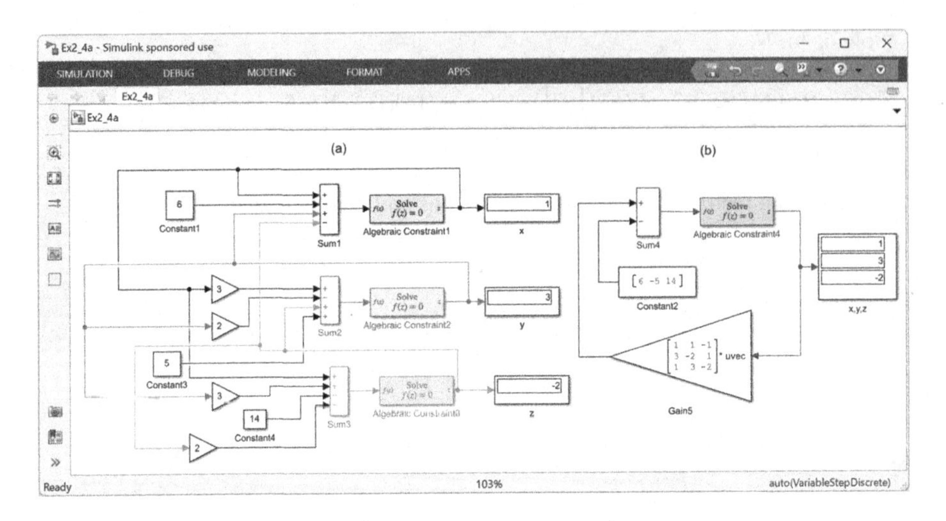

FIGURE 2.5 Simulink block diagram (a) detailed solution (b) using the gain and constants in matrix format to solve the set of equations presented in Example 2.4.

Simulink Solution

We are solving the set of linear equations using Simulink and the Algebraic constraint's function built in the Simulink library. The problem can be solved in two ways: The first method is constructing each algebraic equation using the basic Simulink block diagram using the primary Algebraic constraint function (Figure 2.5a). The second method puts the matrix coefficients in the Gain and the constants of the algebraic equation in the Simulink Constant block, as shown in Figure 2.5b.

Simulink Using Cramer's Rule

In linear algebra, Cramer's Rule is an explicit formula for solving a system of linear equations with as many equations as unknowns. The rule is valid whenever the system has a unique solution. It expresses the solution in terms of the coefficient matrix's determinants and matrices obtained from it by replacing one column with the vector of the right-hand sides of the equations. Figure 2.6 shows the Simulink solution of Example 2.4 using Cramer's Rule. The figure is followed by the MATLAB code that embeds the Simulink function.

```
function Res = Cramer(A,B)
x = ones(3,1);
a_det = det(A);
for i = 1:3
    C = A;
    C(:,i) = B;
    x(i,1) = det(C)/a_det;
end
Res = x';
```

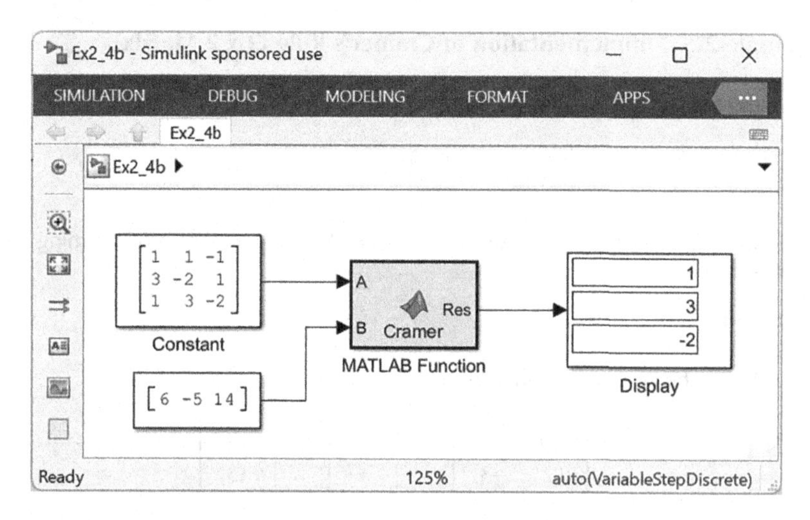

FIGURE 2.6 Simulink block diagram using Cramer's rule encoded in MATLAB and implanted in a MATLAB function indicating the solution to the set of equations defined in Example 2.4.

Python Solution

The following Python code determines the solution of the three linear algebraic equations presented in Example 2.4, followed by the execution results. The program can be used to solve any valid system of three algebraic equations by replacing the matric coefficients (a).

```
# Example 2.4
import numpy as np
a = np.array([[1,1,-1],
              [3,-2,1],
              [1,3,-2]])

b = np.array([6,-5,14])

def cramer(mat,constant):
D = np.linalg.det(mat)
mat1 = np.array([constant,mat[:,1],mat[:,2]])
mat2 = np.array([mat[:,0],constant,mat[:,2]])
mat3 = np.array([mat[:,0],mat[:,1],constant])
Dx = np.linalg.det([mat1,mat2,mat3])
X = Dx/D
print('\n The result,x,y,z')
print(X)

cramer(a,b)

Execution result
The result,x,y,z
[ 1.3.-2.]
```

Example 2.5 Implementation of Cramer's Rule (2 x 2 Matrix)

Solve the following 2×2 system using Cramer's rule manually.

$$12x+3y=15$$

$$2x-3y=13$$

Confirm the manual calculation with Simulink and Python programming of Cramer's rule.

Solution

1. Write the system in matrix form $AX = B$.

$$A=\begin{bmatrix} 12 & 3 \\ 2 & -3 \end{bmatrix}, X=\begin{bmatrix} x \\ y \end{bmatrix}, B=\begin{bmatrix} 15 \\ 13 \end{bmatrix}$$

2. Find D, which is the determinant of A. Also, find the determinants D_x and D_y were:
   D_x = *det* (*A*), where the first column is replaced with *B*
   D_y = *det* (*A*),where the second column is replaced with *B*
3. Find the values of the variables x and y by dividing each of D_x and D_y by D, respectively.

$$x=\frac{D_x}{D}=\frac{\begin{vmatrix} 15 & 3 \\ 13 & -3 \end{vmatrix}}{\begin{vmatrix} 12 & 3 \\ 2 & -3 \end{vmatrix}}$$

$$x=\frac{D_x}{D}=\frac{-45-39}{-36-6}=2$$

4. To solve for *y*, replace the *y* column with the constants (*B*).

$$y=\frac{D_x}{D}=\frac{\begin{vmatrix} 12 & 15 \\ 2 & 13 \end{vmatrix}}{\begin{vmatrix} 12 & 3 \\ 2 & -3 \end{vmatrix}}$$

$$y=\frac{D_x}{D}=\frac{156-30}{-36-6}=-3$$

Simulink Solution

Figure 2.7 describes the Simulink solution of the two linear algebraic equations defined in Example 2.5. The MATLAB embedded code follows the figure.

```
function Res = Cramer(A,B)
x = ones(2,1);
a_det = det(A);
for i = 1:2
```

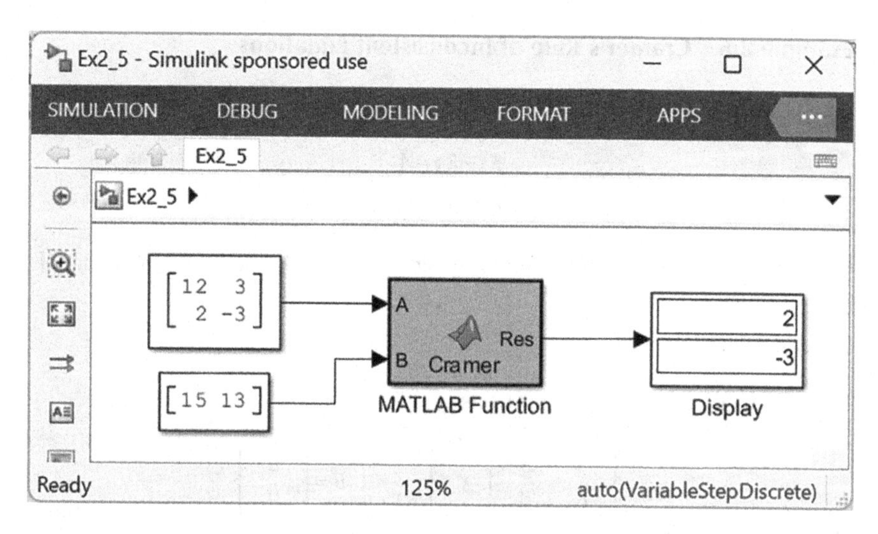

FIGURE 2.7 Simulink block diagram using Cramer's rule encoded in MATLAB and implanted in a MATLAB function representing the solution to the equations defined in Example 2.5.

```
        C = A;
        C(:,i) = B;
        x(i,1) = det(C)/a_det;
    end
    Res = x';
```

Python Solution

The following programming is the Python code, followed by the execution result that determines the solution of the two linear algebraic equations defined in Example 2.5.

```
# Example 2.5
# Implementation f Cramer's rule (2x2)
a = 12
b = 3
c = 2
d = -3
# Right-hand side
e = 15
f = 13
#12x+3y=15
#2x-3y=13

if (a*d - b*c == 0):
print("The equation has no solution")
else:
    x = (e*d-b*f)/(a*d-b*c)
    y = (a*f-e*c)/(a*d-b*c)
    print('\n The result')

    print ("\n x=%s" % x, "y=%s" % y)
The result
x=2.0 y=-3.0
```

Example 2.6 Cramer's Rule of Inconsistent Equations

Using Cramer's rule, solve the following set of equations manually.

$$3x - 2y = 4$$

$$6x - 4y = 0$$

Confirm the manual calculations with Simulink and Python programming of Cramer's rule.

Solution

Write the system in matrix form $AX = B$.

$$A = \begin{bmatrix} 3 & -2 \\ 6 & -4 \end{bmatrix}, X = \begin{bmatrix} x \\ y \end{bmatrix}, B = \begin{bmatrix} 4 \\ 0 \end{bmatrix}$$

The determinant D

$$|D| = \begin{vmatrix} 3 & -2 \\ 6 & -4 \end{vmatrix} = -12 + 12 = 0$$

A determinant of zero means the system has no solution.

Simulink Solution

The Simulink reveals that there is no available solution for the matrix defined in Example 2.6 because the determinant is zero (Figure 2.8).

The following program is the MATLAB code implanted in the Simulink MATLAB function.

```
function Res = fcn(A,B)
x = ones(2,1);
a_det = det(A);
for i = 1:2
    C = A;
    C(:,i) = B;
    x(i,1) = det(C)/a_det;
end
Res = x';
```

Python Solution

The following program is the Python code to find the determinant required in Example 2.6.

```
# Example 2.6
#importing Numpy package
import numpy as np
# creating a 3X3 Numpy matrix
```

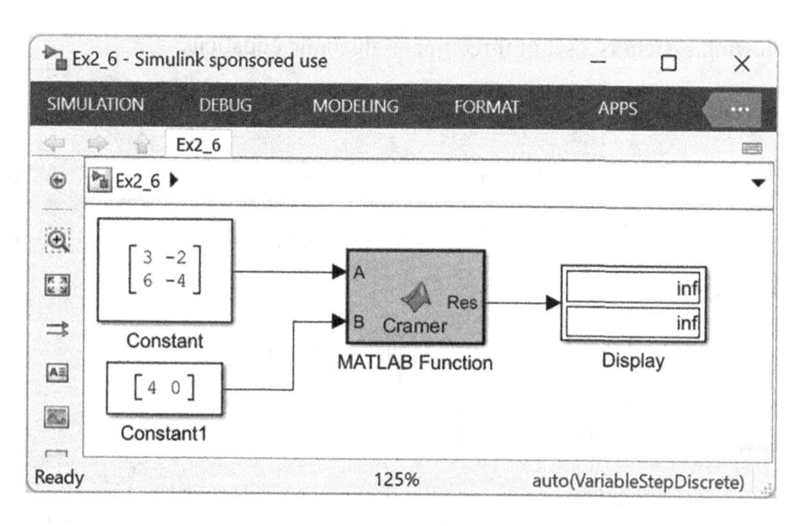

FIGURE 2.8 Simulink solution using Cramer's rule encoded in MATLAB and implanted in a Simulink MATLAB function representing the solution to the linear equations defined in Example 2.6.

```
n_array = np.array([[3,-2],
                    [6,-4]])

# calculating the determinant of matrix
det = np.linalg.det(n_array)
print("\n Determinant of given 3X3 square matrix:",int(det))
```

The Python program executed the result.
Determinant of given 3×3 square matrix: 0

2.3 GAUSS ELIMINATION METHOD

Gauss elimination transforms a system of linear equations into a matrix in the form of an upper triangular matrix, making solving for all variables easy. The steps to performing at Gauss elimination are the following:

1. Create the augmented matrix.
2. Turn every element in the first column to zero except the top one in row 1.
3. Repeat the second step until only one nonzero variable is left in each row.
4. Use backward substitution to solve for all the unknowns.

Gauss elimination is essentially a way to reduce an augmented matrix into an upper triangular matrix to make solving systems of linear equations much more accessible.

The Gauss elimination method has been widely used to solve linear algebraic equations. This method is known as the row reduction algorithm for solving linear equations systems, and it consists of a sequence of operations performed on the corresponding matrix coefficients.

$$Ax = b$$

The following system is a set of three linear algebraic equations:

$$a_{11}x_1 + a_{12}x_2 + a_{13}\ x_3 = b_1 \tag{2.1}$$

$$a_{21}x_1 + a_{22}x_2 + a_{23}\ x_3 = b_2 \tag{2.2}$$

$$a_{31}x_1 + a_{32}x_2 + a_{33}\ x_3 = b_3 \tag{2.3}$$

In matrix form,

$$\begin{bmatrix} a_{11} & a_{12} & a_{13} \\ a_{21} & a_{22} & a_{23} \\ a_{31} & a_{32} & a_{33} \end{bmatrix} \begin{Bmatrix} x_1 \\ x_2 \\ x_3 \end{Bmatrix} = \begin{Bmatrix} b_1 \\ b_2 \\ b_3 \end{Bmatrix}$$

Equation (2.1) is named pivot row, and the first nonzero element (a_{11}) is called a pivotal element. Subsequent operations are performed around the pivot row and pivotal element as the following steps:

Set the equation in matrix form, $AX = B$.

$$\begin{bmatrix} a_{11} & a_{12} & a_{13} \\ 0 & a'_{22} & a'_{23} \\ 0 & a'_{32} & a'_{33} \end{bmatrix} \begin{Bmatrix} x_1 \\ x_2 \\ x_3 \end{Bmatrix} = \begin{Bmatrix} b_1 \\ b'_2 \\ b'_3 \end{Bmatrix}$$

The first equation is divided by $a_{11}(a_{11} \neq 0)$, multiplied by a_{21}, and subtracted from the second equation to yield the modified equation (2.2).

$$a'_{22} = a_{22} - \left(\frac{a_{21}}{a_{11}}\right) \times a_{12}$$

$$a'_{23} = a_{23} - \left(\frac{a_{21}}{a_{11}}\right) \times a_{13}$$

$$b'_2 = b_2 - \left(\frac{a_{21}}{a_{11}}\right) \times b_1$$

Similarly, the first equation is divided by a_{11}, multiplied by a_{31}, and subtracted from the third equation to yield the modified equation (2.3).

$$a'_{32} = a_{32} - \left(\frac{a_{31}}{a_{11}}\right) \times a_{12}$$

$$a'_{33} = a_{33} - \left(\frac{a_{31}}{a_{11}}\right) \times a_{13}$$

$$b'_3 = b_3 - \left(\frac{a_{31}}{a_{11}}\right) \times b_1$$

The new row 2 equals row 2 – row 1 multiplied by (a_{21}/a_{11}), and the new row 3 equals row 3 – row 1 (a_{31}/a_{11}).

Finally, an upper triangular matrix is obtained. The three modified algebraic equations are as follows:

$$a_{11}x_1 + a_{12}x_2 + a_{13}x_3 = b_1$$

$$a'_{22}x_2 + a'_{23}x_3 = b'_2$$

$$a''_{33}x_3 = b''_{33}$$

This is a new set of three linearly independent algebraic equations to solve. Reverse substitution to determine x,

$$x_3 = \frac{b''_3}{a''_{33}}$$

$$x_2 = \frac{b'_2 - a'_{23}x_3}{a'_{22}}$$

$$x_1 = \frac{b_1 - a_{12}x_2 - a_{13}x_3}{a_{11}}$$

In general, for the i variable of n equations

$$x_i = \frac{b_i^{i-1} - \sum_{j=i+1}^{n} a_{ij}^{i-1} x_j}{a_{ii}^{i-1}}$$

Many equations of more than three require a program, such as Simulink or MATLAB programming, to solve.

Example 2.7 Implementation of Gauss Elimination Method

Calculate the determinant of the following matrix:

$$A = \begin{bmatrix} a_{11} & a_{12} \\ a_{21} & a_{22} \end{bmatrix}$$

Solution

The determinant of A is calculated as follows:

$$\det A = (a_{11} \times a_{22}) - (a_{12} \times a_{21})$$

Apply the Gauss elimination method.

$$A' = \begin{bmatrix} a_{11} & a_{12} \\ a_{21} & a'_{22} \end{bmatrix}$$

$$a'_{22} = \left(a_{22} - \left(\frac{a_{21}}{a_{11}} \right) \times a_{12} \right)$$

The determinant of the modified matrix A' =

$$\det A' = (a_{11} \times a'_{22}) - (a_{12} \times a_{21})$$

The value of the determinant is not changed by the forward elimination step of the Gauss elimination, and this must be true because the forward step only modifies the equations.

$$\begin{bmatrix} a_{11} & a_{12} & a_{13} \\ a_{21} & a_{22} & a_{23} \\ a_{31} & a_{32} & a_{33} \end{bmatrix} \rightarrow \text{Gauss elimination} \rightarrow \begin{bmatrix} a_{11} & a_{12} & a_{13} \\ 0 & a'_{22} & a'_{23} \\ 0 & 0 & a'_{33} \end{bmatrix}$$

$$\det[A] = \det[A'\} = a_{11}a'_{22}a''_{33}$$

Example 2.8 Implementation of Gauss elimination Method

Solve the following set of linear algebraic equations utilizing the Gauss elimination method manually.

$$2x + y + z = 2$$

$$2x + 2y + 3z = 5$$

$$2x + 3y + 4z = 11$$

Validate the manual calculations using Simulink and Python programming of the Gauss elimination method.

Solution

First, write these equations in matrix form $AX = B$.

$$\begin{bmatrix} a_{11} & a_{12} & a_{13} \\ a_{21} & a_{22} & a_{23} \\ a_{31} & a_{32} & a_{33} \end{bmatrix} \begin{Bmatrix} x \\ y \\ z \end{Bmatrix} = \begin{Bmatrix} b_1 \\ b_2 \\ b_3 \end{Bmatrix} \rightarrow \begin{bmatrix} 2 & 1 & 1 \\ 2 & 2 & 3 \\ 2 & 3 & 4 \end{bmatrix} \begin{Bmatrix} x \\ y \\ z \end{Bmatrix} = \begin{Bmatrix} 2 \\ 5 \\ 11 \end{Bmatrix}$$

Solving for the new matrix,

$$\begin{bmatrix} a_{11} & a_{12} & a_{13} \\ 0 & a'_{22} & a'_{23} \\ 0 & a'_{32} & a'_{33} \end{bmatrix} \begin{Bmatrix} x_1 \\ x_2 \\ x_3 \end{Bmatrix} = \begin{Bmatrix} b_1 \\ b'_2 \\ b'_3 \end{Bmatrix}$$

Solving for a'_{22}, a'_{23}, a'_{32}, a'_{33}, a'_2 , b'_3,

$$a'_{22} = a_{22} - \left(\frac{a_{21}}{a_{11}}\right) \times a_{12} = 2 - \left(\frac{2}{2}\right) \times 1 = 1$$

$$a'_{23} = a_{23} - \left(\frac{a_{21}}{a_{11}}\right) \times a_{13} = 3 - \left(\frac{2}{2}\right) \times 1 = 2$$

$$b'_2 = b_2 - \left(\frac{a_{21}}{a_{11}}\right) \times b_1 = 5 - \left(\frac{2}{2}\right) \times 2 = 3$$

$$a'_{32} = a_{32} - \left(\frac{a_{31}}{a_{11}}\right) \times a_{12} = 3 - \left(\frac{2}{2}\right) \times 1 = 2$$

$$a'_{33} = a_{33} - \left(\frac{a_{31}}{a_{11}}\right) \times a_{13} = 4 - \left(\frac{2}{2}\right) \times 1 = 3$$

$$b'_3 = b_3 - \left(\frac{a_{31}}{a_{11}}\right) \times b_1 = 11 - \left(\frac{2}{2}\right) \times 2 = 9$$

The new matrix

$$\begin{bmatrix} a_{11} & a_{12} & a_{13} \\ 0 & a'_{22} & a'_{23} \\ 0 & a'_{32} & a'_{33} \end{bmatrix} \begin{Bmatrix} x_1 \\ x_2 \\ x_3 \end{Bmatrix} = \begin{Bmatrix} b_1 \\ b'_2 \\ b'_3 \end{Bmatrix} \rightarrow \begin{bmatrix} 2 & 1 & 1 \\ 0 & 1 & 2 \\ 0 & 2 & 3 \end{bmatrix} \begin{Bmatrix} x_1 \\ x_2 \\ x_3 \end{Bmatrix} = \begin{Bmatrix} 2 \\ 3 \\ 9 \end{Bmatrix}$$

The second step is to follow a similar procedure. In this case, the second row becomes the pivotal equation and a'_{22} abecomes pivotal elements.

$$\begin{bmatrix} a_{11} & a_{12} & a_{13} \\ 0 & a'_{22} & a'_{23} \\ 0 & 0 & a''_{33} \end{bmatrix} \begin{Bmatrix} x_1 \\ x_2 \\ x_3 \end{Bmatrix} = \begin{Bmatrix} b_1 \\ b'_2 \\ b''_3 \end{Bmatrix}$$

where

$$a''_{33} = a'_{33} - \left(\frac{a'_{32}}{a'_{22}}\right) \times a'_{23} = 3 - \left(\frac{2}{1}\right) \times 2 = -1$$

$$b''_3 = b'_3 - \left(\frac{a'_{32}}{a'_{22}}\right) \times b'_{22} = 9 - \left(\frac{2}{1}\right) \times 3 = 3$$

Substituting calculated value,

$$\begin{bmatrix} 2 & 1 & 1 \\ 0 & 1 & 2 \\ 0 & 0 & -1 \end{bmatrix} \begin{Bmatrix} x_1 \\ x_2 \\ x_3 \end{Bmatrix} = \begin{Bmatrix} 2 \\ 3 \\ 3 \end{Bmatrix}$$

Reverse substitution to determine x:

From the last row,

$$x_3 = \frac{b_3''}{a_{33}''} = \frac{3}{-1} = -3$$

From the second last row,

$$x_2 = \frac{b_2' - a_{33}' x_3}{a_{22}'} = \frac{3-(3)(-3)}{1} = 9$$

From the first row,

$$x_1 = \frac{b_1 - a_{12}x_2 - a_{13}x_3}{a_{11}} = \frac{(2-1\times 9-1\times(-3))}{2} = -2$$

Figure 2.9 shows the solution of the set of linear algebraic equations using the 'Algebraic constraints' block available in the Simulink Math functions library. Note that when using the Gain to enter the matrix coefficient [2 1 1;2 2 3;2 3 4], do not forget to select the 'Matrix (*K***u*) (*u* vector)' from the pulldown menu near

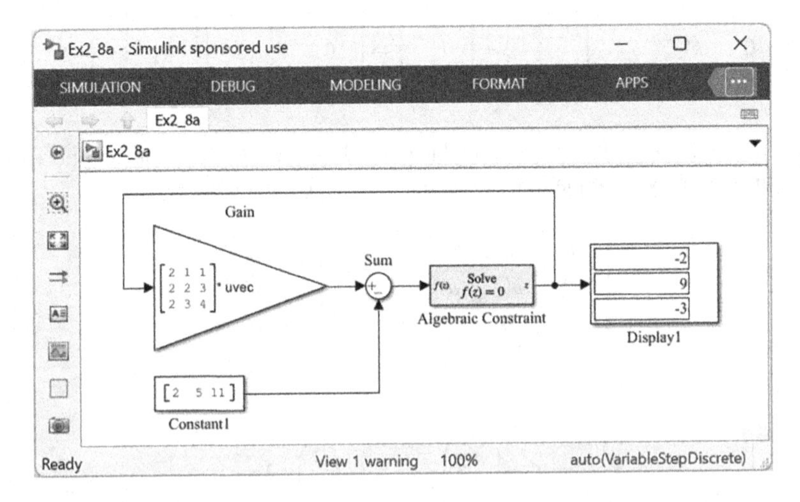

FIGURE 2.9 Simulation calculation using the Simulink algebraic constraint block for the matrix given in Example 2.8.

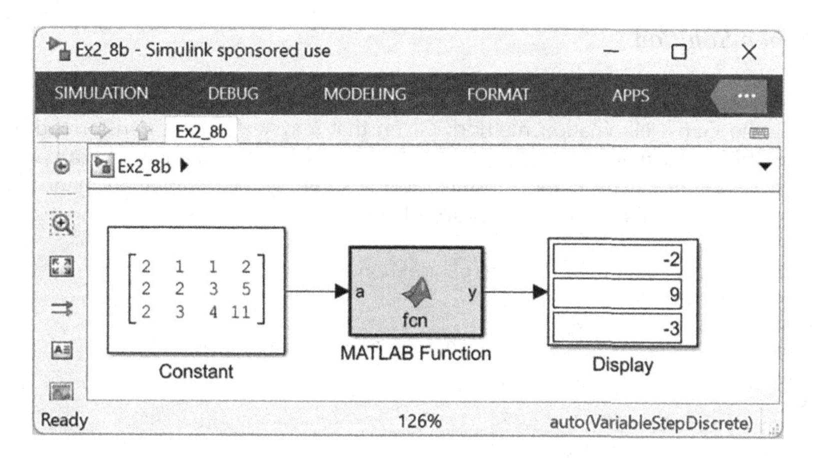

FIGURE 2.10 Simulink solution using the Gauss elimination method coded in MATLAB and implanted in the Simulink MATLAB function of equations specified in Example 2.8.

multiplication. An alternative way is to use the MATLAB function, where the Gauss elimination method is programmed in the MATLAB file implanted in the MATLAB function, followed by Figure 2.10. The Simulink results are the same as those obtained using the Gauss elimination method.

The following program is the embedded MATLAB code utilizing the Gauss elimination methods of the equations defined in Example 2.8.

```
function y = fcn(a)
% Code from "Gauss elimination and Gauss Jordan methods using
MATLAB."
%Gauss elimination method [m,n)=size(a);[m,n]=size(a);
for j=1:m-1
    for z=2:m
        if a(j,j)==0
            t=a(j,:);a(j,:)=a(z,:);a(z,:)=t;
        end
    end
    for i=j+1:m
        a(i,:)=a(i,:)-a(j,:)*(a(i,j)/a(j,j));
end
end
x=zeros(1,m);
for s=m:-1:1
    c=0;
    for k=2:m
        c=c+a(s,k)*x(k);
    end
    x(s)=(a(s,n)-c)/a(s,s);
end
%display the results of the gauss elimination method;
y = x';
```

Python Solution

The following Python program solves systems of linear equations with *n* unknowns using the Gauss elimination method. Given that a system is first transformed to upper triangular matrix row operations, then the solution is obtained by a backward substitution. The Python results are the same as those obtained manually using the Gauss elimination method to solve the equation defined in Example 2.8.

```
#Example 2.8,Gauss elimination method
import numpy as np
# #initial coefficients
a=np.array([[2.0,1.0,1.0],[2.0,2.0,3.0],[2.0,3.0,4.0]])
b=np.array([2.0,5.0,11.0])
def gaussElim(a,b):
    n = len(b)
    # Elimination phase
    for k in range(0,n-1):
        for i in range(k+1,n):
            if a[i,k]!= 0.0:
               #if not null, define λ
               lam = a [i,k]/a[k,k]
               #we calculate the new row of the matrix
               a[i,k+1:n] = a[i,k+1:n]-lam*a[k,k+1:n]
               #we update vector b
               b[i] = b[i]-lam*b[k]
               # backward substitution
    for k in range(n-1,-1,-1):
        b[k] = (b[k]-np.dot(a[k,k+1:n],b[k+1:n]))/a[k,k]
    return b
x = gaussElim(a,b)
#print the result
print('\n Result')
print("x =\n",x)
```

The Python program execution result:

```
Result
x = [-2.9.-3.]
```

2.4 GAUSS-JORDAN ELIMINATION

The Gauss-Jordan elimination method is an algorithm used to solve linear equations systems and find the inverse of any inverted matrix. It is based on three basic class operations that can be used in an array:

- Switch the positions of two rows.
- Multiply a row by a nonzero number.
- Adding or subtracting the standard multiple of one line to another line.

Example 2.9 Implementation of Gauss-Jordan Elimination

Solve the following set of linear algebraic equations manually using the Gauss-Jordan elimination methods.

$$x + y - z = 7$$

$$x - y + 2z = 3$$

$$2x + y + z = 9$$

Verify the manual calculation using the Simulink and Python programming of the Gauss-Jordan elimination method.

Solution

Arrange the equations into matrix form

$$\left[\begin{array}{ccc|c} 1 & 1 & -1 & 7 \\ 1 & -1 & 2 & 3 \\ 2 & 1 & 1 & 9 \end{array}\right]$$

Subtract row 2 from row 1 $(R_1 - R_2)$

$$R_1 - R_2$$

$$1 - 1 = 0$$

$$1 - (-1) = 2$$

$$-1 - (2) = -3$$

$$7 - 3 = 4$$

Replace the second row with 0, 2, –3, and 4.

$$\left[\begin{array}{ccc|c} 1 & 1 & -1 & 7 \\ 0 & 2 & -3 & 4 \\ 2 & 1 & 1 & 9 \end{array}\right]$$

Multiply the row 1 by –2 and add to the third row $(-2R_1 + R_3)$

$$\left[\begin{array}{ccc|c} 1 & 1 & -1 & 7 \\ 0 & 2 & -3 & 4 \\ 0 & -1 & 3 & -5 \end{array}\right]$$

Add the second row to two times the third row ($R_2 + 2R_3$)

$$R_2 + 2R_3$$

$$0 + 2(0) = 0$$

$$2 + 2(-1) = 0$$

$$-3 + 2(3) = 3$$

$$4 + 2(-5) = -6$$

The final matrix is

$$\begin{bmatrix} 1 & 1 & -1 & & 7 \\ 0 & 2 & -3 & | & 4 \\ 0 & 0 & 3 & & -6 \end{bmatrix}$$

Multiply row 1 by 3 minus row 2 ($3R_1 - R_2$)

$$\begin{bmatrix} 3 & 1 & 0 & & 17 \\ 0 & 2 & -3 & | & 4 \\ 0 & 0 & 3 & & -6 \end{bmatrix}$$

Add row 2 and row 3 ($R_2 + R_3$)

$$\begin{bmatrix} 3 & 1 & 0 & & 17 \\ 0 & 2 & 0 & | & -2 \\ 0 & 0 & 3 & & -6 \end{bmatrix}$$

Add two times row 1 minus row 2 ($2R_1 - R_2$)

$$\begin{bmatrix} 6 & 0 & 0 & & 36 \\ 0 & 2 & 0 & | & -2 \\ 0 & 0 & 3 & & -6 \end{bmatrix}$$

Multiply the first row by $1/6$, the second row by $1/2$, and the third row by $1/3$:

$$\begin{bmatrix} 1 & 0 & 0 & & 6 \\ 0 & 1 & 0 & | & -1 \\ 0 & 0 & 1 & & -2 \end{bmatrix}$$

Accordingly,

$$x = 6,\ y = -1,\ z = -2$$

Simulink Solution

The Simulink MATLAB function will return a column vector representing the solution to each variable in the order of appearance in the given matrix (Figure 2.11).

The following program is a MATLAB code associated with MATLAB function programs Gauss-Jordan method to solve the matrix indicated in Example 2.9.

```
function Ans = gauss_jordan (x)
   for n = 1:(length(x)-1)
       % Step 1: make the row N's Nth term 1 by dividing
       % the whole row by it
       A = x(n,:);
       A = A/A(n);
       x(n,:) = A;
       % Step 2: for every other row, add to it -1 * that rows Nth
       term *
       % the Nth row
       for k = 1:(length(x)-1)
           if n~=k
               x(k,:) = A*(-1*x(k,n))+x(k,:);
           end
       end
   end

   y = x(:,length(x))';
Ans = y;
```

Python Solution

Gauss-Jordan elimination method is a structured method of solving a system of linear equations. Thus, it is an algorithm that can easily be programmed in Python to solve linear equations. This function will take a matrix designed to be used by the Gauss-Jordan algorithm and solve it, returning a transposed version of the last

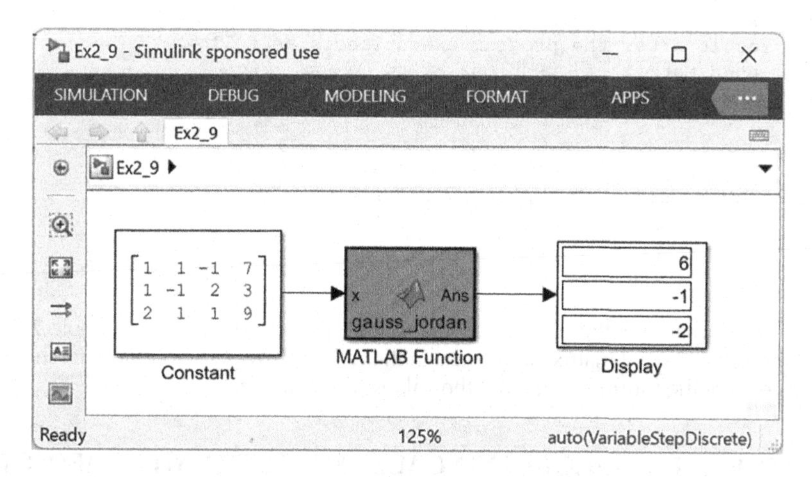

FIGURE 2.11 Simulink's solution uses a Gauss-Jordan elimination method encoded in the MATLAB and implanted in the Simulink MATLAB function to solve the matrix indicated in Example 2.9.

column in the ending matrix representing the solution to the unknown variables. Input: The function takes one matrix of n by $n+1$, where n equals the number of unknown variables. Each row represents the coefficients of the variables in each equation. Furthermore, the last column contains the constants on the right-hand side of each equation.

```
#Example 2.9, Gauss Jordan elimination method
# Matrix 3 x 3
sd = [[1,1,-1,7],[1,-1,2,3],[2,1,1,9]]
print ("\n Augmented matrix of question")
print('=====================================\n')
def Matrix():
for i in sd:
for j in i:
print(j,end="\t\t")
print("\n")
def getone(pp):
for i in range(len(sd[0])):
if sd[pp][pp]!= 1:
q00 = sd[pp][pp]
for j in range(len(sd[0])):
sd[pp][j] = sd[pp][j]/q00
def getzero(r,c):
for i in range(len(sd[0])):
if sd[r][c]!= 0:
q04 = sd[r][c]
for j in range(len(sd[0])):
sd[r][j] = sd[r][j] - ((q04) * sd[c][j])
Matrix()
print('=========== Result ===============\n')
for i in range(len(sd)):
getone(i)
for j in range(len(sd)):
if i!= j:
getzero(j,i)
Matrix()
The result after the program execution is as follows:
Augmented matrix of question
=============================================
1           1           -1          7
1           -1          2           3
2           1           1           9
=================Result====================
1.0         0.0         0.0         6.0
0.0         1.0         0.0         -1.0
0.0         0.0         1.0         -2.0
```

The above matrix is the augmented matrix of equations defined in Example 2.9, whereas the second matrix is the same in the reduced row-echelon form in which elements of the fourth column are the values of x, y, and z.

2.5 THE GAUSS-JACOBI AND GAUSS-SEIDEL ITERATIVE METHODS

The Gauss-Seidel method is an improved version of the Jacobi method, also known as the successive displacement method. The only difference between Jacobi and

Gauss-Seidel methods is that in the Jacobi method, the value of the variables is not modified until the next iteration, while in the Gauss-Seidel method, the values of the variables are modified once a new value is evaluated.

2.5.1 The Gauss-Jacobi Method

The Jacobi method is a method of solving a matrix equation on a matrix that has no zeros along its main diagonal. The Jacobi method is an iterative algorithm for rigorously determining the solutions of a diagonally dominant system of linear equations.

Two assumptions were made on the Jacobi method.

1. The system has a unique solution

$$a_{11}x_1 + a_{12}x_2 + \ldots a_{1n}x_n = b_1$$

$$a_{21}x_1 + a_{22}x_2 + \ldots a_{2n}x_n = b_2$$

$$a_{n1}x_1 + a_{n2}x_2 + \ldots a_{nn}x_n = b_n$$

2. The coefficient matrix A has no zeros on its main diagonal, namely, $a_{11}, a_{22}, \ldots, a_{nn}$, are nonzero. To solve the first equation for x_1, the second equation for x_2, and so on to obtain the rewritten equations

$$x_1 = \frac{1}{a_{11}}(b_1 - a_{12}x_2 - a_{13}x_3 - \ldots a_{1n}x_n)$$

$$x_2 = \frac{1}{a_{22}}(b_2 - a_{21}x_1 - a_{23}x_3 - \ldots a_{2n}x_n)$$

$$x_n = \frac{1}{a_{nn}}(b_n - a_{n1}x_1 - a_{n2}x_2 - \ldots a_{n,n-1}x_{n-1})$$

Make an initial guess of x_1^o, x_2^o, x_n^o. Substitute these values into the right-hand side of the rewritten equation to obtain the first approximation, x_1^1, x_2^1, x_n^1. Substitute the computed approximation, x value, into the right-hand side of the rewritten equations. This accomplished repeated iteration until forming a sequence of approximations.

Example 2.10 Application of the Jacobi Method

Solve the following set of linear algebraic equations using the Jacobi method.

$$5x - 2y + 3z = -1$$

$$-3x + 9y + z = 2$$

$$2x - y - 7z = 3$$

Validate the manual calculations with Simulink and Python programming.

TABLE 2.1
Approximate Solution Using the Jacobi Method of a Set of Equations Defined in Example 2.10

n	0	1	2	3	4	5	6	7
x	0.000	–0.200	0.146	0.192	0.181	0.185	0.186	0.186
y	0.000	0.222	0.203	0.328	0.332	0.329	0.331	0.331
z	0.000	–0.429	–0.517	–0.416	–0.421	–0.424	–0.423	–0.423

Solution

First, put the variables in the following form:

$$x = -\frac{1}{5} + \frac{2}{5}y - \frac{3}{5}z$$

$$y = \frac{2}{9} + \frac{3}{9}x - \frac{1}{9}z$$

$$z = -\frac{3}{7} + \frac{2}{7}x - \frac{1}{7}y$$

Choose the initial guess, $x = y = z = 0$

The first approximation

$$x_1^1 = -0.200$$

$$y^1 = 0.222$$

$$z^1 = -0.429$$

Continue iterations until two successive approximations are identical when rounded to three significant digits (Table 2.1).

Simulink Solution

Jacobi's iterative method is an algorithm for determining the solutions of a diagonally dominant system of linear equations. Each diagonal element is solved, and an approximate value is plugged in. The process is then iterated until it converges (Figure 2.12). The following MATLAB program employs the Gauss-Jacobi method to solve the equation presented in Example 2.10, and the program is embedded into the Simulink MATLAB function block diagram.

```
%Example 2.10
function Res = jacobi(A,b)
Eps=0.001;
%[5 -2 3 ;-3 9 1; 2 -1 -7] equation coefficients
%[-1;2;3] equations right-hand side constants.
```

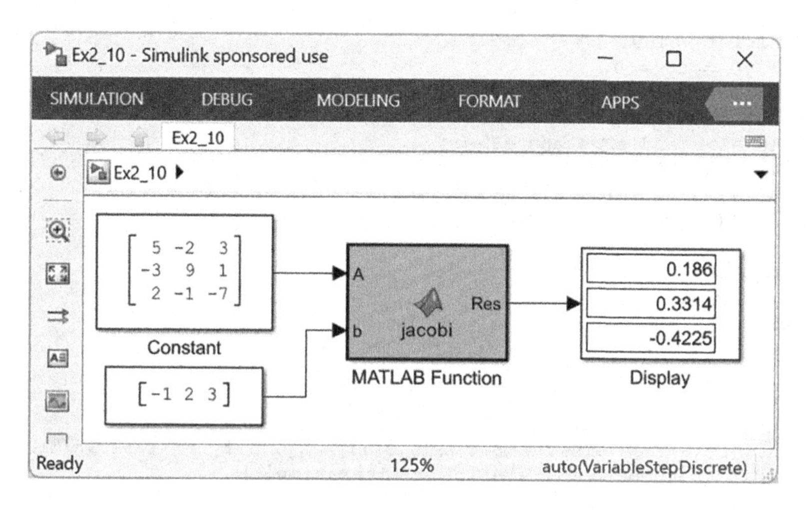

FIGURE 2.12 Simulink simulated block diagram using the Jacobi method coded in MATLAB and inserted in the Simulink MATLAB function to solve the set of algebraic equations given in Example 2.10.

```
n = length(b) ;
x0 = zeros(3,1) ;
x = ones(size(x0)) ;
while norm(x-x0,inf) >= Eps
x0 = x;
for i = 1 : n
x(i) = b(i);
for j = 1 : n
if j ~= i
x(i) = x(i) - A(i, j)*x0(j);
end
end
x(i) = x(i) / A(i, i) ;
end
end
Res =x';
```

Python Solution

The Jacobi method is an iterative matrix used to solve the equation $A\ X = b$ for a known square matrix of size $n \times n$ and known vector or length. The following Python code utilizes the Jacobi method to solve the equations defined in Example 2.10.

```
# Example 2.10
# Jacobi method
f1 = lambda x, y, z : (-1 +2*y - 3*z)/5
f2 = lambda x, y, z : (2 + 3*x - z)/9
f3 = lambda x, y, z : (-3 + 2*x-y)/7
# Tolerance
e = 0.0001
iteration = 0
```

```
# initial values
x0 = y0 = z0 = 0
e1 = e2 = e3 = 1
print("\n n x y z \n")
while e1>e and e2>e and e3>e:
x1 = f1(x0,y0,z0)
y1 = f2(x0,y0,z0)
z1 = f3(x0,y0,z0)
e1 = abs (x0-x1)
e2 = abs (y0-y1)
e3 = abs (z0-z1)
iteration +=1
x0 = x1
y0 = y1
z0 = z1
print(f"{{y1:.3f} {z1:.3f}")
print (f"\n The value of x = {x1:.3f},y= {y1:.3f},z= {z1:.3f}")
print(f"\n number of iteratios = {iteration}")
```

```
Execution result

n        x       y        z
1.0   -0.200   0.222   -0.429
2.0    0.146   0.203   -0.517
3.0    0.192   0.328   -0.416
4.0    0.181   0.332   -0.421
5.0    0.185   0.329   -0.424
6.0    0.186   0.331   -0.423
7.0    0.186   0.331   -0.423

The value of x = 0.186, y= 0.331, z= -0.423
number of iterations = 7
```

2.5.2 Gauss-Seidel Iterative Methods

The Gauss-Seidel method is an iterative numerical method that can aid us in solving linear systems of equations. This method works for linear systems of equations and guarantees convergence if our matrix is diagonally dominant, just like the Jacobi iteration. The steps to solving a system of linear equations using the Gaussian method are as follows:

1. Guess every unknown value in the matrix. Assume a value for as many variables as x, y, and z. Set the initial iteration to 0, as many programming languages like Python have zero indexing, meaning we begin counting from 0. However, if using MATLAB, begin at 1, as it does not have zero indexing. A typical starting guess is just a zero matrix unless we have additional information about the problem that could help our guesses. The benefit of having a better starting guess is that we achieve our final solution in fewer iterations and, therefore, less time.
2. Rewrite all of our independent equations and have them all solve for an individual variable.
3. Once we have rewritten all of our system of linear equations, it is often wise at this point to create a table to summarize all of the relevant information.

Now we are ready to solve for iteration 1 user's latest guesses of y and z year to help us solve for x at iteration 1. Then we must solve for y and use our latest guesses for x and z. Note that this is where the significant difference between the Jacobi iteration method and the Gauss-Seidel method is in the Jacobi method; we would not use the variable values in the same iteration on one another. However, in the Gauss-Seidel method, we do, which almost always leads to finding a solution quicker if there is one. Last, move on to solving z for iteration using x and y that we found in the first iteration.

4. After finding all the variable values for iteration 1, move forward to iteration 2 and solve for x using the y and z values found in iteration 1.
5. Continue iterations until there is an acceptable amount of error. For coding purposes, verify that all equations' right sides are within the error tolerance.

The Gauss-Seidel method is used to solve linear system equations. This method modifies the Gauss-iteration method and the Jacobi method. This modification reduces the number of iterations. The computed value replaces the previous value only at the end of the iteration. Because of this, Gauss-Seidel methods converge much faster than Gauss methods. The number of iteration methods required to get the solution is much less than the Gauss method in Gauss-Seidel methods. The Gauss-Seidel equation

$$x_i = \frac{1}{a_{ii}}(b_i - \sum_{j=1, j\neq i}^{n} a_{ij}x_j)$$

Example 2.11 Application of Gauss-Seidel Method

Use the approximate solution utilizing the Gauss-Seidel iteration method to solve the following set of three linear algebraic equations.

$$5x - 2y + 3z = -1$$

$$-3x + 9y + z = 2$$

$$2x - y - 7z = 3$$

Authenticate the manual calculations using Simulink and Python programming of the Gauss-Seidel iteration method.

Solution

Put the equations in the following form:

$$x = \frac{(-1 + 2y - 3z)}{5}$$

$$y = \frac{(2 + 3x - z)}{9}$$

$$z = \frac{(-3 + 2x - y)}{7}$$

TABLE 2.2
Approximate Solution Using the Gauss-Seidel Method of the Set of Equations Defined in Example 2.11

n	0	1	2	3	4	5
x	0.000	–0.200	0.167	0.191	0.186	0.186
y	0.000	0.156	0.334	0.333	0.331	0.331
z	0.000	–0.508	–0.429	–0.422	–0.423	–0.423

The first calculation is the same as the Jacobi iteration method, as a first initial guess.

$$x = y = z = 0$$

The value of x_1 is based on the initial guess of y and z

$$x_1 = -\frac{1}{5} + \frac{2}{5}(0) - \frac{3}{5}(0) = -0.200$$

The value of y, is obtained by substituting the new value of $x_1 = -0.200$, in the commutation of y

$$y = \frac{2}{9} + \frac{3}{9}(-0.200) - \frac{1}{9}(0) = 0.156$$

In the computation of z, substitute the new values of x, y

$$z = -\frac{3}{7} + \frac{3}{9}(-0.200) - \frac{1}{7}(0.156) = -0.508$$

Continue the iterations until successive values are identical to the three decimal digits. The sequence of approximations is shown in Table 2.2. The solution converges in less number of iterations.

Simulink Solution

The Simulink block diagram using the Simulink MATLAB function is presented in Figure 2.13, followed by the MATLAB code.

The following MATLAB program is the MATLAB code embedded in the Simulink MATLAB function utilizing the Gauss-Seidel iteration method of the set of equations presented in Figure 2.13.

```
% Example 2.11.
% Gauss-Seidel iteration method
function Res =gauss(A,b)
%This function solves the system of algebraic equations using the
%Gaussian method
```

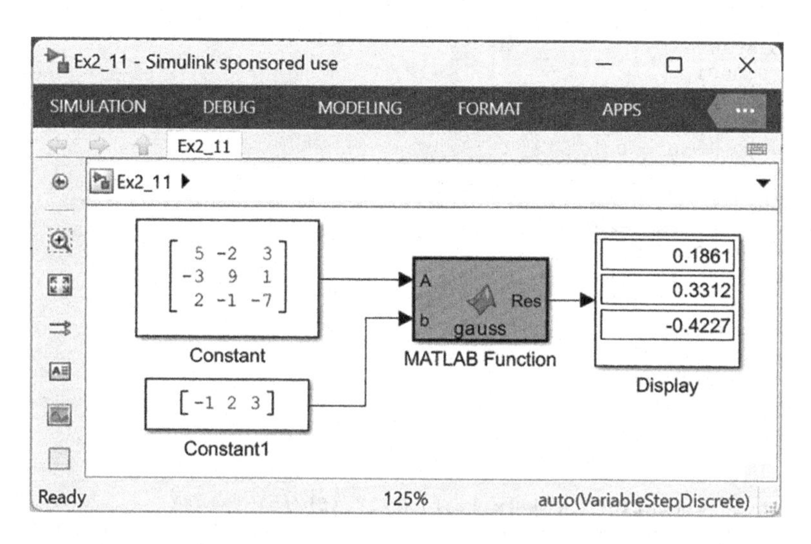

FIGURE 2.13 A simulated block diagram using the MATLAB-encoded Gauss-Seidel method and implanted in the Simulink MATLAB function to solve the set of linear algebraic equations defined in Example 2.11.

```
%Ax=b
%Input: matrix A,column vector b
%Output vector x
n=size(A,1);
b=b(:);%make sure b is a column vector.
nb=n+1;
Ab=[A b];%extended matrix
%Forward process
for i=1:n-1
for j=i+1:n
Ab(j,i:nb)=Ab(j,i:nb)-Ab(j,i)*Ab(i,i:nb)/Ab(i,i);
end
end
%Inverse process
x=zeros(n,1);
x(n)=Ab(n,nb)/Ab(n,n);
for i=n-1:-1:1
x(i)=(Ab(i,nb)-Ab(i,i+1:n)*x(i+1:n))/Ab(i,i);
end
Res = x;
```

Python Solution

The following program is the Python code utilizing the Gauss-Seidel method to solve the equations defined in Example 2.11.

```
# Example 2.11
# Guass Seidel method
f1 = lambda x,y,z:(-1 +2*y - 3*z)/5
f2 = lambda x,y,z:(2 + 3*x - z)/9
f3 = lambda x,y,z:(-3 + 2*x-y)/7
```

```
# Tolerance
e = 0.0001
iteration = 0
# initial values
x0 = y0 = z0 = 0
e1 = e2 = e3 = 1
print("\n n x y z \n")
while e1>e and e2>e and e3>e:
x1 = f1(x0,y0,z0)
y1 = f2(x1,y0,z0)
z1 = f3(x1,y1,z0)
e1 = abs (x0-x1)
e2 = abs (y0-y1)
e3 = abs (z0-z1)
iteration +=1
x0 = x1
y0 = y1
z0 = z1
print(f"{iteration:.1f} {x1:.3f} {y1:.3f} {z1:.3f}")
print (f"\n The value of x = {x1:.3f},y = {y1:.3f},z = {z1:.3f}")
print(f"\n number of iteratios = {iteration}")
```

```
Execution result

n       x       y       z
1.0   -0.200   0.156   -0.508
2.0    0.167   0.334   -0.429
3.0    0.191   0.333   -0.422
4.0    0.186   0.331   -0.423
5.0    0.186   0.331   -0.423

The value of x = 0.186,y = 0.331,z= -0.423
number of iterations = 5
```

2.6 SUMMARY

Solving linear systems is fundamental to numerical computation. Binary and tertiary linear systems are easily solved by successive substitution, and the general linear system can be solved by triangulation and Gaussian elimination. Pivoting is essential for the stable implementation of Gaussian elimination. Algorithms and specialized software are available to solve specific types of linear systems using Simulink and Python.

2.7 PROBLEMS

1. Find the determinant of the following 2×2 matrix:

$$\begin{bmatrix} -4 & 2 \\ -8 & 7 \end{bmatrix}$$

   Answer: (12)

2. Find the determinant of the following 3×3 matrix:

$$\begin{bmatrix} 6 & 2 & -4 \\ 5 & 6 & -2 \\ 5 & 2 & -3 \end{bmatrix}$$

Answer: (6)

3. Find the approximate solution of the system with two variables by Cramer's rule.

$$5x + y = -13$$

$$3x - 2y = 0$$

Answer: (–2, –3)

4. Solve the system with two variables by Cramer's rule.

$$-2x + 3y = -3$$

$$3x - 4y = 5$$

Answer: (3, 1)

5. Solve the following two linear algebraic equations using the Gauss elimination method.

$$-8x - 6y = -2$$

$$2x + 5y = -1$$

Answer: (4/7, –3/7)

6. Solve the system shown below using the Gauss elimination method.

$$-2x + y - z = 8$$

$$-3x - y + 2z = -11$$

$$-2x + y + 2z = -3$$

Answer: (0.133, 4.067, –3.667)

7. Solve the system shown below using the Gauss elimination method.

$$x + y - z = 9$$

$$y + 3z = 3$$

$$-x - 2z = 2$$

Answer: (2/3, 7, –4/3)

8. Solve the system shown below using the Gauss-Jordan elimination method.

$$-x + 2y = -6$$

$$3x - 4y = 14$$

Answer: (2, –2)

9. Solve the system of the following three linear algebraic equations using the Gauss-Jacobi elimination method.

$$2x + y + 2z = 10$$

$$x + 2y + z = 8$$

$$3x + y - z = 2$$

Answer: (1, 2, 3)

10. Solve the system of the below three linear algebraic equations using the Gauss-Seidel elimination method.

$$x + y + z = 4$$

$$2x - 3y + z = 2$$

$$-x + 2y - z = -1$$

Answer: (2, 1, 1)

REFERENCE

1. Kong Q., Siauw T., Bayen A., 2020. *Python Programming and Numerical Methods, A Guide for Engineers and Scientists*, 1st ed., London: Academic Press Inc.

3 Bracketing Numerical Methods for Solving Systems of Nonlinear Equations

Root-finding methods of nonlinear algebraic equations are categorized into bracketed and open methods. This chapter covers the graphical and bracketing methods such as the bisection method, the false-position method, and the Ridders method. Programming these methods using Simulink and the open-source Python language is explained extensively.

LEARNING OBJECTIVES

1. Determine the approximate roots of nonlinear equations using the graphical method.
2. Calculate the roots of nonlinear equations using the bisection method.
3. Estimate the approximate roots of nonlinear equations using the false-position method.
4. Find the roots of nonlinear equations using the Ridders method.

3.1 INTRODUCTION

Most numerical methods for finding roots of linear and nonlinear algebraic equations use repetition, resulting in a series of numbers that expectantly approach the root as a bound. It requires one or more initial guesses of the root as initial values, and then each iteration of the algorithm results in a more accurate approximation of the root successively.

Nonlinear equations involve multiple variables. To recognize a nonlinear equation is to observe that the x is not alone as in ax, but involves a product with itself, such as in $x^3+3x^2-9=0$. In this case, x^3 and $2x^2$ are nonlinear terms. The differences between 'closed' and 'open' methods are that the closed method uses a bounded interval. This method usually converges slowly and always finds a root if it exists. The open method usually converges quickly; many do not find a root if it exists. The available methods require a single or two starting values that do not necessarily bracket a root. Closed methods generate at least one root. Depending on the initial guess and type of function, the available methods sometimes result in divergence. By contrast, once converged, they are faster than closed methods and converged quickly. The bracketing methods

DOI: 10.1201/9781003360544-3

include graphical, bisection, and false-position methods. The following equations are a set of the nonlinear algebraic equations:

$$\left.\begin{array}{l} f_1(x_1 \ldots\ldots\ldots x_n) = 0 \\ f_2(x_1 \ldots\ldots\ldots x_n) = 0 \\ f_n(x_1 \ldots\ldots\ldots x_n) = 0 \end{array}\right\} = \bar{f}(\bar{x}) \tag{3.1}$$

or

$$\bar{F}(\bar{X}) = \left\{\begin{array}{c} f(\bar{x}) \\ \vdots \\ f_n(\bar{x}) \end{array}\right\} = 0 \tag{3.2}$$

where $F_1(\bar{X})$ takes the form such as $(X_1 - \sin X_1) = 0$, or

$$x - \sin x = 0, \text{ and } x_1 = e^{x2} \tag{3.3}$$

Solving a set of nonlinear algebraic equations usually is difficult. Solving one nonlinear equation $f(\alpha) = 0$, finding the root (α).

3.2 GRAPHICAL METHOD

The graphical method allows for solving simple problems intuitively and visually. This method is limited to two or three problem decision variables since it is impossible to illustrate more than 3D graphically. For any cartesian equation in one variable, the root of the given equation is where the graph intersects the x-axis. Now, if it is practically not easy to plot a graph of a given equation directly, for such an equation, we can split the given equation into two parts as follows. If the given equation is

$$f(x) = 0$$

We split the function f into f_1 and f_2 such that

$$f(x) = f_1(x) - f_2(x)$$

Now to find x such that $f(x) = 0$, we can find x such that $f_1(x) = f_2(x)$.

Steps for solving equations by graphical method:

1. First, find the suitable range x such that the root of the given equation lies in that range.
2. For this selected range, plot the graph of f_1 and f_2 using the appropriate scale.
3. Find the approximate point on the x-axis corresponding to the intersection of these two graphs, which will be the required solution.

Example 3.1 Root Finding Using Graphical Method

Using the graphical method, find the roots of the following equation:

$$f(x) = x^3 - 3x^2 - 3x + 1 = 0$$

Validate the manual calculations with Simulink and Python programming of the root finding using a graphical method.

Solution

Split the equation into two such as:

$$f_1(x) = x^3$$

$$f_2(x) = 3x^2 + 3x - 1$$

So that

$$f(x) = f_1(x) - f_2(x) = (3x^3) - (3x^2 + 3x - 1)$$

Take appropriate values of x as shown in Table 3.1.

The root is at $x = -1$, because both functions give the same value at that intersection point.

Simulink Solution

Figure 3.1 symbolizes the Simulink block diagram utilizing the graphical method to solve the equation defined in Example 3.1.

To enter the X and Y label in the scope, select the X-Y graph generated by Simulink, and then run the below commands in the MATLAB command window:

```
set(0,'ShowHiddenHandles', 'on')
set(gcf, 'menubar', 'figure')
```

These commands will enable the toolbar in the X – Y graph figure window. Individuals can add axes names from Insert >> X Label and Insert >> Y Label (Figure 3.2).

TABLE 3.1
Values of *f*(*x*) at Various *x* Values

x	**–4**	**–3**	**–2**	**–1**	**0**	**1**	**2**	**3**	**4**
$f_1(x)$	–64	–27	–8	–1	0	1	8	27	64
$f_2(x)$	35	17	5	–1	–1	5	17	35	59

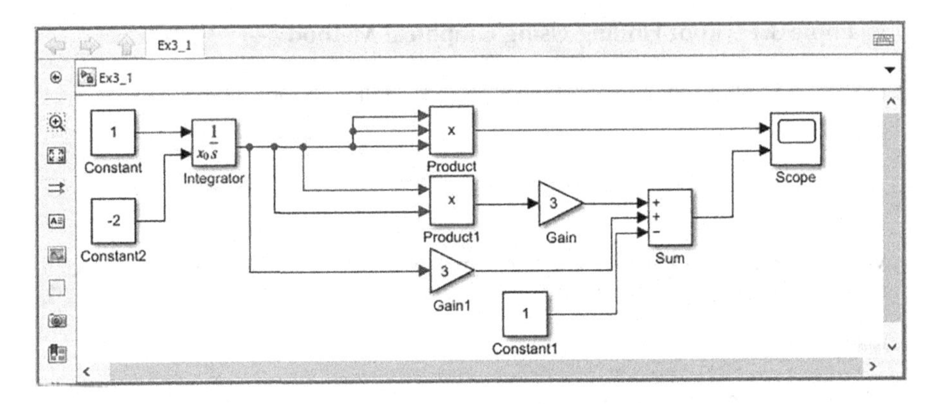

FIGURE 3.1 Simulink programming utilizes a graphical method to solve the equation defined in Example 3.1.

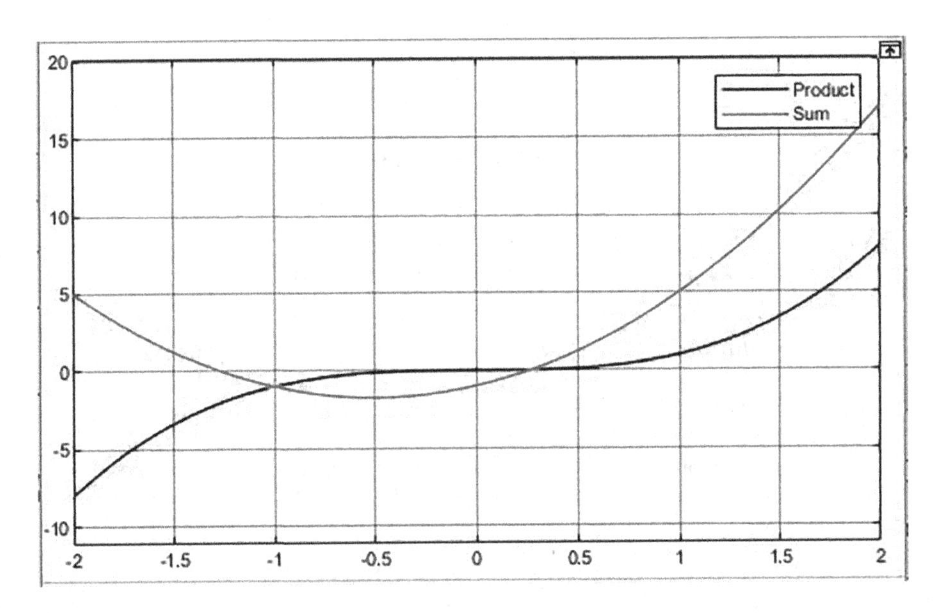

FIGURE 3.2 Graphical representation using the graphical root finding of the equation defined in Example 3.1.

Python Solution

The following program is the Python code that generates the plot shown in Figure 3.3. The figure represents the graphical root-finding method of the equation defined in Example 3.1.

```
# Example 3.1
# Bisection method
import numpy as np
import matplotlib.pyplot as plt
def f(x):
    y= x**3 - 3*x**2 - 3*x + 1
    return y
def g(x):
    y1 = x**3
    return y1
def h(x):
    y2 = 3*x**2 + 3*x - 1
    return y2
a = -5
b = 5
def bisection(a, b):
 if f(a)*f(b)>0:
     print("no root found")
     return
 c = a
 while ((b-a)>=0.01):
     c = (a +b)/2
     if f(c)==0:
         exit()
     if f(c)*f(a)<0:
         b = c
     else:
         a = c
```

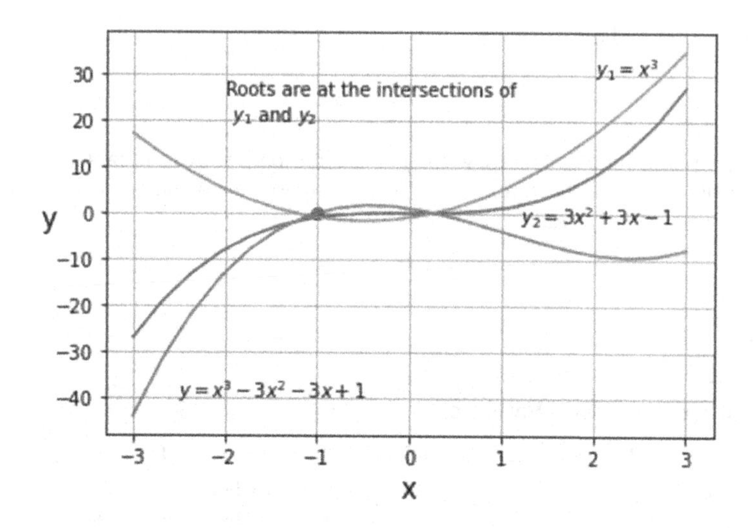

FIGURE 3.3 Find the root plot generated by Python programming for the graphical method of the given equation in Example 3.1.

```
 print("root of of i=", c)
 plt.plot(c, f(c), 'ro')
bisection(a, b)
x=np.linspace(-3,3,20)
plt.plot (x, g(x), x, h(x), x, f(x))
plt.grid()
plt.xlabel("x", fontsize = 15)
plt.ylabel("y", fontsize = 15, rotation = 0)
plt.text(-2,20, 'Roots are at the intersections of
\n $y_1$ and $y_2$')
plt.text(-2.5,-40, '$y=x^3 - 3x^2 - 3x + 1$')
plt.text(2,30, '$y_1= x^3$')
plt.text(1.2,-2, '$y_2= 3x^2 + 3x - 1$')
```

Example 3.2 Root Finding of High-Order Equations

Find the root of the following equation using the graphical method.

$$f(x) = x^4 - 11x^2 + 2x + 1$$

Validate the manual calculation of the root using Simulink and Python programming of the graphical method.

Solution

The function can be arranged in the following form:

$$f_1(x) = x^4$$

$$f_2(x) = 11x^2 - 2x - 1$$

$$f(x) = f_1(x) - f_2(x) = \left(x^4\right) - \left(11x^2 - 2x - 1\right)$$

The root is at x where

$$f_1(x) = f_2(x)$$

Taking an appropriate value of x to generate the data shown in Table 3.2. Two roots exist at $x = -0.22437$, and $x = 0.40987$. At these values, differences between $f_1(x)$ and $f_2(x)$ equal zero.

TABLE 3.2
Values of *f*(*x*) at Various Values of *x*

x	**–4**	**–3**	**–2**	**–1**	**–0.22437**	**0**	**0.40987**	**1**	**2**	**3**	**4**
$f_1(x)$	256	81	16	1	0.0025	0	0.0282	1	16	81	256
$f_2(x)$	183	104	47	12	0.0025	–1	0.0282	8	39	92	167

Simulink Solution

The Simulink solution is described by the Simulink block diagram shown in Figure 3.4. By double-clicking the scope released in Figure 3.5.

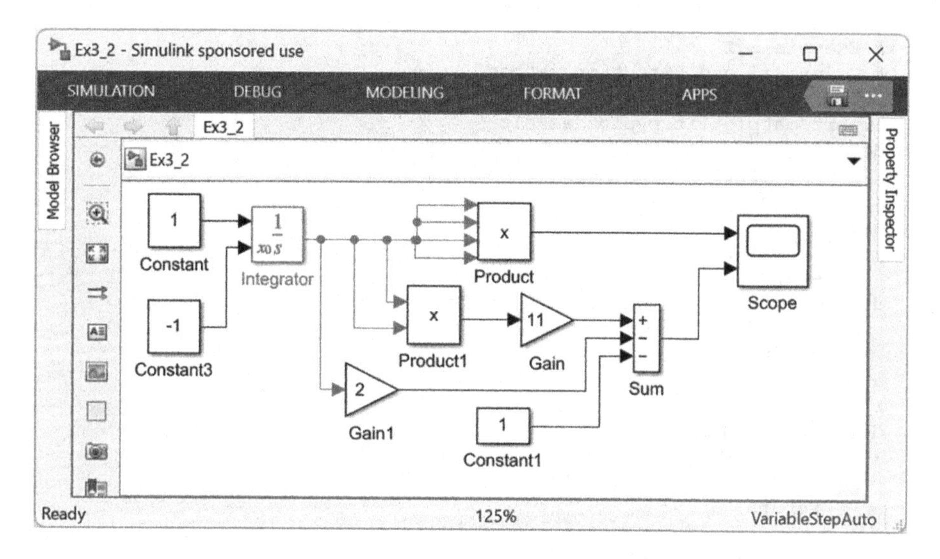

FIGURE 3.4 The Simulink block diagram represents the method for finding the root by the graphical approach of the given equation in Example 3.2.

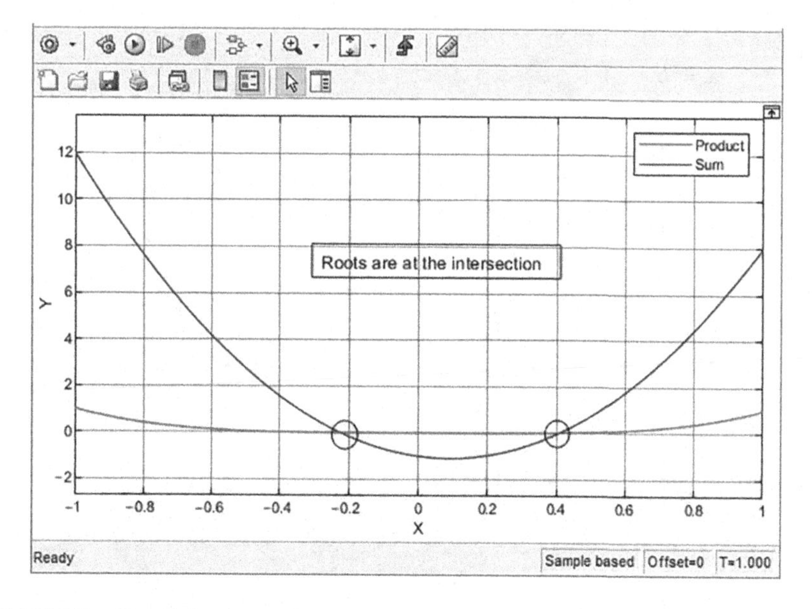

FIGURE 3.5 Simulink generated a plot of the graphical root-finding representation of the equation identified in Example 3.2.

Python Solution

The following Python code uses the graphical root-finding method to solve the equation defined in Example 3.2. Figure 3.6 shows the plot with the intersections where the root lies.

```
# Example 3.2
# Graphical and Bisection method
import numpy as np
import matplotlib.pyplot as plt
def f(x):
    y= x**4-11*x**2+2*x+1
    return y
def g(x):
    y1 = x**4
    return y1
def h(x):
    y2 = 11*x**2-2*x-1
    return y2
a = 0
b = 2
def bisection(a, b):
 if f(a)*f(b)>0:
     print("no root found")
     return
c = a
while ((b-a)>=0.0001):
    c = (a +b)/2
    if f(c)==0:
        exit()
    if f(c)*f(a)<0:
        b = c
else:
        a = c
```

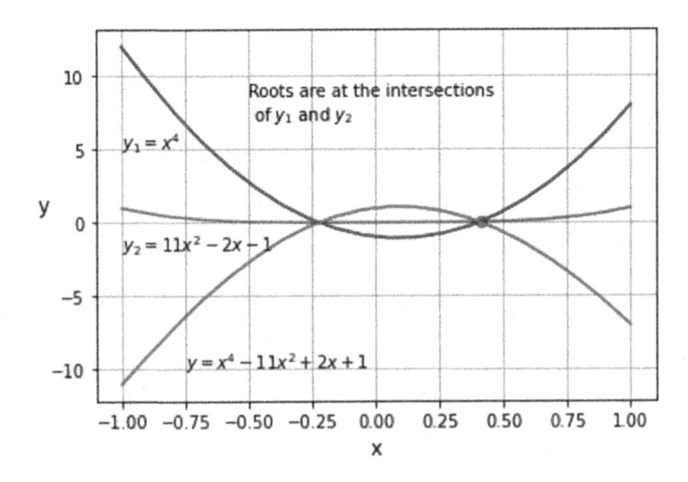

FIGURE 3.6 Python-generated plot of the root finding using the graphical technique of the equation defined in Example 3.2.

```
  print("root of of i=", c)
  plt.plot(c, f(c), 'ro')
bisection(a, b)
x=np.linspace(-1,1,20)
plt.plot (x, g(x), 'g', x, h(x), 'b', x, f(x), 'r')
plt.grid()
plt.xlabel("x", fontsize = 12)
plt.ylabel("y", fontsize = 12, rotation=0)
plt.text(-0.5,7, 'Roots are at the intersections of
\n $y_1$ and $y_2$')
plt.text(-0.75,-10, '$y=x^4-11x^2+2x+1$')
plt.text(-1,5, '$y_1= x^4$')
plt.text(-1,-2, '$y_2= 11x^2-2x-1$')
```

Example 3.3 Graphical Methods for the Root of Quadratic Equations

Find the roots of the following quadratic equation using manual calculations, Python programming, and Simuling graphical programming.

$$y = x^2 - 4x + 3 = 0$$

Confirm the manual calculation using the graphical root-finding method utilizing Simulink and Python programming.

Manual Solution

The equation can be arranged in the following form.

$$x^2 - 4x + 3 = 0$$

This can be analyzed as follows:

$$(x-3)(x-1) = 0$$

Accordingly, the roots are 1 and 3.

Simulink Solution

Figure 3.7 shows the Simulink block diagram representing the graphical root-finding method of the equation defined in Example 3.3. The intersection points represent the roots of the quadratic equations (Figure 3.8).

Python Solution

The following Python code describes the graphical root-finding method of the equation defined in Example 3.3. After execution, the results are plotted in Figure 3.9.

```
# Example 3.3
# Graphical & Bisection methods
import numpy as np
import matplotlib.pyplot as plt
def f(x):
```

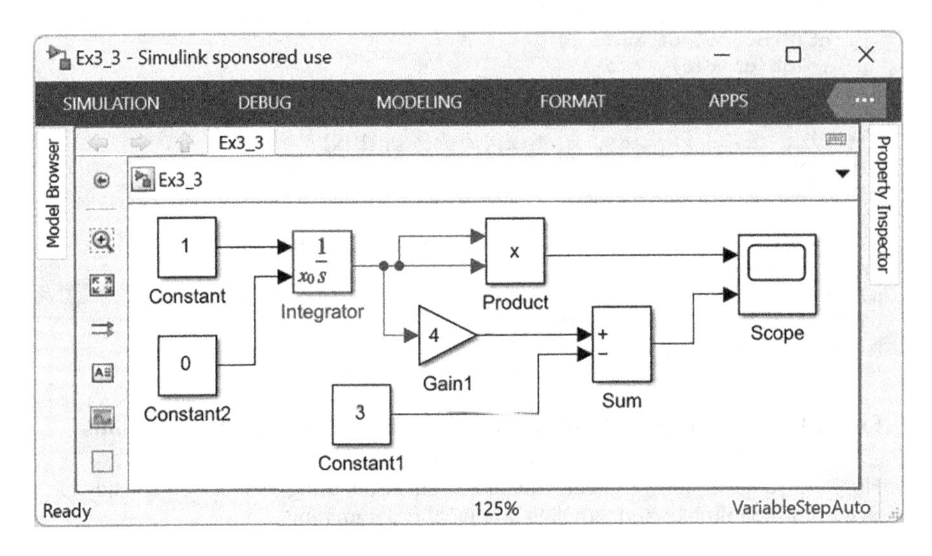

FIGURE 3.7 Simulink block diagram utilizing the graphical method to solve the equation declared in Example 3.3.

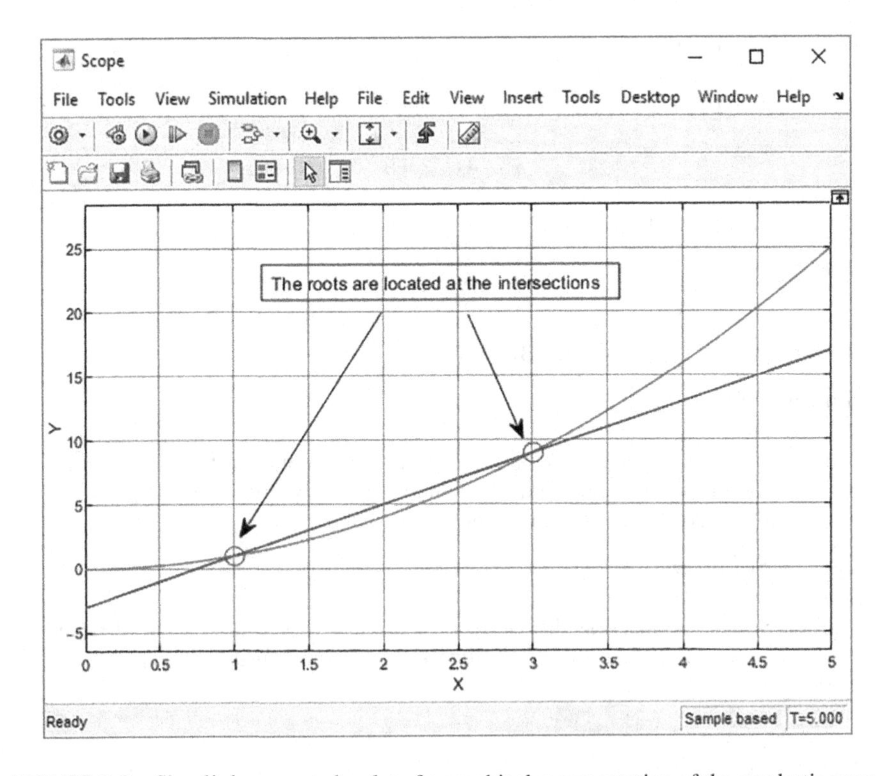

FIGURE 3.8 Simulink generated a plot of a graphical representation of the quadratic equation mentioned in Example 3.3.

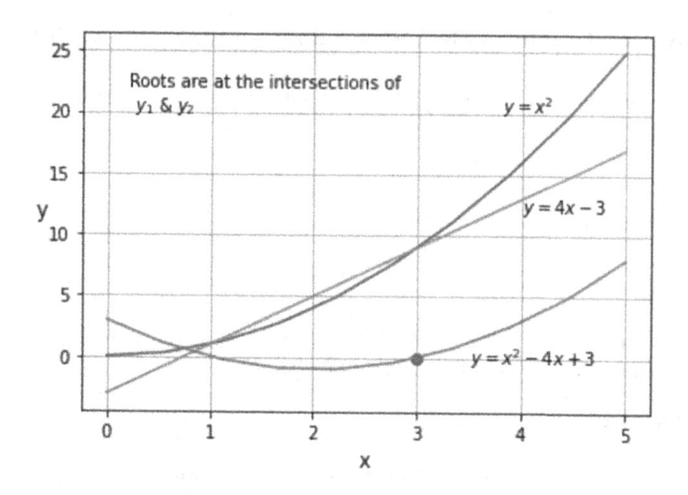

FIGURE 3.9 Root finding using the graphical method in Python of the equation described in Example 3.3.

```
        y = x**2 - 4*x + 3
        return y
    def g(x):
        y1 = x**2
        return y1
    def h(x):
        y2=4*x - 3
        return y2
    a = 2
    b = 5
    def bisection(a, b):
     if f(a)*f(b)>0:
         print("no root found")
         return
    c = a
    while ((b-a)>=0.01):
        c = (a +b)/2
        if f(c)==0:
            exit()
        if f(c)*f(a)<0:
            b = c
        else:
            a = c
     print("root of of i=", c)
     plt.plot(c, f(c), 'ro')
    bisection(a, b)
    x=np.linspace(0,5,10)
    plt.plot (x, g(x), x, h(x), x, f(x))
    plt.grid()
    plt.xlabel("x", fontsize = 12)
    plt.ylabel("y", fontsize = 12, rotation=0)
    plt.text(0.2,20, 'Roots are at the intersections')
    plt.text(3.5,-0.5, '$y=x^2-4x+3$')
    plt.text(3.8,20, '$y= x^2$')
    plt.text(4,12, '$y=4x-3$')
```

Example 3.4 Graphical Methods for Roots of Quadratic Equation

Find the roots of the following equation using the graphical method.

$$y = e^x + x - 3$$

The equation can be written as follows:

$$e^x = 3 - x$$

There is one root which is around 0.8. Validate the root found manually with the graphical method using Simulink and Python programming.

Simulink Solution

Figure 3.10 is a Simulink block diagram describing a graphical method for finding the root of the equation given in Example 3.4, and the output is displayed in Figure 3.11.

Python Solution

The following Python code uses the graphical root search method programmed to find one of the roots of the equation given in Example 3.4.

```
# Example 3.4
import numpy as np
import matplotlib.pyplot as plt
def f(x):
y = np.exp(x) + x - 3
```

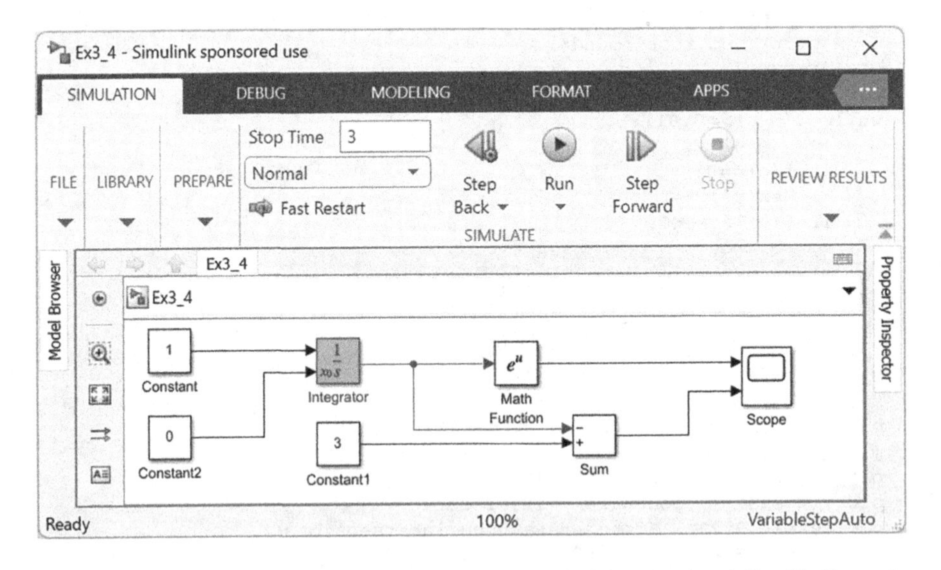

FIGURE 3.10 Simulink graphical root-finding method of the equation defined in Example 3.4.

```
    return y
def g(x):
    y1 = np.exp(x)
    return y1
def h(x):
    y2= 3-x
    return y2
# interval [a, b]
a = 0
b = 3
def bisection(a, b):
 if f(a)*f(b)>0:
     print("no root found")
     return
 c = a
 while ((b-a)>=0.01):
     c = (a +b)/2
     if f(c)==0:
         exit()
     if f(c)*f(a)<0:
         b = c
else:
     a = c
 print("The root of the function :", c)
```

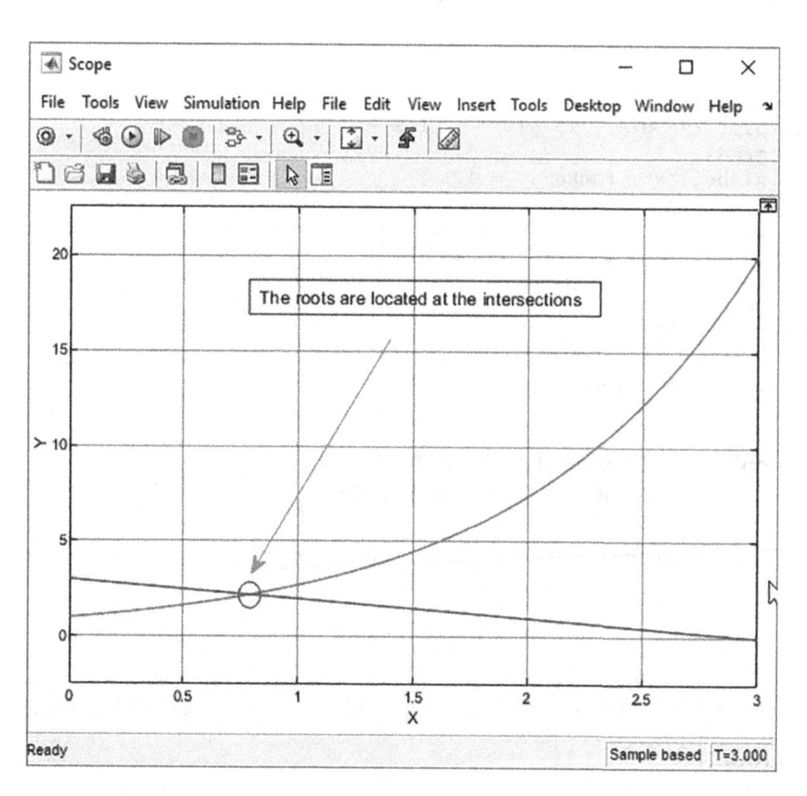

FIGURE 3.11 Simulink generated a plot of the graphical root-finding method to solve the equation defined in Example 3.4.

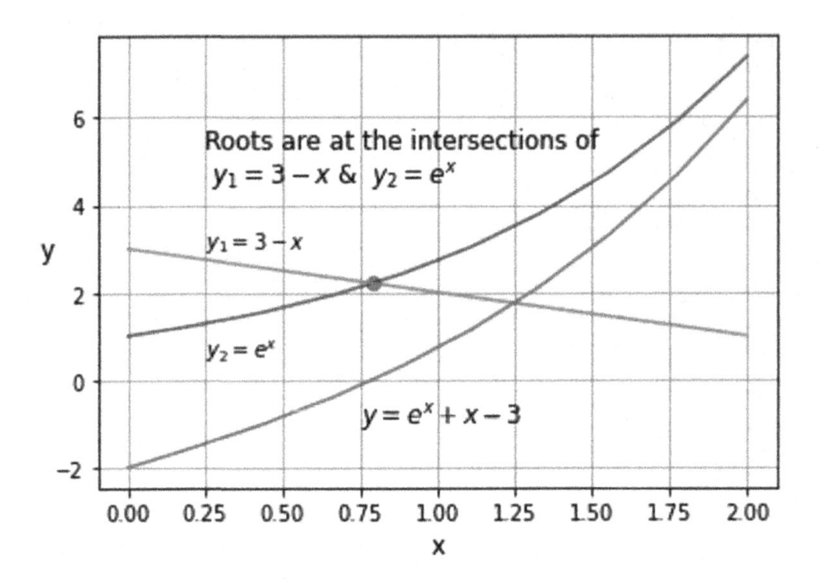

FIGURE 3.12 Graphical method for finding the root of the equation defined in Example 3.4 using Python.

```
plt.plot(c, g(c), 'ro')
bisection(a, b)
x=np.linspace(0,2,10)
plt.plot (x, g(x), x, h(x), x, f(x))
plt.grid()
plt.xlabel("x", fontsize = 12)
plt.ylabel("y", fontsize = 12, rotation=0)
plt.text(0.25,4.5,'Roots are at the intersections of'
'\n $y_1=3-x$ & $y_2=e^x$', fontsize = 12)
plt.text(0.75,-1,'$y=e^x+x-3$', fontsize = 12)
plt.text(0.25,0.5, '$y_2=e^x$', fontsize = 10)
plt.text(0.25,3, '$y_1=3-x$', fontsize = 10)
```

The program-executed plot is shown in Figure 3.12.

Example 3.5 Graphical Methods for Finding the Root of a Quadratic Equation

Find the roots of the following equation using the graphical root-finding method.

$$y = \ln(x) + x - 3$$

Using Simulink and Python programming to find the root of the defined equation.

Solution

The root is around 2.208.

Simulink Solution

The Simulink solution utilizing the graphical root-finding method of the equation present in Example 3.5 is shown in Figures 3.13 and 3.14.

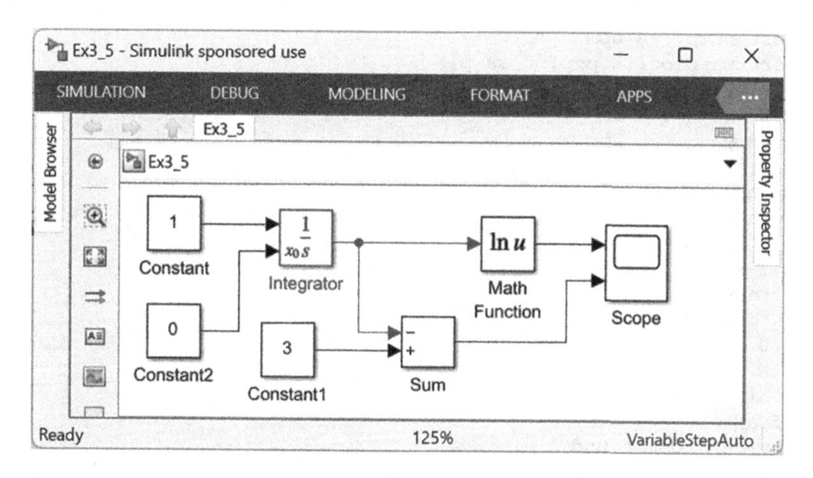

FIGURE 3.13 Simulink graphical simulation method of finding the root with a cut-off time is five to solve the given equation in Example 3.5.

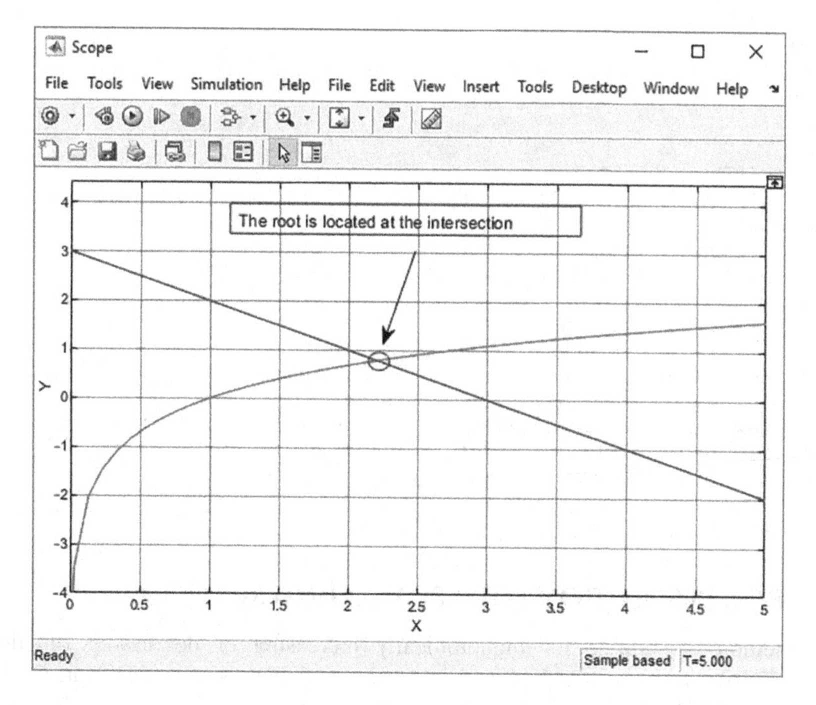

FIGURE 3.14 A graphically simulated root-finding method generated by Simulink to determine the root of the equation given in Example 3.5.

Python solution

The following Python code programmed the graphical root-finding method to explore the root of the equation needed in Example 3.5.

```
# Example 3.5
import numpy as np
import matplotlib.pyplot as plt
def f(x):
y = np.log(x) + x - 3
return y
def g(x):
y1 = np.log(x)
return y1
def h(x):
y2= 3-x
return y2
# interval [a, b]
a = 0
b = 5
def bisection(a, b):
if f(a)*f(b)>0:
print("no root found")
return
c = a
while ((b-a)>=0.01):
c = (a +b)/2
if f(c)==0:
exit()
if f(c)*f(a)<0:
b = c
else:
a = c
print("The root is:", c)
plt.plot(c, g(c), 'ro')
bisection(a, b)
x=np.linspace(0,5,30)
plt.plot (x, g(x), x, h(x), x, f(x))
plt.grid()
plt.xlabel("x", fontsize = 13)
plt.ylabel("y", fontsize = 13, rotation=0)
plt.text(0.65,2.5,'Roots are at the intersections of'
'\n $y_1=3-x$ & $y_2=ln(x)$', fontsize = 12)
plt.text(1.3,-2, '$y=ln(x)+x-3$', fontsize = 13)
plt.text(4,0.5, '$y_2=ln(x)$', fontsize = 12)
plt.text(0.1,1.5, '$y_1=3-x$', fontsize = 12)
```

The result is shown in Figure 3.15.

3.3 ROOTS SOLUTION WITH BRACKETING METHOD

The method is suitable for monotonically increasing or decreasing functions (Figure 3.16). The closed method is an interval where there is a root inside it. At each step, the method produces intervals that contain the root and steadily shrink. This method is guaranteed to find a root within the interval [a, b]. This technique is called the bracketing method. The following steps are taken to resolve the root.

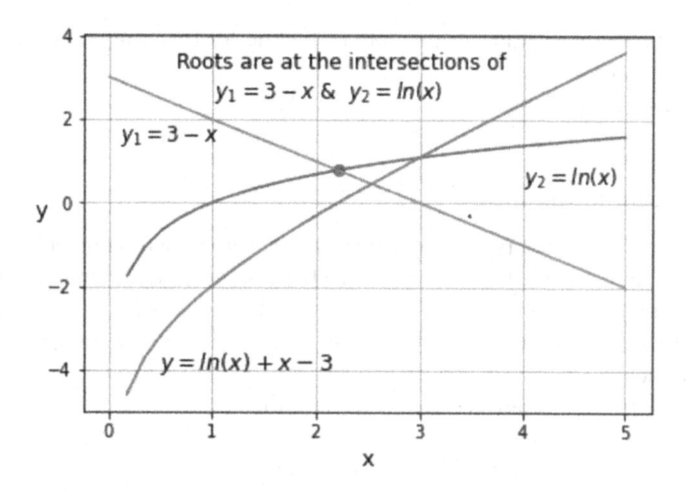

FIGURE 3.15 Python-generated plot represents the graphical root-finding method of the equation present in Example 3.5.

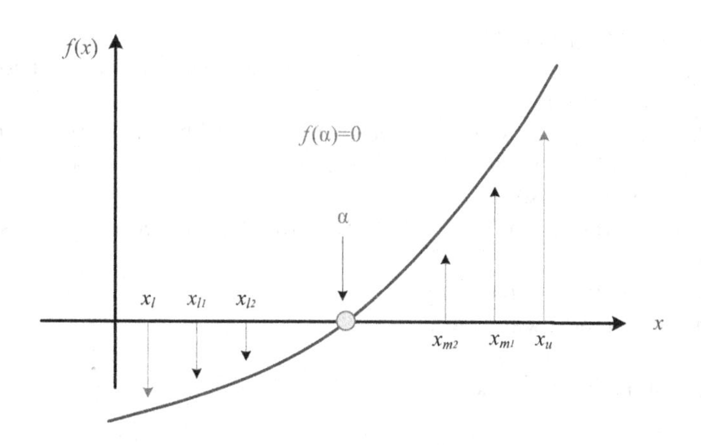

FIGURE 3.16 Graphical representation of the bracketing method of numerical analysis.

1. Find x_l and x_u such that

$$f(x_l) \times f(x_u) < 0$$

   This means that one root was trapped or enclosed in parentheses. Some functions may have multiple roots.
2. Check, $f(x_m) = 0$, if true, then this is the root or $x_m = \alpha$, if not, then check if $f(x_m) f(x_1) < 0$, if true, then

$$x_U = x_m \text{ (Replace } x_U \text{ with } x_m \text{), else } x_1 = x_m \text{ and replace } x_1 \text{ with } x_m$$

3. The bracketing method is a simple technique for obtaining an estimate of the root for the equation $f(x)=0$.
4. It gives a rough approximation of the root solution.

3.3.1 Bisection Method

The bisection method is a type of root-finding bracketing method in which the interval is always divided in half. If the function changes signs over an interval, the value of the function will be evaluated at the midpoint. The root location is then identified as within the subinterval where the tag change occurs. The bisection method requires two initial guesses:

$a = x_0$ and $b = x_1$, to satisfy the bracket condition

$$f(x_0)\cdot f(x_1) < 0$$

To check for a root of some function, $f(x)$, select an interval $[x_o, x_1]$ in which the root lies in between x_o and x_1.

1. Calculate $f(x_0)$ and $f(x_1)$.
2. Check to see if they have the same sign. If so, there may be no root between them, and then quit.
3. If they have opposite signs, then pick a point, $f(x_i)$ between the two ends.
4. Check the sign of the $f(x_i)$, if it is signed the same as $f(x_1)$, the root lies between x_i and x_2, so set $x_1 = x_i$.
5. If it is signed the same as $f(x_1)$, then the root lies between x_i and x_1, so set x_i as x_1.
6. If $f(x_i)=0$, then x_i is the root.
7. Otherwise, repeat with new x_i.

Advantages of the bisection method:

1. The root needs to be inside the range.
2. It is always converging.
3. It only requires evaluation of function, not derivative.

Disadvantage of the bisection method:

1. It converges slowly compared to other methods

Example 3.6 Implementation of Bisection Technique

Use the bisectional method to find the root of the following equation on the interval [1, 2].

$$f(x)=x^2-3$$

Stop when the relative fraction change is 1×10^{-5}. The exact root is $\mp\sqrt{3}$. Verify the exact root with Simulink and Python programming of the bisection method.

Solution

Now, find the value of $f(x)$ at $a = 1.0$ and $b = 2$.

$$f(x_0 = 1) = 12 - 3 = 1 - 3 = -2 < 0$$

$$f(x_1 = 2) = 22 - 3 = 4 - 3 = 1 > 0$$

The given function is continuous, and the root lies in the interval [1, 2]. The value of the lower limit is negative, and the upper interval is positive. Let 'x_i' be the midpoint of the interval. We calculate the $f(x_1 = 1)$ and $f(x_2 = 2)$ the first guess is the midpoint of the interval, $x_i = 1.5$, and then we evaluate the function $f(x_i = 1.5)$.

$$x_i = \frac{(1+2)}{2} = 1.5$$

Therefore, the value of the function at x_i is

$$f(x_i) = f(1.5) = (1.5)2 - 3 = 2.25 - 3 = -0.75 < 0$$

If $f(x_i) < 0$, assume $a = x_i$, and if $f(x_i) > 0$, assume $b = x_i$.

$f(x_i)$ is negative, so a is replaced with $x_i = 1.5$ for successive iterations. The rest of the iterations for the given functions are tabulated in Table 3.3.

So, we get the final interval [1.7266, 1.7344]. Hence, 1.7344 is the approximated solution.

The Simulink solution is shown in Figure 3.17.

An alternative way to use Simulink is the MATLAB function from the Simulink user-defined functions library (Figure 3.18).

The following program is the MATLAB code that embeds the Simulink MATLAB function.

```
% Example 3.6
% Bisection Method in MATLAB
function y = fcn(a, b)
```

TABLE 3.3

Solution of Equation Present in Example 3.6 Using the Bisectional Method

Iterations	a	b	x_i	$f(a)$	$f(b)$	$f(x_i)$
1	1	2	1.5	–2	1	-0.75
2	1.5	2	1.75	–0.75	1	0.062
3	1.5	1.75	1.625	–0.75	0.0625	–0.359
4	1.625	1.75	1.6875	–0.3594	0.0625	–0.1523
5	1.6875	1.75	1.7188	–1523	0.0625	–0.0457
6	1.7188	1.75	1.7344	–0.0457	0.0625	0.0081
7	1.7188	1.7344	1.7266	–0.0457	0.0081	–0.0189

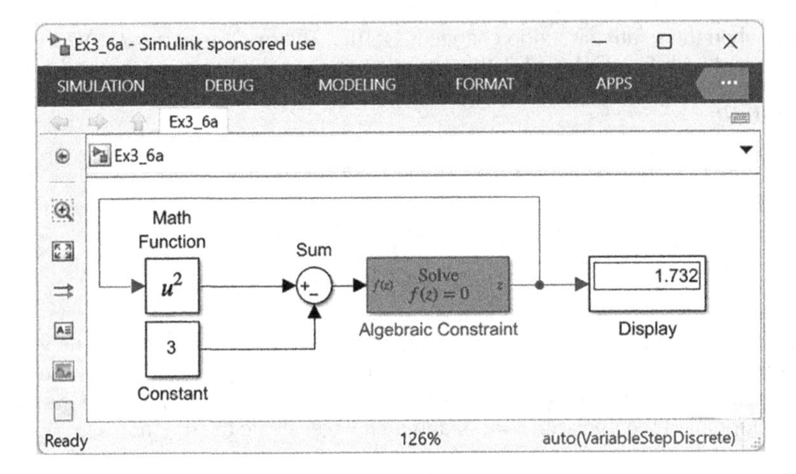

FIGURE 3.17 Simulink solution using the 'algebraic constraint' of the equation defined in Example 3.6.

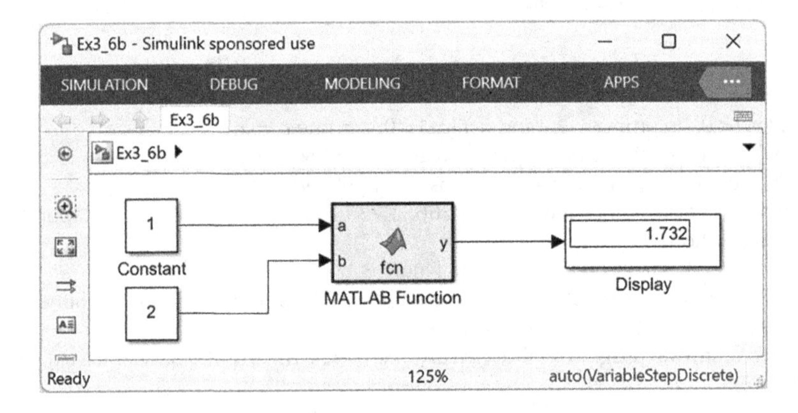

FIGURE 3.18 Simulink solution using the MATLAB function and the associated MATLAB coding of the bisection method programmed in MATLAB of the equation defined in Example 3.6.

```
f = @(x) (x^2-3);
%a = 1;
%b = 2;
eps = 0.001;
m = (a+b)/2;
fprintf('\n The value of, after bisection method, m
is %f\n', m);
while abs(b-a)>eps
    if (f(a)*f(m))<0
        b=m;
    else
        a=m;
    end
    m = (a+b)/2;
end

y = m;
```

Python solution

The following program is the Python code for finding one of the roots of the equation exhibited in Example 3.6.

```
#Example 3.6
# Our f(x) function
def f(x):
    y = x**2-3
    return y
# Function to update R
def UpdateBisec(L, U):
    R = (L + U)/2
    return R
# The Bisection method (BSM) function
def BSM(f, L, U):
    R = UpdateBisec(L, U)
    while abs(f(R)) > 1e-3:
        if f(L) * f(R) < 0:
            U = R
        elif f(U) * f(R) < 0:
            L = R
        elif f(R) == 0:
            return R
        R = UpdateBisec(L, U)
    return R
Root= BSM(f, 0, 5)
print ("The calculated root is =", root)
```

The results:

```
The calculated root is = 1.732177734375
```

3.3.2 False-Position Method

False-position methods or regula-falsi is a root-finding technique that can be considered an advanced version of the bisection method. The regula-falsi method is a numerical method for estimating the roots of a polynomial. The value x replaces the midpoint in the bisection method and serves as the new approximation of a root $f(x)$. (Figure 3.19).

$$x_{i+1} = \frac{x_i f(x_{i-1}) - x_{i-1} f(x_i)}{f(x_{i-1}) - f(x_i)}$$

Example 3.7 Regula False Method

Solve the following equation using the regula-falsi method up to three decimal places within the interval [0,1].

$$f(x) = x^3 - 4x + 1$$

Verify the manual calculations with Simulink and Python programming of the regula-falsi method.

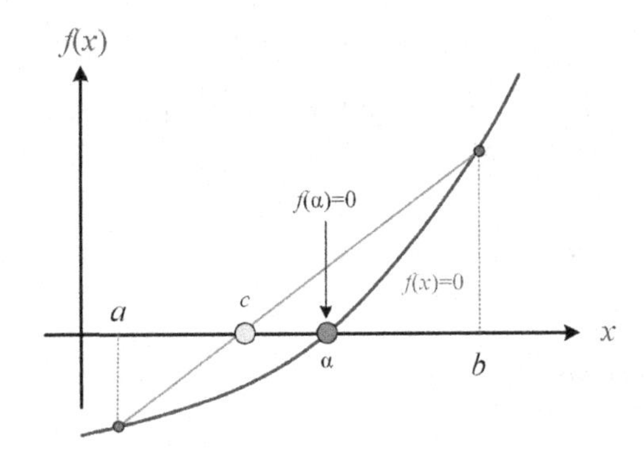

FIGURE 3.19 Graphical representation of regula-falsi method.

Solution

Start with finding the function at 0 and 1.

$$f(0) = 0^3 - 4(0) + 1 = 1$$

$$f(1) = (1)^3 - 4(1) + 1 = -2$$

Because the value of $f(0)$ is positive and the value of $f(1)$ is negative, accordingly, the root lies between 0 and 1.

$$x_1 = \frac{af(b) - bf(a)}{f(b) - f(a)} = \frac{0(-2) - 1(1)}{-2 - 1} = 0.333$$

$$f(x_1) = (0.333)^3 - 4(0.333) + 1 = -0.2962$$

Since the value is negative, the root lies between a and x_1, and it will replace b

$$x_2 = \frac{af(x_1) - x_1 f(a)}{(f(x_1) - f(a))}$$

Substitute the values

$$x_2 = \frac{0(-0.2962) - 0.333(1)}{-0.296 - 1} = 0.2571$$

$$f(x_2) = (0.2571)^3 - 4(0.2571) + 1 = -0.0115$$

Since the value of $f(x_2)$ is negative; the root lies between a and x_2, accordingly,

$$x_3 = \frac{af(x_2) - x_2 f(a)}{(f(x_2) - f(a))}$$

Substitution

$$x_3 = \frac{0(-0.0115) - 0.2571(1)}{(-0.0115 - 1)} = 0.2541$$

Find $f(x_3)$

$$f(x_3) = (0.2541)^3 - 4(0.2541) + 1 = 0.0001$$

Because the value is positive, the root lies between x_2 and x_3

$$x_4 = \frac{x_3 f(x_2) - x_2 f(x_3)}{f(x_2) - f(x_3)}$$

Substitute and find x_4

$$x_4 = \frac{0.2541(-0.0115) - 0.2571(0.0001)}{-0.0115 - 0.0001} = 0.2541$$

Figure 3.20 represents the Simulink block diagram using the algebraic constraints for finding the root of the equation required in Example 3.7.

An alternative way is to program the regula-falsi method using the MATLAB function built in the Simulink library. The Simulink block diagram shown in Figure 3.21 describes the solution of Example 3.7.

The following program is the MATLAB code associated with the MATLAB function using the regula-falsi method to solve the equation described in Example 3.7.

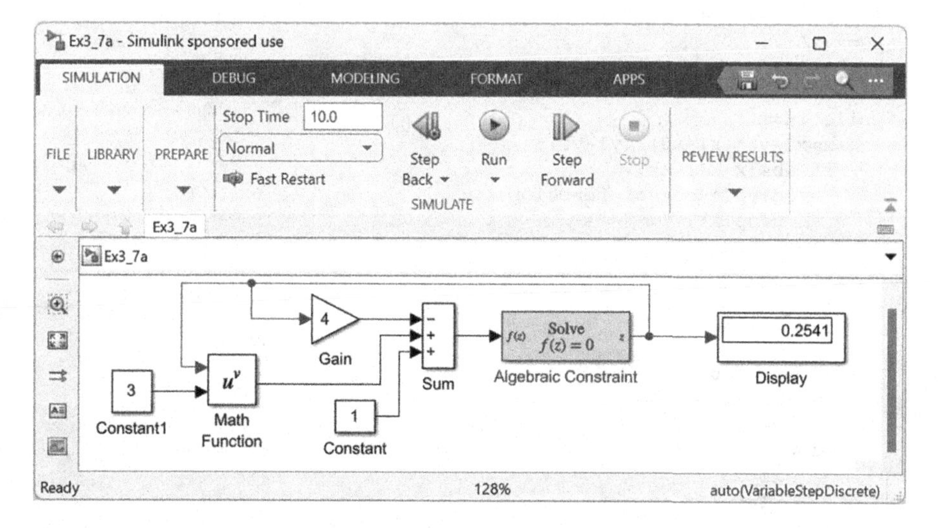

FIGURE 3.20 Simulink block diagram using the 'algebraic constraints' represents the solution of the equation required to find its root in Example 3.7.

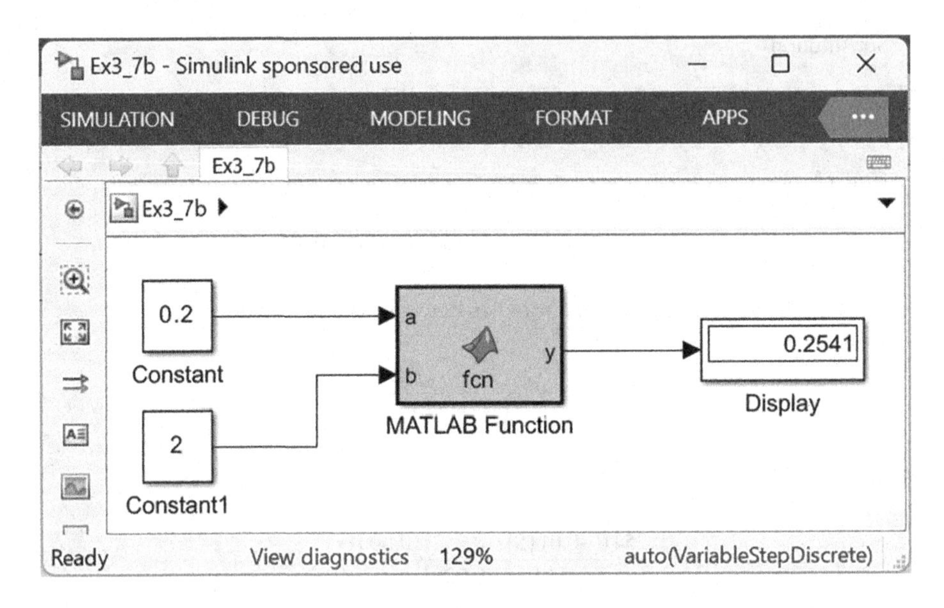

FIGURE 3.21 Regula-falsi method for solving the equation required in Example 3.7.

```
%Example 3.7
%Using False Position Method
% Find out the root of a function
function y = fcn(a, b)
% within the interval [a, b]
x0=a;
x1=b;
tol=0.001;
n=100;
f=@(x)x^3-4*x + 1;
y0=f(x0);
y1=f(x1);
i=1;
while i<=n
    x=x1-y1*(x1-x0)/(y1-y0);
    if abs(x-x1)<tol
        disp('Root of function:');
        disp(x);
        break
    end
    i=i+1;
    y=x^3-4*x + 1;

    if y0*y1<0
        x0=x;
        y0=y;
    else
        x1=x;
        y1=y;
    end
end
y = x1;
```

Python solution

The following program is the Python code that uses the false-position method, representing the equation's solution defined in Example 3.5.

```
# Example 3.7
# False Position method (FPM)
# import numpy as np
# Our f(x) function
def f(x):
y = x**3 - 4*x +1
return y
# Function to update R
def UpdateFalseP(f, L, U):
R = U - (f(U) * (L - U))/(f(L) - f(U))
return R
# The False Position method function
def FPM(f, L, U):
R = UpdateFalseP(f, L, U)
while abs(f(R)) > 1e-3:
if f(L) * f(R) < 0:
U = R
elif f(U) * f(R) < 0:
L = R
elif f(R) == 0:
return R
R = UpdateFalseP(f, L, U)
return R
Root = FPM(f, 0, 1)
print ("\n The calculated root = ", Root)
```

Output

```
The calculated root = 0.2542021166217058
```

3.3.3 Ridders' Method

In numerical analysis, Ridders' method is a root-finding algorithm proposed by Ridders in 1979. The Ridders method is a process for finding the root based on the regula-falsi method that uses an exponential function to fit a given function bracketed between x_0 and x_1. The Ridders algorithm for the determination of the root of the equation $f(x)=0$ is as follows:

1. Assume suitable guessed values for x_0 and x_1 as a proper bracketing point where the function must have different signs.

$$f(x_o)\times f(x_1)<0$$

2. Find the midpoint, u, such that

$$u=\frac{x_0+x_1}{2}$$

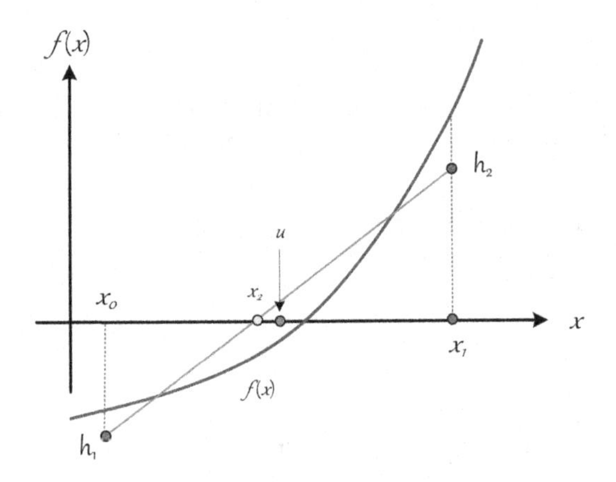

FIGURE 3.22 Graphical illustration of Ridders' method.

3. Find a new approximation for the root

$$x_2 = \frac{x_0 + x_1}{2} + \frac{x_1 - x_0}{2} \cdot \frac{\sin\left|f(x_o) - f(x_1)\right| f(u)}{\sqrt{f^2(u) - f(x_o) f(x_1)}}$$

4. If x_2 reached the convergence conditions, stop.
5. Otherwise, assign new x_o and x_1 using the old values, where the end of the bracket is x_2.
6. Repeat the loop until it converges (Figure 3.22).

Example 3.8 Applying Ridders Numerical Method

Solve the following equation using the Ridders method within the interval [0.0,0.5].

$$f(x) = x^3 - 4x + 1$$

Verify the manual calculations with Simulink and Python programming of the Ridders method.

Solution

1. Assume suitable guessed values for $x_0 = 0$ and $x_1 = 0.5$ as a proper bracketing point where the function must have different signs.

$$f(x_o) \times f(x_1) < 0$$

$$f(x_o) = f(0) = 1$$

$$f(x_1) = f(0.5) = -0.875$$

Accordingly,

$$f(x_o)\times f(x_1)=1\times(-0.875)=-0.875<0$$

2. Find the midpoint, u, such that

$$u_1=\frac{x_0+x_1}{2}=\frac{0+0.5}{2}=0.25$$

3. Find a new approximation for the root

$$x_2=\frac{x_0+x_1}{2}+\frac{x_1-x_0}{2}\cdot\frac{\sin\left|f(x_o)-f(x_1)\right|f(u_1)}{\sqrt{f^2(u)-f(x_o)f(x_1)}}$$

Substitute the values,

$$x_2=\frac{0+0.5}{2}+\frac{0.5-0}{2}\cdot\frac{\sin\left|1+0.875\right|0.0156}{\sqrt{0.0156^2-1(-0.875)}}=0.254175$$

The new interval should depend on $f(x_1)f(x_2)<0$ an x_2 should be the end of the interval,

$$f(x_0)=f(0)=1$$

$$f(x_1)=f(0.5)=-0.875$$

$$f(x_2)=f(0.25417)=-0.00028$$

Since the new value (x_2) is close to zero, it is considered one of the roots of the equation.

Simulink Solution

Figure 3.23 represents the Simulink graphical programming of the Ridders numerical method utilizing the Simulink MATLAB function for solving the equation defined in Example 3.8. The following MATLAB code programs the Ridders technique.

The following program describes the MATLAB code of the Ridders method associated with the MATLAB function to solve the equation specified in Example 3.6.

```
%Example 3.8
%Using Ridders Method
% Find out the root of a function
function root = riddler(a, b)
    xl=a;
    xh=b;
    func=@(x)x^3-4*x + 1;
    fl=func(a);
```

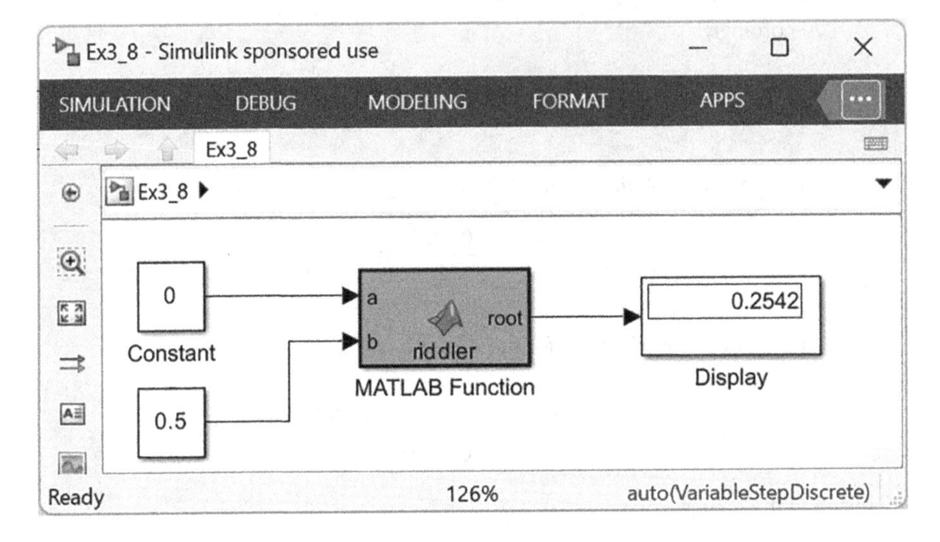

FIGURE 3.23 Simulink solution utilizing Ridder's method to solve the equation defined in Example 3.8.

```
        fh=func(b);
        tol=0.001;
        N=100;
        root = 0;

    for i = 1:N
        xm = 0.5*(xl+xh);
        fm = func(xm);
        s = sqrt(fm*fm - fl*fh);
        if s == 0
          return;
        end
        xnew = xm + (xm - xl)*sign(fl - fh)*fm/s ;
        root =xnew;
        if (xnew-xm) <tol
            return
        end

        end % end for
        root = xnew;
```

Python Solution

The following program deals with the Ridders numerical method using Python language to solve the equation specified in Example 3.8.

```
# Example 3.8
''' Finds a root of f(x) = 0 with Ridder's method.
The root must be bracketed in (a, b).
'''
from numpy import sign
from numpy import sqrt
```

```
xl=0.0
xh=0.5
def func(x):
    return x**3-4*x + 1
fl=func(xl)
fh=func(xh)
tol=0.001
N=100
root = 0;
N=100
for i in range(0, N):
    xm = 0.5*(xl+xh)
    fm = func(xm)
    s = sqrt(fm*fm - fl*fh)
    if s == 0:
        exit()
    xnew = xm + (xm - xl)*sign(fl - fh)*fm/s
    if (xnew-xm) <tol:
        exit()
# print the root
print ('The Root is', xnew)
```

The executed result

```
The Root is: 0.2541753744434931
```

3.4 SUMMARY

Graphical methods are a straightforward way to estimate the root of equation, $f(x)=0$. We can approximate the root if we plot the function $f(x)=0$ and note where it intersects the x-axis. The bracketing methods are based on the mean value theory, and the bisection method is a bracketing type of the root-finding method. The bisection method decreases the error by a constant factor by each iteration, referred to as linear convergence. The Ridders method is a root-finding algorithm based on the false-position method and uses an exponential function to approximate the root of a continuous function successively.

3.5 PROBLEMS

1. Using the graphical method, find the roots of the following second-order equation.

$$y = x^2 - 4x + 3$$

   Answer: (1, 3)

2. Determine the root of the given equation using the bisectional method within the interval [1, 2].

$$f(x) = x^2 - 3$$

   Answer: (1.7344)

3. Use the bisectional method to solve the following nonlinear equation within the interval [1, 4].

$$f(x) = x^2 + 2x - 8$$

Answer: (1.75)

4. Using the false-position method, find the solution of the following nonlinear equation within the interval [0, 2].

$$f(x) = x^3 - x - 1$$

Answer: (1.325)

5. Find the roots of the following nonlinear equation using the bisection method within the interval [1, 2].

$$f(x) = x^3 - 5$$

Answer: (1.71)

6. Consider finding the root of the following nonlinear equation using the false-position method in the interval [3, 4].

$$f(x) = e^{-x}\left(3.2\sin(x) - 0.5\cos(x)\right)$$

Answer: (3.2969)

7. Use the false-position method to find the roots of the nonlinear equation. Conduct three iterations to estimate the root within the interval [2.5, 4].

$$x^3 - 6x^2 + 11x - 6 = 0$$

Answer: (3)

8. Find the roots of the following equation using the graphical method.

$$f(x) = e^x - \sin(x) = 0$$

9. Find the roots of the following equation using the graphical method.

$$f(x) = x + \ln(x) = 0$$

10. Find the roots of the following equation using the graphical method within the interval [–1 1].

$$f(x) = x^2 - \tan(x) = 0$$

REFERENCE

1. William, H. Press, Saul A. Teukolsky, William T. Vetterling, and Brian P. Flannery, 2007. *Numerical Recipes 3rd Edition: The Art of Scientific Computing*. London: Cambridge University Press.

4 Open Numerical Methods for Solving Systems of Nonlinear Equations

An equation is said to be nonlinear when it involves terms of a degree higher than 1 in an unknown quantity. Several methods can be used to numerically solve a nonlinear system of equations, and Newton's method is the most feasible method to solve systems of nonlinear equations. This chapter studied open root-finding methods such as the method of substitution, fixed-point iteration method, Newton-Raphson (NR) method, secant method, and Muller's method.

LEARNING OBJECTIVES

1. Use the method of substitution for solving simultaneous equations.
2. Utilize the fixed-point iteration method to solve equations of the form $f(x)=0$.
3. Employ the NR method for solving equations numerically.
4. Apply the secant method for root finding.
5. Apply Muller's method to solve equations of the form $f(x)=0$.

4.1 INTRODUCTION

Nonlinear algebraic equations are one of the subjects in mathematics. A system of nonlinear equations is a system of two or more equations in two or more variables containing at least one equation that is not linear. Any equation that cannot be written in the following form is considered nonlinear [1].

$$Ax + By + C = 0 \tag{4.1}$$

Numerical methods can solve root exploration in complicated nonlinear equations, and different methods are available to solve the nonlinear equations. This chapter presents the open numerical methods, including the method of substitution, the secant method, the fixed-point iteration method, the NR method, and Muller's method.

DOI: 10.1201/9781003360544-4

Examples of nonlinear equations:

1. One-dimensional equations

$$x^2 - 6x + 9 = 0 \tag{4.2}$$

$$x - \sin(x) = 0 \tag{4.3}$$

2. Two-dimensional equations

$$(1 - x)y^2 - x^3 = 0 \tag{4.4}$$

$$y^2 + x^2 - 1 = 0 \tag{4.5}$$

The open methods do not restrict the root from remaining trapped in a closed interval. Accordingly, they are not as robust as bracketing methods and can divert, and they are more difficult to obtain compared to closed methods. The most popular are Newton's method, the secant method, and Muller's method.

4.2 METHOD OF SUBSTITUTION

The substitution method used previously for solving linear equations can still be used for nonlinear systems. We solve one equation for one variable and then substitute the result into the second equation to solve for another variable. The method of substitution involves three steps:

1. Solve one equation for one of the variables.
2. Substitute this expression into the other equation and solve.
3. Resubstitute the value into the original equation to find the corresponding variable.

Suppose we have the following algebraic equations and we want to use the substitution method to solve the two algebraic equations:

$$y - 2x = 8 \tag{4.6}$$

$$y = 5 - x \tag{4.7}$$

1. Solve the first equation for y.

$$y = 8 + 2x \tag{4.8}$$

2. Substitute this expression into the second equation.

$$(8 + 2x) = 5 - x \tag{4.9}$$

3. Rearrange.

$$8 - 5 = -2x - x = -3x \tag{4.10}$$

$$\begin{aligned} 3 &= -3x \\ x &= -1 \end{aligned} \tag{4.11}$$

4. Resubstitute the value of $x = -1$ into the first equation to find the corresponding y value.

$$y - 2x = 8 \tag{4.12}$$

$$\begin{aligned} y - 2(-1) &= 8 \\ y &= 6 \end{aligned} \tag{4.13}$$

Example 4.1 Applying Method of Substitution

Solve the following two equations using the method of substitution.

$$y = x + 1$$

$$y = x^2 + 1$$

Support the manual calculation with Simulink and Python programming.

Solution

Solve for x

$$x = y - 1$$

Substitute in the second equation to solve for y

$$y = (y - 1)^2 + 1$$

Simplify

$$y = y^2 - 2y + 2$$

Rearrange

$$0 = y^2 - 3y + 2$$

Factorize

$$(y-2)(y-1) = 0$$

Solving for y, gives $y = 1$ and $y = 2$.

Substitute each value of y in the first equation to find the value of x.

For $y = 1$,

$$x - 1 = -1$$

$$x = 0$$

For $y = 2$,

$$x - 2 = -1$$

$$x = 1$$

Accordingly, the solutions are (0,1) and (1,2).

Simulink Solution

Figure 4.1 represents the Simulink block diagram using the MATLAB function dragged from the Simulink/User-Defined Functions library [2]. The MATLAB code is associated with the MATLAB function. The MATLAB function is attached to two

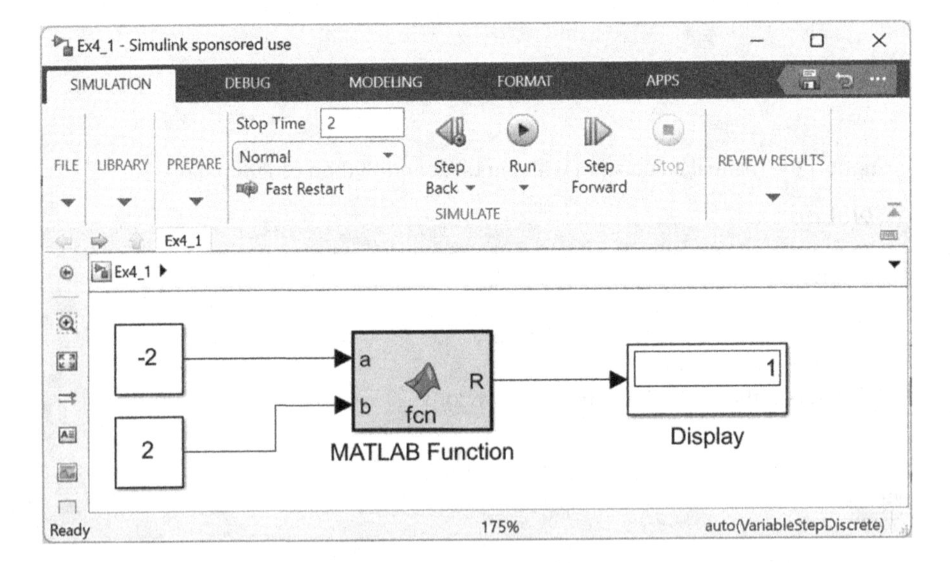

FIGURE 4.1 Graphical demonstration of the solution of equation present in Example 4.1.

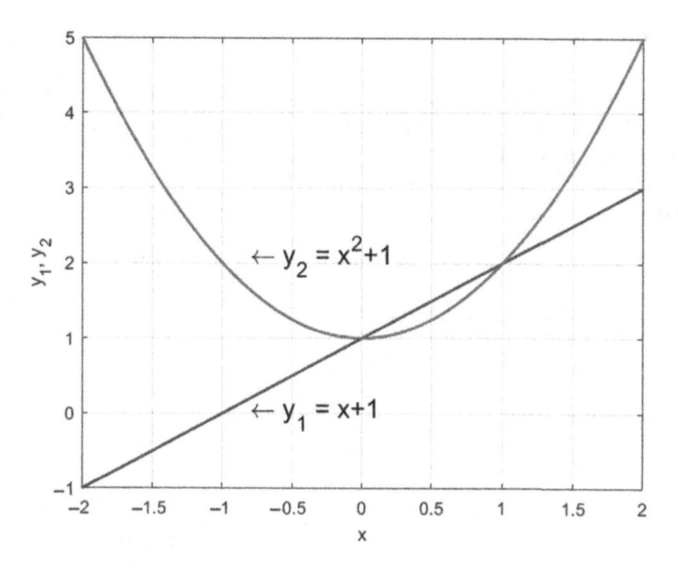

FIGURE 4.2 Plot of $y = x + 1$ and $y = x^2 + 1$ demonstrating the graphical root finding of the equation defined in Example 4.1.

ports representing the lower and upper root boundaries, a, and b, respectively. The exit port is attached to a Display block showing the value of the root within the selected interval. While Figure 4.2 illustrates the plot of the two equations in Example 4.1, the intersection of the two curves represents the root of equations ($x = 1$).

The MATLAB code embeds the Simulink function of Example 4.1 as follows:

```
% Example 4.1 Graphical method
function R = fcn(a, b)
R1=0;
for x = a:b
   y1 = x+1;
   y2 = x.^2+1;
   if abs(y2-y1)<0.001
   R1= x;
   end
end
x= linspace(a, b);
y1 = x+1;
y2 = x.^2+1;
plot (x, y1,'b', x, y2,'r', 'LineWidth', 1.5)
txt1 = '\leftarrow y_1 = x+1';
txt2 = '\leftarrow y_2 = x^2+1';
text(-0.8,0,txt1,'FontSize', 15);
text(-0.8,2.1,txt2,'FontSize', 15);
ylabel('y_1, y_2');
xlabel ('x');
grid on;
R=R1;
```

Python Solution

The following program is the Python code used to solve the two nonlinear algebraic equations defined in Example 4.1. Running the program yields the following values of the two roots [3].

```
# Example 4.1
# importing library sympy
from sympy import symbols, Eq, solve

# defining symbols used in equations
# or unknown variables
x, y = symbols('x, y')

# defining equations
eq1 = Eq((x-y), -1) # x-y =1
eq2 = Eq((y-x**2), 1) # y-x^2=1

# solving the equation
print("Values of 2 unknown variables are as follows:")

print(solve((eq1, eq2), (x, y)))
```

The results are as follows:

```
Values of 2 unknown variables are as follows:
[(0, 1), (1, 2)]
```

4.3 FIXED-POINT ITERATION METHOD

The fixed-point iteration method is a numerical analysis technique for finding roots with fixed-point iteration. This method frequently uses fixed-point concepts to calculate the solution to a given equation. The fixed point is a point in the function f as $f(x)=x$. The given function is algebraically transformed into the form $f(x)=x$; for example, assume a function $f(x)=0$, rewrite x as of $x_{n+1}=x_n$. Label the left side as x_{n+1} and right side with x_n, choose x_1, and plug it into the equation. Repeat the iteration until it converges, as shown in the graphical presentation of Figure 4.3.

The fixed-point iteration method employs the principle of a fixed point in a repeated approach to compute the solution of the desired equation. It is a point in the function $g(x)$ domain. The desired function in the fixed-point iteration technique is algebraically converted in the form of $g(x)=x$.

Consider solving the equation: $f(x)=0$. Set the equation as $x=g(x)$. Choose $g(x)$ such as:

$$|g'(x)|<1 \text{ at } x=x_o \tag{4.14}$$

where x_o is the initial guess called the fixed-point iteration approach. Then the iterative method is applied by successive approximation given by

$$x_n=g(x_n-1) \tag{4.15}$$

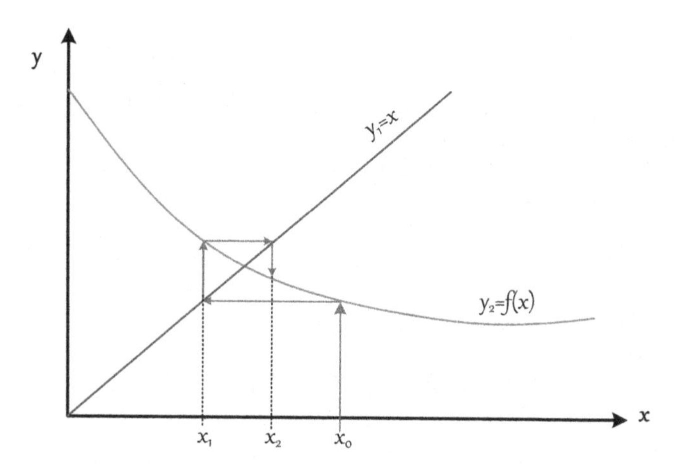

FIGURE 4.3 Graphical representation of the fixed-point iteration method.

That is

$$x_1 = g(x_o),\ x_2 = g(x_1) \tag{4.16}$$

Adhere to the following approach:

1. Choose the initial value x_o. Find the value $x = a$ and $x = b$, where $f(a) < 0$ and $f(b) > 0$. The value of x_o, is the average of a and b values.
2. Express the given equation in the form $x = g(x)$ such that:

$$|g'(x)| < 1\ at\ x = x_o \tag{4.17}$$

3. Apply the successive approximations

$$x_n = g(x_n - 1) \tag{4.18}$$

For the continuous function, $f(x) = 0$, there will be a sequence of x_n, that converges to the approximate solution.

Example 4.2 Application of the Fixed-Point Iteration Method

Using the fixed-point iteration method, find the first approximate root of the following equation up to four decimal places.

$$f(x) = 2x^3 - 2x - 5 = 0$$

Support the manual calculation with Simulink/MATLAB and Python programming of the fixed-point iteration method.

Solution

1. Follow the solution approach to finding the value of x_o, for which we must find a and b, such that $f(a) < 0$ and $f(b) > 0$.
   Let $a = 0$, $f(0) = -5$, $f(0) < 0$, and $b = 2$, $f(2) = 7$, $f(b) > 0$
2. Find $g(x)$ such that

$$|g'(x)| < 1 \text{ at } x = x_o$$

From the given function $2x^3 - 2x - 5$,

$$x^3 = \frac{2x+5}{2} \rightarrow x = \left(\frac{2x+5}{2}\right)^{1/3} = g(x)$$

To find x_o,

$$g'(x) = \frac{1}{3}\left(\frac{2x+5}{2}\right)^{-2/3} < 1$$

Solve

$$|g'(x)| = \left|\frac{1}{3}\left(\frac{2x+5}{2}\right)^{-\frac{2}{3}}\right| < 1$$

$x = 1.5$ satisfies $|g'(x)| < 1$
Applying the iterative method (Table 4.1),
$x_n = g(x_n - 1)$, for $n = 1$ to 6
Start with $x_o = 1.5$
The approximate root of $f(x)$ by the fixed-point iteration method is 1.6006.

TABLE 4.1
The Fixed-Point Iteration Method to Solve the Equation in Example 4.2

n	$g(x)$	x
1	$\left[\frac{2(1.5)+5}{2}\right]^{1/3}$	1.5874
2	$\left[\frac{2(1.5874)+5}{2}\right]^{1/3}$	1.5989
3	$\left[\frac{2(1.5989)+5}{2}\right]^{1/3}$	1.60037
4	$\left[\frac{2(1.60037)+5}{2}\right]^{1/3}$	1.60057
5	$\left[\frac{2(1.60057)+5}{2}\right]^{1/3}$	1.60059
6	$\left[\frac{2(1.60059)+5}{2}\right]^{1/3}$	1.600597

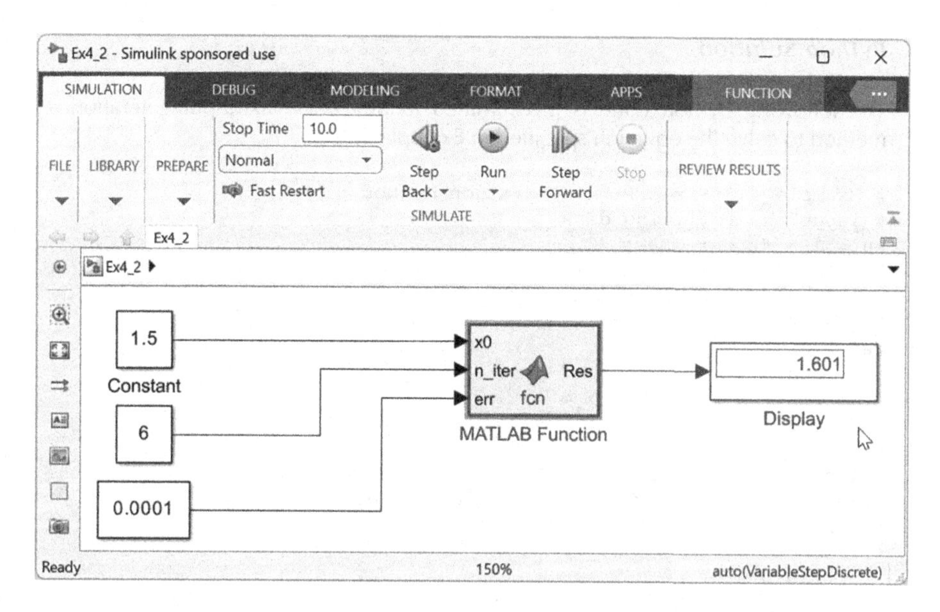

FIGURE 4.4 Simulink program uses the fixed-point iteration numerical analysis method to solve the desired equation in Example 4.2.

Simulink Solution

The Simulink block diagram demonstrating the solution of Example 4.2 is shown in Figure 4.4. The MATLAB code associated with the MATLAB function (available in the Simulink/User-Defined Functions) relates to three ports, start point (x_o), number of iterations (n_{iter}), and the allowable error (err). The output port of the function relates to the Display block showing the root of the desired equation.

The following program is MATLAB code that describes the solution of the nonlinear algebraic equation using the fixed-point iteration method. The following MATLAB code is implanted in the Simulink MATLAB function.

```
% Example 4.2
function Res = fcn(x0,n_iter, err)
g = @(x) ((2*x + 5)/2)^(1/3);
p0 = x0; %enter initial approximation
n = n_iter; % enter no. of ierations
tol = err; %tolerance
i = 1;
p=0;
while i <= n
   p = g(p0);
   if abs(p-p0) < tol
      break;
else
   i = i+1;
   p0 = p;
   end
end
Res = p;
```

Python Solution

The following Python code is programmed to utilize the fixed-point iteration method to solve the equation specified in Example 4.2.

```
# Example 4.2 Fixed Point Iteration Method
# input the following data
x0 = 1.5 #Enter Guess
e = 0.0001 #Tolerable Error
N = 6      #Maximum Number of Steps
# The required function
def f(x):
    return 2*x**3-2*x -5
# Re-writing f(x)=0 to x = g(x)
def g(x):
return ((2*x + 5)/2)**(1/3)
# Implementing Fixed Point Iteration Method
print('\n\n=== Fixed Point Iteration Method ===\n')
def fp(x0, e, N):
    i = 1
    flag = 1
    condition = True
    while condition:
        x1 = g(x0)
        print('%d x1 = %0.6f f(x1) = %0.6f' % (i, x1, f(x1)))
        x0 = x1
        i = i + 1
        if i > N:
           flag=0
           break
        condition = abs(f(x1)) > e
if flag==1:
        print('\nThe required root is: %0.8f' % x1)
else:
        print('\nNot Convergent.')

# Starting Fixed point iteration method
res= fp(x0,e, N)
```

Execution

Running the Python program produces the results shown below, where the value of $f(x_1)$ is close to zero and the discrepancy between two consecutive values is too low.

```
=== Fixed Point Iteration Method ===
1 x1 = 1.587401 f(x1) = -0.174802
2 x1 = 1.598880 f(x1) = -0.022957
3 x1 = 1.600375 f(x1) = -0.002991
4 x1 = 1.600569 f(x1) = -0.000389
5 x1 = 1.600595 f(x1) = -0.000051
The required root is: 1.60059476
```

Example 4.3 Application of Fixed-Point Iteration

Find the root of the following nonlinear algebraic equation using the fixed-point iteration method.

$$x^2 - x - 1 = 0$$

Support the manual calculations with the Simulink and Python programming of the fixed-point iteration method.

Solution

Rearrange the equation such that $x = \ldots$

$$x = 1 + \frac{1}{x}$$

Label the left side as x_{n+1} and right side with x_n

$$x_{n+1} = 1 + \frac{1}{x_n}$$

Choose

$$x_1 = 2$$

$$x_1 = 1 + \frac{1}{2} = 1.5$$

$$x_2 = 1 + \frac{1}{x_1} = 1 + \frac{1}{1.5} = 1.666$$

Repeat

$$x_3 = 1 + \frac{1}{x_2} = 1 + \frac{1}{1.5} = 1.666$$

$$x_4 = 1 + \frac{1}{x_3} = 1 + \frac{1}{1.666} = 1.6$$

$$x_5 = 1 + \frac{1}{x_4} = 1 + \frac{1}{1.6} = 1.625$$

$$x_6 = 1 + \frac{1}{x_5} = 1 + \frac{1}{1.625} = 1.6125$$

Simulink Solution

The fixed-point iteration method is coded with MATLAB and linked to the Simulink MATLAB function available under the Simulink/User-Defined Functions library. Figure 4.5 demonstrates the Simulink graphical programming using the MATLAB function techniques. The inlet port of the MATLAB function is connected to a

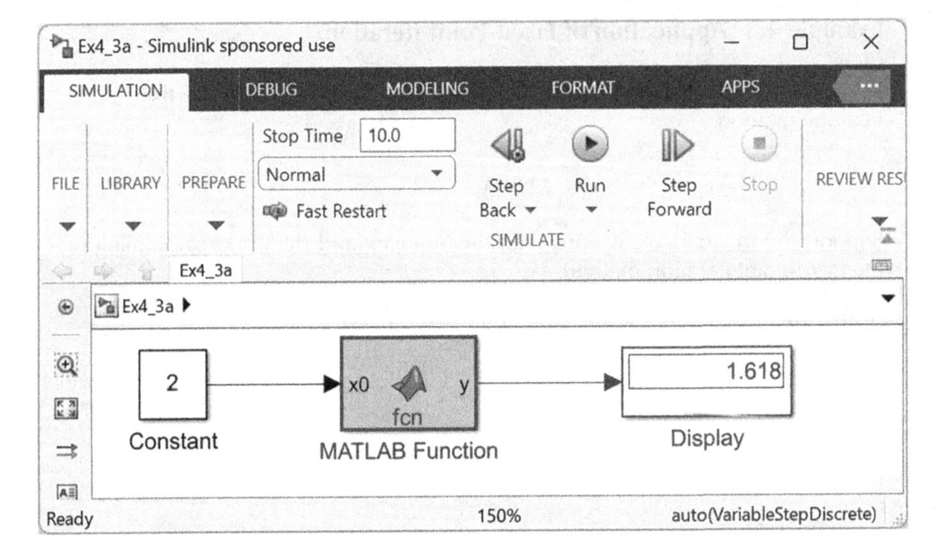

FIGURE 4.5 Simulink block diagram utilizing the fixed-point iteration method to solve the equation specified in Example 4.3.

constant block that shows the initial guess (x_o), and the exit port is connected to a Display block showing the numerical value of the calculated root.

The following program is the MATLAB code programmed for the fixed-point iteration method in MATLAB. The code is placed in the Simulink MATLAB function.

```
% Exmaple 4.3
% Fixed point iteration method
function y =fcn(x0)
x=x0;
TOL = 0.001;
N = 100;
% set the iteration index
i=1;
err = 1;
% Start the iteration
while (i<N && err>TOL)
      i = i +1;
      x = 1 + 1/x;
      err = abs (x-1-1/x);
end
% display the results
y=x;
```

An alternative way of using Simulink is utilizing the Algebraic Constraint block from the Simulink/Math operation library and the other needed blocks, as demonstrated in Figure 4.6. After connecting the blocks, the figure represents the solution of the equation specified in Example 4.3. The initial guess must be changed; accordingly, double-click on the "Algebraic Constraint" block, and set the initial guess to 2 to get the root value of 1.618.

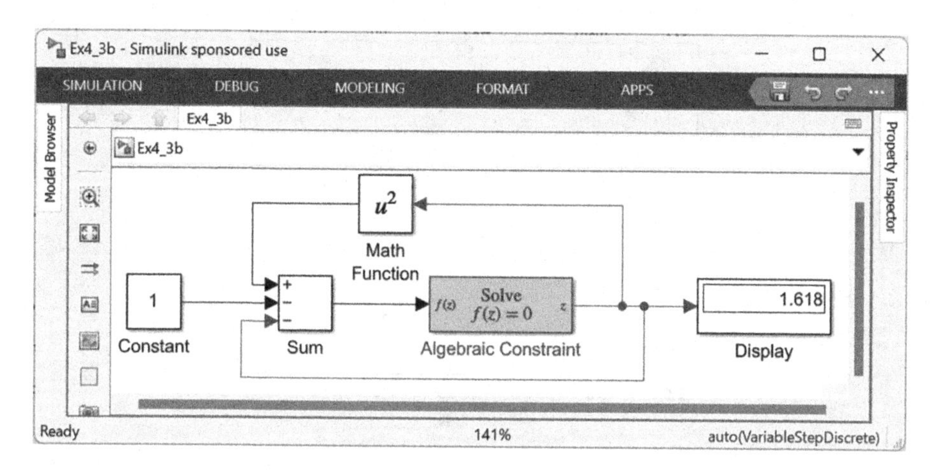

FIGURE 4.6 Simulink solution using the Algebraic constraint block to solve the equation specified in Example 4.3.

Python Solution

The following Python program uses the fixed-point iteration method to solve Example 4.3. The executed results follow the program.

```
# Example 4.3 Fixed Point Iteration Method
import numpy as np
import matplotlib.pyplot as plt
def f(x):
    return x*x - x -1
# Input Section
x0 =1.5 # Enter Guess
e = 0.0001 # Tolerable Error
N = 100 #Maximum Step

# Re-writing f(x)=0 to x = g(x)
def g(x):
return (x+1)/x
# Implementing Fixed Point Iteration(FIP) Method
def FIP(x0, e, N):
    print('\n\n*** FIXED POINT ITERATION ***')
    step = 1
    flag = 1
    condition = True
    while condition:
        x1 = g(x0)
        print('Iteration-%d, x1 = %0.6f and f(x1) = %0.6f' %
        (step, x1, f(x1)))
x0 = x1

step = step + 1
if step > N:
flag=0
break
condition = abs(f(x1)) > e

if flag==1:
print('\nRequired root is: %0.8f' % x1)
```

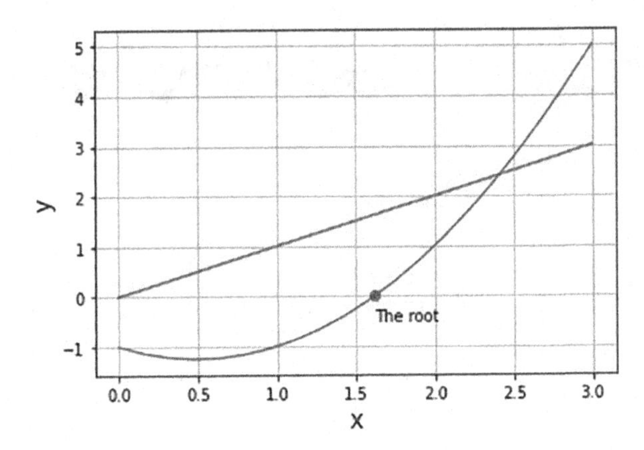

FIGURE 4.7 Python programming of the fixed-point iteration method to solve the equation present in Example 4.3.

```
plt.plot (x1,0, 'ro')
plt.text(x1,-0.5, 'The root')
else:
    print('\nNot Convergent.')

# Starting the Fixed Point Iteration mehtod
res=FIP(x0,e, N)
x=np.linspace(0,3,20)
plt.plot(x, f(x), x, x, 'r')
plt.xlabel('x', fontsize = 15)
plt.ylabel ('y', fontsize = 15)
plt.grid(True)
```

After running the Python program, you should get the following results and the plot shown in Figure 4.7.

```
*** FIXED POINT ITERATION ***

Iteration-1, x1 = 1.666667 and f(x1) = 0.111111
Iteration-2, x1 = 1.600000 and f(x1) = -0.040000
Iteration-3, x1 = 1.625000 and f(x1) = 0.015625
Iteration-4, x1 = 1.615385 and f(x1) = -0.005917
Iteration-5, x1 = 1.619048 and f(x1) = 0.002268
Iteration-6, x1 = 1.617647 and f(x1) = -0.000865
Iteration-7, x1 = 1.618182 and f(x1) = 0.000331
Iteration-8, x1 = 1.617978 and f(x1) = -0.000126
Iteration-9, x1 = 1.618056 and f(x1) = 0.000048

Required root is: 1.61805556
```

4.4 NR METHOD

The NR method is utilized to find an approximate root of a real-valued function ($f(x)=0$). The method uses a continuous and differentiable function that can be approximated by a straight-line tangent, as graphically demonstrated in Figure 4.8. We make an initial guess for the desired root.

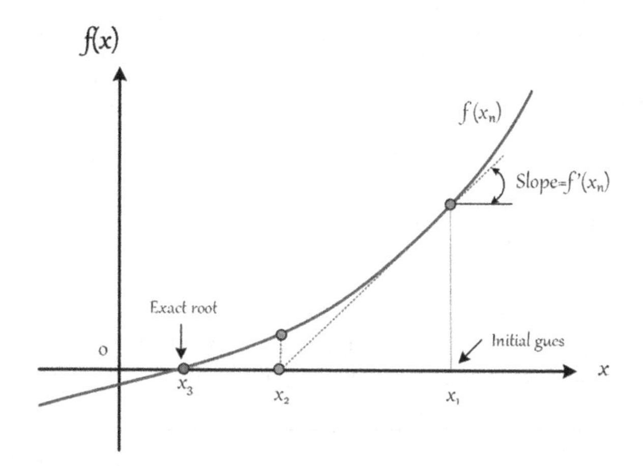

FIGURE 4.8 Graphical demonstration of NR method for root finding.

Find points a and b, such that $a < b$, and $f(a) \cdot f(b) < 0$. The NR method finds the tangent to the function $f(x)$ at $x = x_0$ and extrapolates it to intersect the x-axis to get x_1. The intersection point represents the new approximation to the root, and the procedure is repeated until convergence is obtained.

$$x_{n+1} = x_n - \frac{f(x_n)}{f'(x_n)} \tag{4.19}$$

The procedure of finding the root using the NR method:

1. Find points a and b, such that $a < b$ and $f(a) \cdot f(b) < 0$.
2. Select an interval $[a, b]$ where the root exists in between.
3. Find $f(x_o)$ and $f'(x_o)$.
4. Find x_1, such that

$$x_1 = x_o - \frac{f(x_o)}{f'(x_o)} \tag{4.20}$$

5. If $f(x_1) = 0$, then x_1 is an exact root, else $x_0 = x_1$.
6. Repeat steps 3 and 4 until $f(x_i) = 0$, or $f(x_i)$ desired accuracy.

Example 4.4 NR Method for Single Polynomial Equation

Use the NR method to find the root of the following quadratic equation.

$$f(x) = x^3 - x - 1 = 0$$

Validate the manual root finding with Simulink and Python programming of the NR method.

Solution

The function $f(x)$,

$$f(x) = x^3 - x - 1$$

The slope of $f(x)$ is the $f'(x)$,

$$f'(x) = 3x^2 - 1$$

Find the interval where the root lies between the intervals. Try different values of x until the function $f(x)$ gets consecutive negative and positive signs.

x	0	1	2
$f(x)$	–1	–1	5

The root lies within intervals 1 and 2. The first initial guess is

$$x_o = \frac{1+2}{2} = 1.5$$

The first iteration

$$x_1 = x_o - \frac{f(x_0)}{f'(x_o)} = 1.5 - \frac{1.5^3 - 1.5 - 1}{3*(1.5)^2 - 1} = 1.5 - \frac{0.87}{5.7} = 1.35$$

Repeat the iterations until the value of x_i is almost constant, as shown in Table 4.2. The approximate root is 1.3247.

Simulink Solution

With Simulink, the root can be found in two ways: the first is by using the "Algebraic Constraint" block along with other required blocks such as Multiplication, Constant, Sum, and Display, as illustrated in Figure 4.9. The second method uses the Simulink "Matlab function" block available in the user-defined functions library (Figure 4.10). The MATLAB function requires writing a MATLAB code encrypting the NR method. The associated MATLAB program with the MATLAB function utilizes the NR method.

TABLE 4.2
NR Method for Finding the Root of the Equation Given in Example 4.4

S. No.	x_0	$f(x_0)$	$f'(x_o)$	x_i
1	1.5	0.875	5.75	1.3478
2	1.3478	0.10068	4.44991	1.3252
3	1.3252	0.00206	4.26847	1.3247
4	1.3247	0	4.26463	1.3247

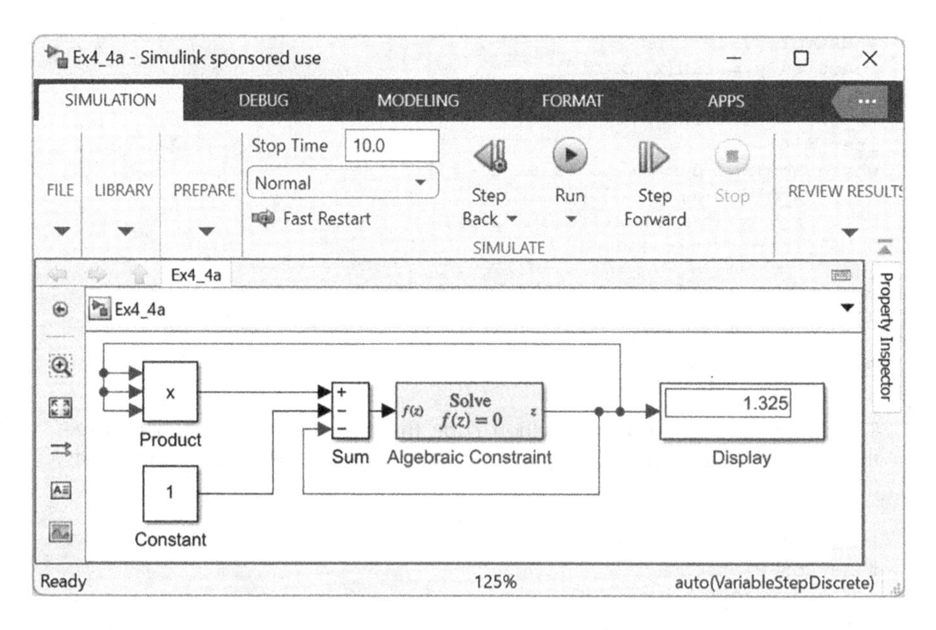

FIGURE 4.9 Root finding using the Simulink Algebraic constraint block to solve the quadratics equations defined in Example 4.4.

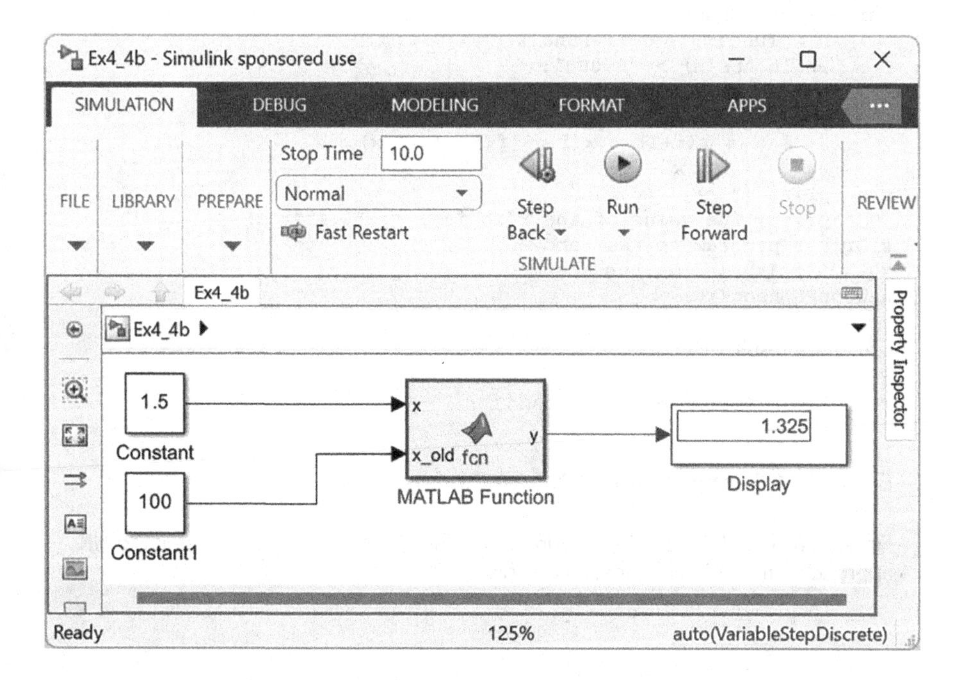

FIGURE 4.10 Simulink block diagram utilizing the MATLAB function and NR method representing the solution of the equation defined in Example 4.4.

```
% Example 4.4
function y = fcn(x, x_old)
%x = initial guess;
%x_old = previous x;
iter = 0;
while abs(x_old-x) > 10^-3 && x ~= 0
        x_old = x;
        x = x - (x^3- x-1)/(3*x^2-1);
        iter = iter + 1;
end
y=x;
```

Python Solution

The following program is the Python code that uses the NR method to solve the quadratic equation in Example 4.4. The program is followed by the results implemented.

```
# Example 4.4
# Newton Raphson
# The function is x^3- x -1
def func(x):
      return x**3- x -1
# Derivative of the above function
def derivFunc(x):
      return 3 * x**2-1
# Function to find the root
def newtonRaphson(x):
      h = func(x) / derivFunc(x)
      while abs(h) >= 0.0001:
               h = func(x)/derivFunc(x)

               # x(i+1) = x(i) - f(x) / f'(x)
               x = x - h
      print (" ")
      print("The value of the root is : ", "%.4f"% x)
# Driver program to test above
x0 = 5 # Initial values assumed
newtonRaphson(x0)
```

Execution result:

```
The value of the root is: 1.3247
```

Example 4.5 Application of NR Method

Find the root of the following quadratic nonlinear algebraic equation using the NR approach, and an initial guess value equals 4.

$$f(x)=x^2-7$$

Confirm the manual root finding with Simulink and Python programming of the NR method.

TABLE 4.3
Solution of Example 4.5 Using the NR Method

S. No.	x_o	$f(x_0)$	$f'(x_o)$	x_i
1	4.00000	9.00000	8.00000	2.87500
2	2.87500	1.26560	5.75000	2.65489
3	2.65489	0.04844	5.30978	2.64577
4	2.64577	0.0000832	5.29153	2.64575

Solution

Derivation of a nonlinear algebraic equation:

$$f'(x) = 2x$$

Using the NR method, the first iteration (starting with $x=4$):

$$x_1 = x_o - \frac{f(x_0)}{f'(x_o)} = 4 - \frac{4^2 - 7}{2(4)} = 4 - \frac{9}{8} = 2.875$$

The second iteration uses the new $x_1 = 2.875$

$$x_2 = x_1 - \frac{f(x_1)}{f'(x_1)} = 2.875 - \frac{2.875^2 - 7}{2(2.875)} = 2.875 - \frac{1.266}{5.75} = 2.655$$

Repeat the iterations until the value of x_i is almost constant, as shown in Table 4.3. The table shows that after four iterations, the value of 2.64575 is the approximate value of one of the roots.

Simulink Solution

Figure 4.11 is the Simulink block diagram utilizing NR's method for solving the equation required in Example 4.5. The MATLAB code incorporated the MATLAB function block as follows.

The following MATLAB code built into the MATLAB function block is programmed in the NR method to solve the equation specified in Example 4.5.

```
%Example 4.5
function y = fcn(x, x_old)
%x = initial guess;
%x_old = previous x;
iter = 0;
while abs(x_old-x) > 10^-3 && x ~= 0
    x_old = x;
    x = x - (x^2-7)/(2*x);
    iter = iter + 1;
end
y=x;
```

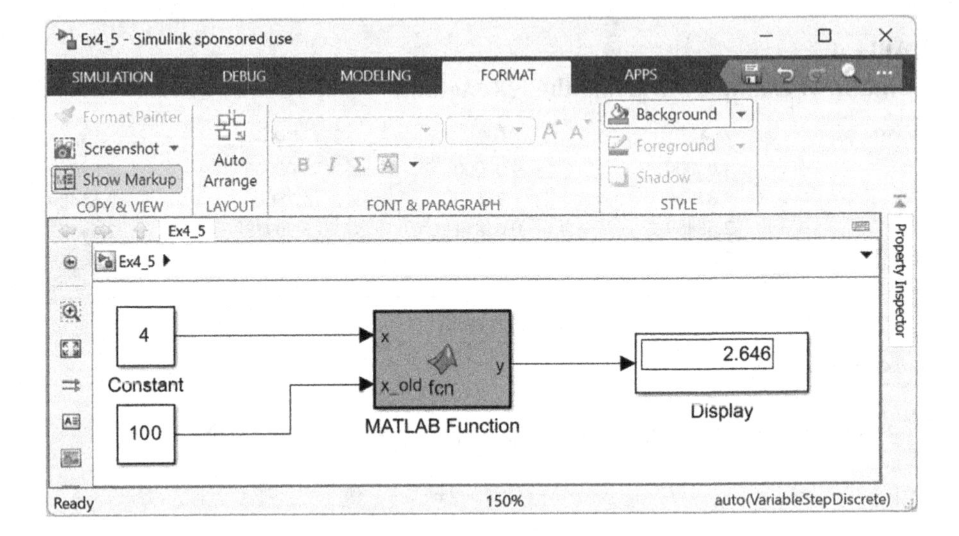

FIGURE 4.11 Simulink block diagram utilizing the NR method to solve of equation defined in Example 4.5.

Python Solution

The below program is written in Python, employing the NR method to solve the nonlinear algebraic equation defined in Example 4.5. The implementation result follows the program.

```
# Example 4.5
# Newton Raphson
# The function is x^3- x -1
def func(x):
     return x**2-7
# Derivative of the above function
def derivFunc(x):
     return 2 * x
# Function to find the root
def newtonRaphson(x):
     h = func(x) / derivFunc(x)
     while abs(h) >= 0.0001:
          h = func(x)/derivFunc(x)

          # x(i+1) = x(i) - f(x) / f'(x)
          x = x - h
     print (" ")
     print("The value of the root is : ", "%.4f"% x)
# Driver program to test above
x0 = 7 # Initial values assumed
newtonRaphson(x0)
```

Execution results:

```
The value of the root is: 2.6458
```

4.5 NR MULTIVARIABLE

The multivariate NR method directly extends the single variable method. While the single variable method solves a single equation of the form $f(x)=0$, the multivariate version solves a system of more than one equation of the form

$$f_1(x_1, x_2, x_3, \ldots, x_{n-1}, x_n)=0$$

$$f_2(x_1, x_2, x_3, \ldots, x_{n-1}, x_n)=0$$

$$f_3(x_1, x_2, x_3, \ldots, x_{n-1}, x_n)=0$$

$$f_n(x_1, x_2, x_3, \ldots, x_{n-1}, x_n)=0$$

The multivariable method formula is as follows:

$$\begin{Bmatrix} x \\ y \end{Bmatrix}^{k+1} = \begin{Bmatrix} x \\ y \end{Bmatrix}^{k} - \begin{bmatrix} \dfrac{\partial f_1}{\partial x} & \dfrac{\partial f_1}{\partial y} \\ \dfrac{\partial f_2}{\partial x} & \dfrac{\partial f_2}{\partial y} \end{bmatrix}^{-1} \begin{Bmatrix} f_1 \\ f_2 \end{Bmatrix}^{k} \tag{4.21}$$

Example 4.6 Multivariate NR Method

Consider the system of two nonlinear algebraic equations. Find the values of the unknown variables using the multivariate NR method.

$$2x^3+8x-4y=4$$

$$4x-3y-y^2=1$$

Validate the manual root finding with Simulink and Python programming of the multivariate NR method.

Solution

Rewrite the equations in the following form:

$$f_1(x,y)=4-8x-2x^3+4y=0$$

$$f_2(x,y)=1-4x+3y+y^2=0$$

Take initial guess $(x,y)=(0.5,0.5)$

$$\begin{Bmatrix} x \\ y \end{Bmatrix}^{k+1} = \begin{Bmatrix} x \\ y \end{Bmatrix}^{k} - \begin{bmatrix} \dfrac{\partial f_1}{\partial x} & \dfrac{\partial f_1}{\partial y} \\ \dfrac{\partial f_2}{\partial x} & \dfrac{\partial f_2}{\partial y} \end{bmatrix}^{-1} \begin{Bmatrix} f_1 \\ f_2 \end{Bmatrix}^{k}$$

Take the derivatives of functions 1 and 2

$$\begin{Bmatrix} x \\ y \end{Bmatrix}^{k+1} = \begin{Bmatrix} 0.5 \\ 0.5 \end{Bmatrix}^{k} - \begin{bmatrix} -8-6x^2 & 4 \\ -4 & 3+2y \end{bmatrix}^{-1} \begin{Bmatrix} f_1 \\ f_2 \end{Bmatrix}^{k}_{0.5,0.5}$$

Find the values of f_1 and f_2

$$\begin{Bmatrix} x \\ y \end{Bmatrix} = \begin{Bmatrix} 0.5 \\ 0.5 \end{Bmatrix} - \begin{bmatrix} -8-6x^2 & 4 \\ -4 & 3+2y \end{bmatrix}^{-1} \begin{Bmatrix} 1.75 \\ 0.75 \end{Bmatrix}$$

Substitute values of x and y

$$\begin{Bmatrix} x \\ y \end{Bmatrix} = \begin{Bmatrix} 0.5 \\ 0.5 \end{Bmatrix} - \begin{bmatrix} -9.5 & 4 \\ -4 & 4 \end{bmatrix}^{-1} \begin{Bmatrix} 1.75 \\ 0.75 \end{Bmatrix}$$

Find the inverse of the matrix

$$\begin{Bmatrix} x \\ y \end{Bmatrix} = \begin{Bmatrix} 0.5 \\ 0.5 \end{Bmatrix} - \frac{1}{22} \begin{bmatrix} 4 & -4 \\ 4 & -9.5 \end{bmatrix} \begin{Bmatrix} 1.75 \\ 0.75 \end{Bmatrix}$$

$$x = 0.5 + \frac{1}{22}(4 \times 1.75 - 4 \times 0.75) = 0.6818$$

$$y = 0.5 + \frac{1}{22}(4 \times 1.75 - 9.5 \times 0.75) = 0.4943$$

Continue the iteration as before till there is a convergence.

Simulink Solution

Figure 4.12 presents the Simulink block diagram for solving Example 4.6 using the Simulink Algebraic constraint block.

Python Solution

The following Python program shows how to use the NR algorithm to find the roots of a multivariate system of equations. The NR algorithm is relatively simple and very powerful for finding the roots of a real-valued function that has a continuous derivative within a given range

```
# Example 4.6 Multivariant Newton-Raphson
import numpy as np
# differentiate (Jacobian)
def jacobian_Ex(xy):
    x, y = xy
    return [[6*x**2 + 8, -4],
            [4, -3-2*y]]
# add your equations
def func_Ex(xy):
x, y = xy
return [2*x**3+8*x-4*y-4, 4*x-3*y-y**2-1]
```

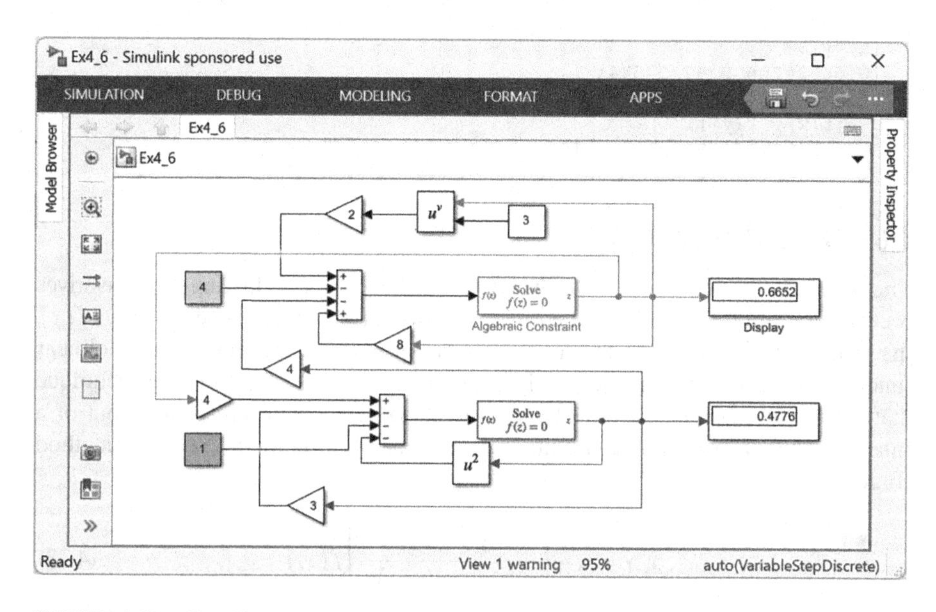

FIGURE 4.12 Simulink solution equation defined in Example 4.6.

```
# From the exercise:
def iter_newton(fun, x_init, jacobian):
    max_iter = 50
    epsilon = 1e-8

x_last = x_init
for k in range(max_iter):

# Solve J(xn)*(xn+1- xn) = -F(xn):
J = np.array(jacobian(x_last))
F = np.array(fun(x_last))

diff = np.linalg.solve(J, -F)
x_last = x_last + diff

# Stop condition:
if np.linalg.norm(diff) < epsilon:
   print('\nNumber of iterations:', k)
   break
else: # only if the for loop end 'naturally.'
   print('not converged')
return x_last

# Results:
xy_sol = iter_newton(func_Ex, [1.0,2.0], jacobian_Ex)
print('\nSolution (roots)\n [x, y]:\n', xy_sol)
print('\nresubstitution of the roots\n [f(x), f(y)]:\n',
func_Ex(xy_sol))
```

Execution results:

```
Number of iterations: 5

Solution (roots)
```

```
[x, y]:
[0.66520106 0.47757534]
resubstitution of the roots
   [f(x), f(y)]:
   [-8.881784197001252e-16, -1.1102230246251565e-16]
```

4.6 SECANT METHOD

The secant method is a variation of Newton's method when evaluating the derivatives is complex. Imitated locally by the linear function $g(x)$, which is the secant to $f(x)$, the root of $g(x)$ is taken as an improved approximation to the root of the nonlinear function $f(x)$. The secant method is numerical analysis, a root-finding technique that uses a succession of roots of secant lines to better approximate the root of a function $f(x)$. The method is a finite difference approximation of Newton's method (Figure 4.13).

$$x_{i+1} = x_i - \left[\frac{x_i - x_{i-1}}{f(x_i) - f(x_{i-1})}\right] f(x_i) \tag{4.22}$$

As shown in Figure 4.2 ($i = 1$),

$$x_2 = x_1 - \left[\frac{x_1 - x_0}{f(x_1) - f(x_o)}\right] f(x_1) \tag{4.23}$$

In general, applying a backward approximation, the secant algorithm to find the root becomes

$$x_n = x_{n-1} - \frac{(x_{n-1} - x_{n-2})}{f(x_{n-1}) - f(x_{n-2})} f(x_{n-1}) \tag{4.24}$$

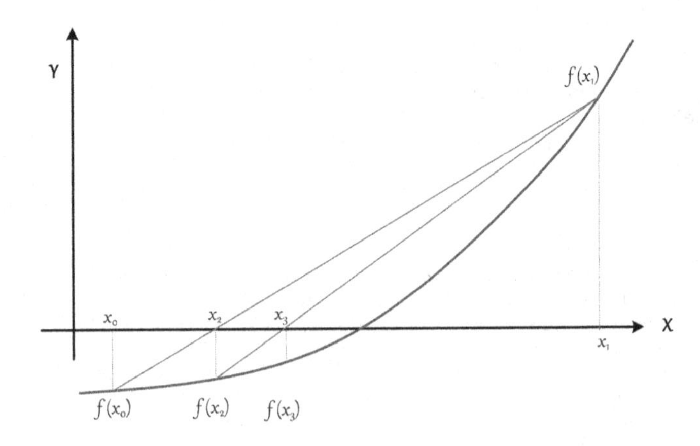

FIGURE 4.13 Graphical explanation of the numerical root finding using the secant method.

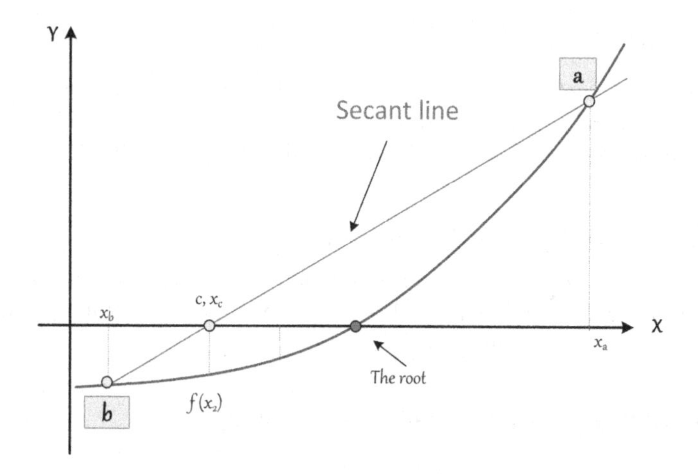

FIGURE 4.14 Graphical description of the secant method.

for $i = 2$,

$$x_2 = x_1 - \frac{(x_1 - x_0)}{f(x_1) - f(x_0)} f(x_1) \tag{4.25}$$

Figure 4.13 is a graphical demonstration of the secant method for root finding. The method is like the bisectional method, which divides each interval by choosing the midpoint; in contrast, the secant method divides each interval by the secant line connecting the endpoints. Providing that $f(x)$ is continuous on $[a, b]$ and $f(a)f(b) < 0$, the secant method always converges to a root for $f(x) = 0$. Considering the connection of the endpoint values $(a, f(a))$ and $(b, f(b))$. The line connecting these two points is called the secant line (Figure 4.14) and is given by the following formula:

$$y = \frac{f(b) - f(a)}{b - a}(x - a) + f(a) \tag{4.26}$$

The point at which the secant line crosses the x-axis is

$$0 = \frac{f(b) - f(a)}{b - a}(x - a) + f(a) \tag{4.27}$$

Solving for x

$$x = a - f(a)\frac{b - a}{f(b) - f(a)} \tag{4.28}$$

4.6.1 Secant Method Advantages

1. It uses the recent approximation roots to find new approximations instead of being bounded by the interval to enclose the root.
2. The secant method converges faster than the bisection method.

4.6.2 Secant Method Drawbacks

1. The secant technique does not always guarantee convergence.
2. Since convergence is not assured, a maximum number of iterations is implemented on computer programs.

Example 4.7 Application of the Secant Method

Compute the following function in two iterations using the secant method, in which the real roots of the equation $f(x)$ lie in the interval (0,1).

$$f(x) = x^3 - 5x + 1 = 0$$

Confirm the manual rooting find process using Simulink and Python programming of the secant method.

Solution

Using the given interval

$$x_o = 0,\ f(x_0) = f(0) = 1$$

$$x_1 = 1,\ f(x_1) = f(1) = -3$$

Hence,

$$f(a)f(b) < 0$$

Using the secant method formula:

$$x_2 = x_1 - \frac{(x_1 - x_o)}{f(x_1) - f(x_o)} f(x_1) = 1 - \frac{(1-0)}{f(1) - f(0)} f(1)$$

$$x_2 = 1 - \frac{(1-0)}{f(1) - f(0)} f(1) = 1 - \frac{1}{-3-1}(-3) = 0.25$$

The second approximation

$$x_3 = x_2 - \frac{(x_2 - x_1)}{f(x_2) - f(x_1)} f(x_2) = 0.25 - \frac{(0.25-1)}{f(0.25) - f(1)} f(0.25)$$

$$x_3 = 0.25 - \frac{(0.25-1)}{-0.234-(-3)}(-0.234) = 0.1866$$

Since $f(x_3) = 0.073$, more trials are required until $f(x_i)$ is close to zero.

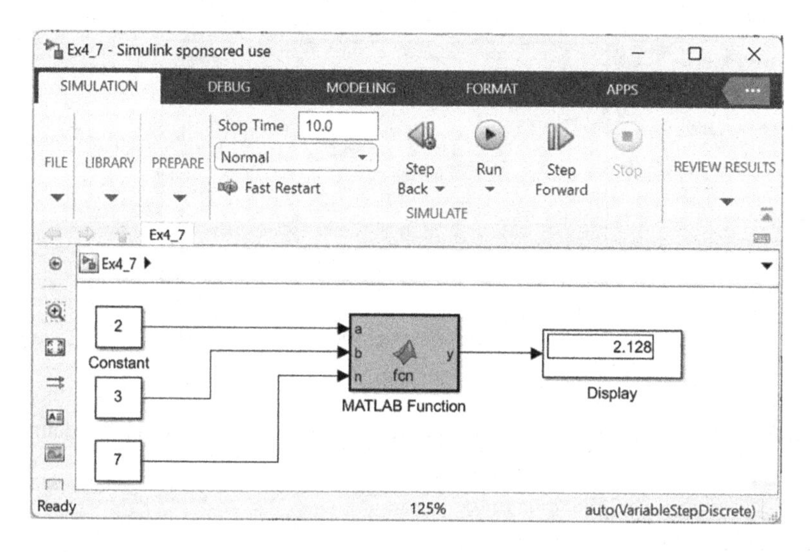

FIGURE 4.15 Simulink block diagram using the secant method describing the solution of the equation defined in Example 4.7.

Simulink Solution

Figure 4.15 is the Simulink block diagram representing the solution of the equation specified in Example 4.7. The MATLAB code follows the diagram representing the Secant method implants in the Simulink MATLAB function.

The embedded MATLAB code programmed the secant method to solve the equation defined in Example 4.7.

```
% Example 4.7
function y = fcn(a, b, n)
f=@(x)(x^3-5*x + 1);
xn1=a;
xn2=b;
xn = (xn2*f(xn1) - xn1*f(xn2))/(f(xn1) - f(xn2));
k = 0;
while (k < n)
   k = k + 1;
   xn2 = xn1;
   xn1 = xn;
   xn = (xn2*f(xn1) - xn1*f(xn2))/(f(xn1) - f(xn2));
end
y = xn;
```

Python Solution

The following is the Python code utilizing the secant method in solving the equation specified in Example 4.7.

```
# Example 4.7
# The Secant method function (SM)
```

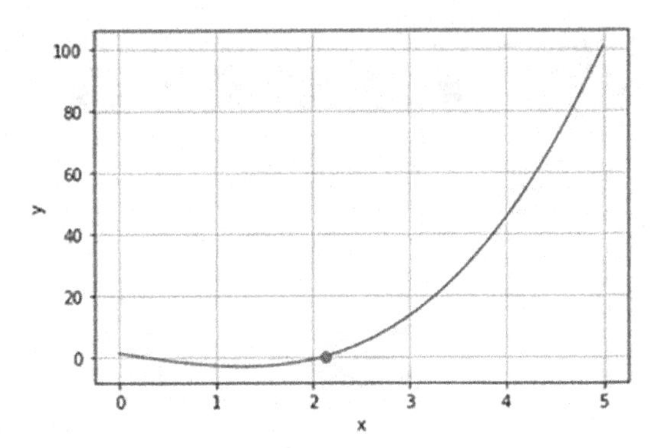

FIGURE 4.16 Python solution using the secant method of the equation defined in Example 4.7.

```
import numpy as np
import matplotlib.pyplot as plt
# Our f(x) function
def f(x):
y = x**3-5*x + 1
return y

def SM(f, x1, x2):

while abs(f(x1)) > 1e-6:
      fx1 = f(x1)
      fx2 = f(x2)
      xtemp = x1
      x1 = x1- (x1- x2) * fx1/ (fx1- fx2)
      x2 = xtemp
   return x1
#call the SM function with inital guess, x1, x2
Root = SM(f, 2, 4)
print (Root)
x =np.linspace (0, 5)
plt.plot(x, f(x), Root, f(Root), 'ro', Root)
plt.xlabel ("x")
plt.ylabel("y")
plt.grid()
```

The data generated from running the Python code is plotted in Figure 4.16.

Example 4.8 Secant Method to Solve a Cubic Equation

Using the secant method, compute the root of the equation in the interval [0, 1]. Make sure that the root should be correct to three decimal places.

$$f(x) = x^3 - 5x + 1 = 0$$

Validate the manual root finding with Simulink and Python programming of the secant method for a cubic equation.

Solution

The equation is bounded by the interval [0,1].

$$x_o = 0,\ f(x_o) = 1,\ x_1 = 1,\ f(x_1) = -3$$

$$x_2 = x_1 - \left[\frac{x_0 - x_1}{f(x_o) - f(x_1)}\right] f(x_1)$$

Substitute

$$x_2 = 1 - \left[\frac{0-1}{1-(-3)}\right](-3) = 0.25$$

$$f(x_2) = -0.234375$$

The second approximation is

$$x_3 = x_2 - \left[\frac{x_1 - x_2}{f(x_1) - f(x_2)}\right] f(x_2)$$

$$x_3 = (-0.234375) - \left[\frac{1-0.25}{-3-(-0.234375)}\right](-0.234375) = 0.186441$$

The third approximation is

$$x_4 = x_3 - \left[\frac{x_2 - x_3}{f(x_2) - f(x_3)}\right] f(x_3)$$

Substitute previous x_3 and x_2

$$x_4 = 0.186441 - \left[\frac{0.25 - 0.186441}{(-0.234375) - (0.074276)}\right](-0.234375) = 0.201736$$

The fourth approximation is

$$x_5 = x_4 - \left[\frac{x_4 - x_5}{f(x_4) - f(x_5)}\right] f(x_4)$$

Substitute the old x_4

$$x_5 = 0.201736 - \left[\frac{0.186441 - 0.201736}{0.074276 - -(-0.00470)}\right](-0.00470) = 0.201640$$

Simulink Solution

The Secant method followed by the implanted MATLAB coded in MATLAB function is presented in the Simulink graphical programming shown in Figure 4.17.

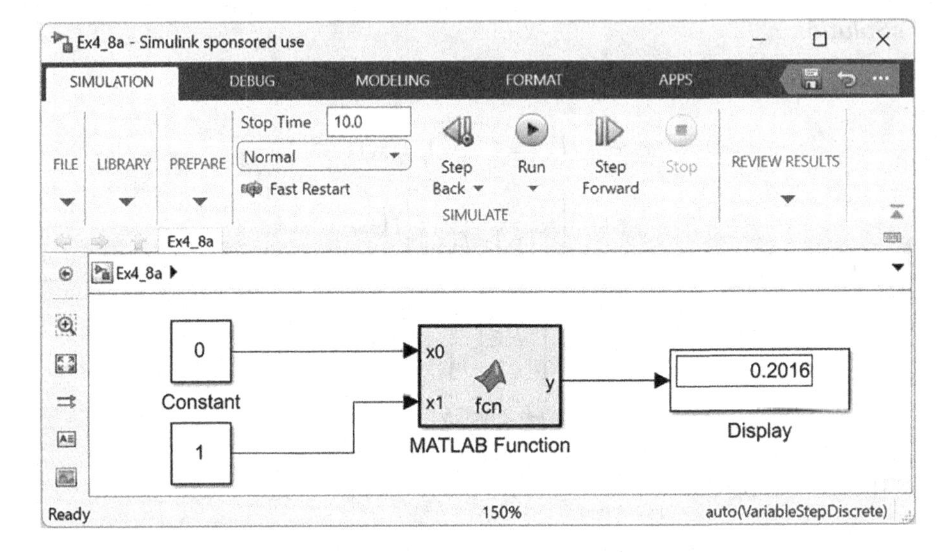

FIGURE 4.17 Secant method for the solution of the equation defined in Example 4.8.

The following program is the MATLAB code encrypting the secant method and is associated with the Simulink MATLAB function block.

```
function y = fcn(x0,x1)
% Secant Method in MATLAB
f=@(x)x^3-5*x+1; % Roots of the function;xA3-5*x+1
tol = 0.001; % tolerance
itr=1000;%nurnber of iteration
for i=1:itr
    x2=(x0*f(x1)-x1*f(x0))/(f(x1)-f(x0));
    if abs(x2-x1)<tol
        break;
else
    x0=x1;
    x1=x2;
    end
end
y=x2;
```

The following root is obtained using the Math operation/Algebraic constraint function from the Simulink library with an initial guess of zero (Figure 4.18).

Python Solution

The following Python program utilizes the secant method coded to solve the equation defined in Example 4.8.

```
# Example 4.8
# Implementation of SECANT METHOD
# Defining Function
```

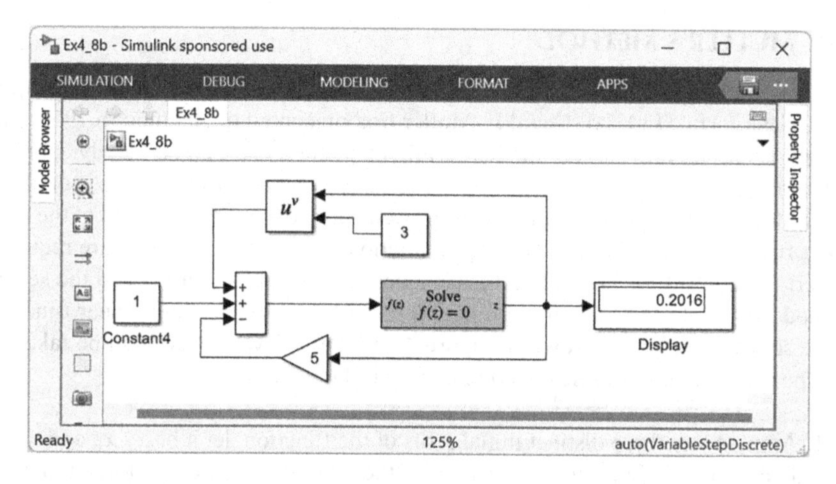

FIGURE 4.18 Simulink solution using the Algebraic constraint block of the equation defined in Example 4.8.

```
def f(x):
return x**3-5*x + 1

# Implementing Secant Method
def secant(x0,x1,e, N):
    print('\n\n Secant Method ')
    print(' =================================== ')
    step = 1
    condition = True
    while condition:
        if f(x0) == f(x1):
        print('Divide by zero error!')
break
x2 = x0- (x1-x0)*f(x0)/(f(x1) - f(x0))
x0 = x1
x1 = x2
step = step + 1
if step > N:
print('Not Convergent!')
break

condition = abs(f(x2)) > e
print('\n Required root is: %0.8f' % x2)
# Input Section
x0 = 0 # First Guess
x1 = 5 # Second Guess
ET = 0.0001 # Error Tolerable (ET)
N = 100 # Maximum Step

# Starting Secant Method
secant(x0,x1,ET, N)
```

Execution results:

```
Secant Method
===================================
The required root is: 0.20162848
```

4.7 MULLER'S METHOD

Muller's method is a root-finding algorithm, a numerical method for solving equations of the form $f(x) = 0$. David E. Muller first presented the method in 1956. The method is based on the secant method. Muller's method is based on locally approximating the nonlinear function $f(x)$ by a quadratic function $g(x)$, and the root of the quadratic function $g(x)$ is taken as an improved approximation to the root of the nonlinear function $f(x)$. Three initial approximations x_0, x_1, and x_2, which are required to start the algorithm. The only difference between Muller's method and the secant method is that $g(x)$ is a quadratic function in Muller's method but a linear function in the secant method, as shown in Figure 4.19. The following steps can be taken to find the roots of an algebraic equation using Muller's method:

1. Assume any three distinct initial roots of the function; let it be x_o, x_1, and x_2.
2. Draw a parabola through the values of the function $f(x)$ for the three initial guess points x_0, x_1, and x_2.

   The equation of the second-degree polynomial through the three points is:

$$g(x) = c + b(x - x_2) + a(x - x_2)^2 \tag{4.29}$$

   where a, b, and c are constants.
3. After drawing the parabola, then find the intersection of this parabola with the x-axis; let us say x_3.
   Calculate δ_o

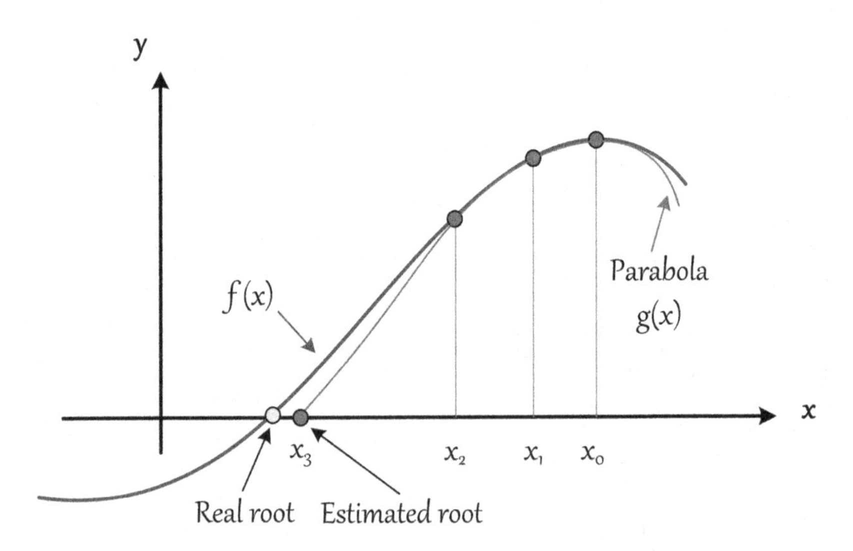

FIGURE 4.19 Graphical presentation of Muller's method.

$$\delta_o = \frac{f(x_1) - f(x_o)}{x_1 - x_o} \tag{4.30}$$

Calculate δ_1

$$\delta_1 = \frac{f(x_2) - f(x_1)}{x_2 - x_1} \tag{4.31}$$

Calculate h_o

$$h_o = x_2 - x_o \tag{4.32}$$

Calculate h_1

$$h_1 = x_1 - x_2 \tag{4.33}$$

Calculate a, b, and c

$$a = \frac{\delta_1 - \delta_o}{h_1 + h_o} \tag{4.34}$$

$$b = ah_1 + \delta_1 \tag{4.35}$$

$$c = f(x_2) \tag{4.36}$$

Calculate x_3 using the above-calculated values

$$x_3 = x_1 + \frac{-2c}{b \mp \sqrt{b^2 - 4ac}} \tag{4.37}$$

The percent error

$$\varepsilon = \left|\frac{x_3 - x_2}{x_3}\right| 100\% \tag{4.38}$$

Example 4.9 Application of Muller's Method

Use Muller's method to find the roots of

$$x^3 - 13x - 12 = 0$$

The three initial guesses are $x_0 = 4.5$, $x_1 = 5.5$, $x_2 = 5.0$

Confirm the manual root findings with Simulink and Python programming of Muller's method.

Solution

Following Muller's method,

$$f(x_0) = f(4.5) = 20.626$$

$$f(x_1) = f(5.5) = 82.875$$

$$f(x_2) = f(5) = 48.0$$

$$h_o = x_1 - x_0 = 5.5 - 4.5 = 1$$

$$h_1 = x_2 - x_1 = 5 - 5.5 = -0.5$$

$$\delta_o = \frac{f(x_1) - f(x_o)}{(x_1 - x_o)} = \frac{82.875 - 20.625}{5.5 - 4.5} = 62.25$$

$$\delta_1 = \frac{f(x_1) - f(x_2)}{(x_1 - x_2)} = \frac{82.875 - 48}{5.5 - 5.0} = 69.75$$

$$a = \frac{\delta_1 - \delta_o}{h_1 + h_o} = \frac{69.75 - 62.25}{1 + (-0.5)} = 15$$

$$b = ah_1 + \delta_1 = 15(-0.5) + 69.75 = 62.25$$

$$c = f(x_2) = 5^3 - 13(5) - 12 = 48$$

$$x_3 = x_1 + \frac{-2c}{b \mp \sqrt{b^2 - 4ac}} = 5 + \frac{-2(48)}{62.25 \mp \sqrt{(62.25)^2 - 4(15)(48)}} = 3.976$$

The percentage error is:

$$\varepsilon = \left| \frac{3.976 - 5}{3.976} \right| 100\% = 25.79\%$$

Simulink Solution

Using Muller's method to build the Simulink block diagram representing the solution of the equation defined in Example 4.9 is shown in Figure 4.20. The MATLAB code follows the Simulink graphical solution implanted in the MATLAB function programs, the Muller method. The MATLAB function is connected to three input ports representing the three constants containing the three initial guesses.

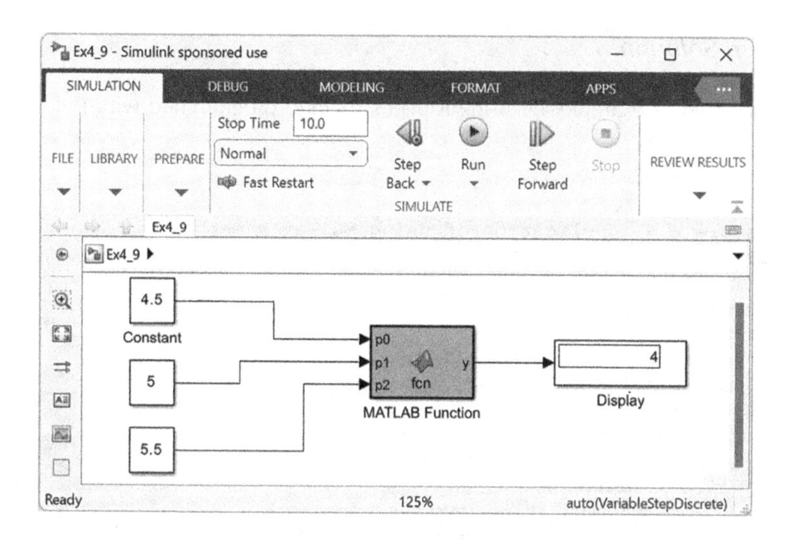

FIGURE 4.20 Simulink block diagram using the Muller method to solve the equation defined in Example 4.9.

The following are the MATLAB codes implanted in the MATLAB function of Example 4.9.

```
% Example 4.9
function y = fcn (p0, p1, p2)
TOL = 10^-5;
NO = 100;
f =@ (x) x^3-13*x-12; % the required function
h1 = p1- p0; h2 = p2- p1;
DELTA1 = (f(p1) - f(p0))/h1;
DELTA2=(f(p2) - f(p1))/h2;
d = (DELTA2- DELTA1)/(h2 + h1);
i=3;
p=0;
while i <= NO
       b = DELTA2 + h2*d;
       D = (b^2-4*f(p2)*d)^(1/2);
       if abs(b-D) < abs(b+D)
E = b + D;

else
E = b - D;
end
       h = -2*f(p2)/E; p=p2+h;
       if abs(h) < TOL
break
end
p0 = p1; p1= p2; p2 = p;
h1 = p1- p0; h2 = p2- p1;
DELTA1 = (f(p1) - f(p0))/h1;
DELTA2 = (f(p2) - f(p1))/h2;
d = (DELTA2- DELTA1)/(h2 + h1); i = i + 1;
end
y = p;
```

Python Solution

The solution of Example 4.9 using Muller's method programmed with Python is shown as follows:

```
# Example 4.9: Application of using Muller's method
import math;
MAX_ITERATIONS = 100
# Function to calculate f(x)
def f(x):
      # Taking f(x) = x ^ 3-13x -12
      return (x**3-13 * x -12)
def Muller(a, b, c):
      res = 0
      i = 0
      while (True):
               # Calculating various constants
               f1 = f(a); f2 = f(b); f3 = f(c)
               d1 = f1- f3
               d2 = f2- f3
               h1 = a - c
               h2 = b - c
               a0 = f3
               a1 = (((d2 * pow(h1, 2)) -
                     (d1 * pow(h2, 2))) /
                     ((h1 * h2) * (h1- h2)))
               a2 = (((d1 * h2) - (d2 * h1)) /
                     ((h1 * h2) * (h1- h2)))
               x = ((-2 * a0) / (a1 +
                     abs(math.sqrt(a1 * a1-4 * a0 * a2))))
               y = ((-2 * a0) / (a1-
                     abs(math.sqrt(a1 * a1-4 * a0 * a2))))
               # Taking the root, which is
               # closer to x2
               if (x >= y):
                     res = x + c;
               else:
                     res = y + c;
               # checking for the resemblance of x3
               m = res * 100;
               n = c * 100;
               m = math.floor(m)
               n = math.floor(n)
               if (m == n):
                     break
               a = b
               b = c
               c = res
               if (i > MAX_ITERATIONS):
                     print("Root cannot be found ")
                     break
               i =i+ 1
      if (i <= MAX_ITERATIONS):
               print("\n using Muller's method, the root is:",
round(res, 4))
```

```
# Initial guesses
a = 4.5
b = 5.5
c = 5
Muller(a, b, c)
```

Execution result:

```
using Muller's method, the root is: 4.0
```

4.8 SUMMARY

There are various methods available to find the roots of equations. The simplest method is the bisection method. The method is based on the changes in sign of a function in the vicinity of a root. NR is based on using an initial guess for the root and finding the intersection with the straight-line axis representing the slope at the initial guess. Based on a reasonable estimate of the initial guess, the method works and converges extremely fast.

4.9 PROBLEMS

1. Solving the following two equations using the substitution method.

$$x - 2y = 8$$

$$x + y = 5$$

   Answer: ($x = 6$, $y = -1$)
2. Solve the following two equations using the method of substitution.

$$2x + 3y + 15 = 0$$

$$x + 4y + 2 = 0$$

   Answer: ($x = -54/5$ and $y = 11/5$)
3. Solve the following two linear algebraic equations using the method of substitution.

$$y = 2x + 8$$

$$y = 5x - 1$$

   Answer: (x = 3,y = 14)

4. Solve the following two linear algebraic equations using the substitution method.

$$y = \frac{19}{5} - \frac{2}{5}x$$

$$y = 3(x-1)$$

Answer: ($x = 2$, $y = 3$)

5. Find the roots of the following nonlinear algebraic equation using the fixed-point iteration method.

$$x^2 - 2x - 3 = 0$$

Answer: ($x = 3$)

6. Determine the root of the following equation using the fixed-point iteration method within the interval [0 1].

$$2x^3 - 11.7x^2 + 17.7x - 5 = 0$$

Answer: ($x = 0.365$)

7. Using NR method, find the root of the following equation:

$$x^3 - 3x - 5 = 0$$

Answer: ($x = 2.279$)

8. Using NR method, find the roots of the following function:

$$x^3 - 2x^2 + x - 3 = 0$$

As an initial guess use, $x_o = 4$

Answer: ($x = 2.174$)

9. Use the NR method to find the values of x and y. Use an initial guess of $x_0 = 1.5$ and $y_o = 3.5$

$$x^2 + xy - 10 = 0$$

$$y + 3xy^2 - 57 = 0$$

Answer: $x_1 = 2.036$, $y_1 = 2.843$

10. Find the roots of the following equation using Muller's method, starting with the initial approximations as $x_o = 2$, $x_1 = 2.5$, $x_2 = 3$

$$x^3 - 3x - 5 = 0$$

Answer: 2.279

11. Find the roots of the following equation using Muller's method, starting with the initial guesses as $x_o = 0$, $x_1 = 1$, $x_2 = 2$

$$f(x) = x^3 + 2x^2 + 10x - 20 = 0$$

Answer: 1.368

12. Find the roots of the following equation using Muller's method, starting with three initial guesses $x_o = 0$, $x_1 = 1$, $x_2 = 2$

$$x^5 - 5x + 2 = 0$$

Answer: 1.372

REFERENCES

1. Hoffman, J.D. and Frankel, S., 2018. *Numerical* Methods *for* Engineers *and* Scientists. London: CRC Press.
2. Kiusalaas, J., 2005. *Numerical Methods in Engineering with MATLAB*. Cambridge: Cambridge University Press,
3. Kong, Q., Siauw, T. and Bayen, A., 2020. *Python Programming and Numerical Methods: A Guide for Engineers and Scientists*. Cambridge, MA: Academic Press.

5 Initial Value Problem Differential Equations

An initial value problem (IVP) is an ordinary differential equation (ODE) with an initial condition that specifies the value of the unknown function at a given point in the domain. Numerical methods are essential for solving system IVPs in ODEs, especially in complicated cases. This chapter introduces eigenvalue, eigenvector, and elimination techniques to solve systems of more than one initial value ODEs.

LEARNING OBJECTIVES

1. Solve IVP using eigenvalue and eigenvector technique.
2. Apply elimination techniques to solve initial value problems (IVPs).
3. Master the use of the Simulink Integrator block in designing and simulating dynamic systems.
4. Utilize Python programming language to address an IVP.

5.1 INTRODUCTION

In mathematics, an ODE is a differential equation that contains one or more functions of one independent variable and the derivatives of those functions. The term ordinary is used in contrast to the partial differential equation, which may concern more than one independent variable. ODEs may represent system dynamics in transferring heat and mass and related problems in chemical engineering and other engineering disciplines. The IVP describes the behavior of a system due to certain disturbances that cause the system to deviate from its steady-state value. Differential equations can arise naturally from physical modeling. The following IVP describes the concentration profile of three chemicals in a reactor as a function of time. The following set of ODEs represents certain dynamic equations [1].

$$\frac{dc_1}{dt} = f_1(c_1 \ldots\ldots c_3)$$

$$\frac{dc_2}{dt} = f_2(c_1 \ldots\ldots c_3)$$

$$\frac{dc_3}{dt} = f_n(c_1 \ldots\ldots c_3)$$

DOI: 10.1201/9781003360544-5

The set of equations can be written in the following form:

$$\frac{dC}{dt} = f(c)$$

Under steady-state conditions, it can be written as

$$0 = f_1(c_1 \ldots\ldots c_3)$$

$$0 = f_2(c_1 \ldots\ldots c_3)$$

$$0 = f_n(c_1 \ldots\ldots c_3)$$

The solution method is the same as the single first-order ODE solution method and can be extended to multiple ODEs system.

5.2 EIGENVALUE AND EIGENVECTOR

Eigenvalues and eigenvectors are used to solve linear systems of ODEs. Each eigenvalue is associated with a specific eigenvector. IVPs can be presented in terms of an eigenvalue technique as follows:

$$\boldsymbol{Ax} = \lambda \boldsymbol{x}$$

where $\boldsymbol{A}$ is an $n \times n$ matrix (the coefficients of the ODEs), $\boldsymbol{x}$ is a nonzero $n \times 1$ (vector), and λ is the eigenvalue (scaler) of matrix $\boldsymbol{A}$. The eigenvalue problem is rewritten as follows:

$$\boldsymbol{Ax} - \lambda \boldsymbol{x} = 0$$

$$\boldsymbol{Ax} - \lambda \boldsymbol{Ix} = 0$$

where (I) is the identity matrix, rearranging

$$(\boldsymbol{A} - \lambda \boldsymbol{I})\boldsymbol{x} = 0$$

Since $\boldsymbol{x}$ is nonzero, the characteristic equation would be

$$|\boldsymbol{A} - \lambda \boldsymbol{I}| = 0$$

The roots (λ) are called the eigenvalues of $\boldsymbol{A}$.

Example 5.1 Application of Eigenvalue and Eigenvector

Solve the following two ODEs simultaneously using the eigenvalue and eigenvector technique.

$$\frac{dy_1}{dt} = -3y_1 + 2y_2, \quad y_1(0) = 1$$

$$\frac{dy_2}{dt} = y_1 - 2y_2, \quad y_2(0) = 0$$

Find the values of y_1 and y_2 when $t = 2$.

Manual Solution

First, convert the system into matrix form. The equations can be written as follows:

$$\frac{dY}{dt} = AY$$

where

$$A = \begin{bmatrix} -3 & 2 \\ 1 & -2 \end{bmatrix}$$

First determine eigenvalue λ

$$|A - \lambda I| = 0$$

$$\begin{bmatrix} -3 & 2 \\ 1 & -2 \end{bmatrix} - \begin{bmatrix} \lambda & 0 \\ 0 & \lambda \end{bmatrix} = \begin{bmatrix} -\lambda - 3 & 2 \\ 1 & -2 - \lambda \end{bmatrix} = 0$$

Find the determinant

$$\det \begin{bmatrix} -\lambda - 3 & 2 \\ 1 & -2 - \lambda \end{bmatrix} = (-\lambda - 3)(-2 - \lambda) - (1)(2) = 0$$

Rearrange

$$\lambda^2 + 5\lambda + 6 - 2 = \lambda^2 + 5\lambda + 4 = 0$$

Factorize

$$(\lambda + 4)(\lambda + 1) = 0$$

The two eigenvalues: $\lambda_1 = -1$, $\lambda_2 = -4$

Determine corresponding eigenvectors

$$\lambda_1 = -1 : \begin{bmatrix} -3 & 2 \\ 1 & -2 \end{bmatrix} \begin{Bmatrix} x_{11} \\ x_{12} \end{Bmatrix} = -1 \begin{Bmatrix} x_{11} \\ x_{12} \end{Bmatrix}$$

Solving the matrix,

$$-3x_{11} + 2x_{12} = -x_{11}$$

$$x_{11} - 2x_{12} = -x_{12}$$

Rearranging and simplifying,

$$x_{11} = x_{12}$$

Accordingly,

$$X_1 = \begin{Bmatrix} 1 \\ 1 \end{Bmatrix}$$

Finding the eigenvector of λ_2,

$$\lambda_2 = -4 : \begin{bmatrix} -3 & 2 \\ 1 & -2 \end{bmatrix} \begin{Bmatrix} x_{21} \\ x_{22} \end{Bmatrix} = -4 \begin{Bmatrix} x_{21} \\ x_{22} \end{Bmatrix}$$

Solving the matrix,

$$-3x_{21} + 2x_{22} = -4x_{21}$$

$$x_{21} - 2x_{22} = -4x_{22}$$

Rearranging and simplifying,

$$-2x_{21} = -4x_{21} - 4x_{22}$$

Simplify

$$2x_{21} = -4x_{22}$$

Hence,

$$x_{21} = -2x_{22}$$

$$X_2 = \begin{Bmatrix} -2 \\ 1 \end{Bmatrix}$$

The general solution is

$$Y = c_1 e^{(\lambda_1 t)} X_1 + c_2 e^{\lambda_2 t} X_2$$

In another format,

$$\begin{Bmatrix} y_1 \\ y_2 \end{Bmatrix} = c_1 \begin{Bmatrix} 1 \\ 1 \end{Bmatrix} e^{(-1t)} + c_2 \begin{Bmatrix} -2 \\ 1 \end{Bmatrix} e^{-4t}$$

The solution is

$$y_1 = c_1 e^{-t} - 2c_2 e^{-4t}$$

$$y_2 = c_1 e^{-t} + c_2 e^{-4t}$$

c_1 and c_2 are determined from the initial conditions.

$$1 = c_1 - 2c_2$$

$$0 = c_1 + c_2$$

Solving for c_1 and c_2 , the values are as follows:

$$c_1 = \frac{1}{3}, \ c_2 = -\frac{1}{3}$$

When $t = 2$ min, the values of y_1 and y_2 are as follows:

$$y_1 = \frac{1}{3}e^{-2} - 2\left(-\frac{1}{3}\right)e^{-8} = 0.0453$$

$$y_2 = \frac{1}{3}e^{-2} - \frac{1}{3}e^{-8} = 0.045$$

Simulink Solution

Figure 5.1 represents the Simulink block diagram using the Simulink Integrator block for solving the two ODEs defined in Example 5.1. The integrator's default numerical integration method is the fourth-order Runge-Kutta method.

An alternative way is to use the MATLAB function block available in the User-Defined Functions of the Simulink library. The MATLAB function is associated with a program written in MATLAB programming language. The solution using the eigenvalue and eigenvector approaches of the two ODEs specified in Example 5.1 is presented by the Simulink block diagram shown in Figure 5.2. The Simulink block diagram is followed by the MATLAB code associated with the MATLAB function. The predicted approximate results are identical to the case using the Integrator block and the default RK4. The MATLAB function input port relates to a constant (*n*), representing the independent variable's simulation stop time (*t*). The exit ports are connected with two Display blocks for the output of y_1 and y_2 at the simulation stop time. The change in the dependent variable (y_1, y_2) versus the independent time variable (t) is presented in Figure 5.3.

The following program represents the MATLAB code implanted in the MATLAB function shown in Figure 5.2. The results are plotted in Figure 5.3.

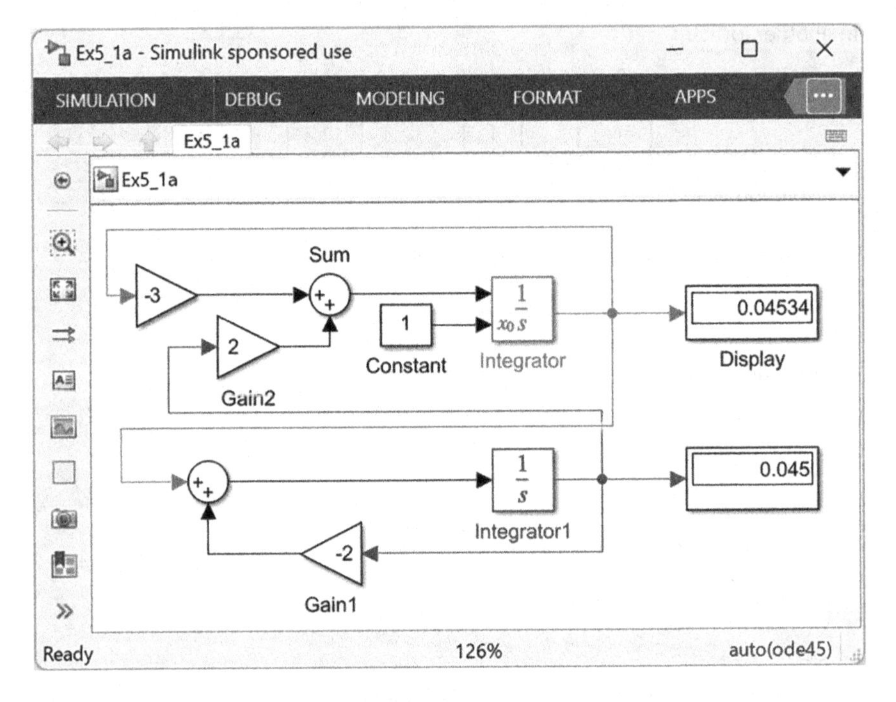

FIGURE 5.1 Solution using the Simulink integrator block and the fourth-order Runge-Kutta approach of equations defined in Example 5.1.

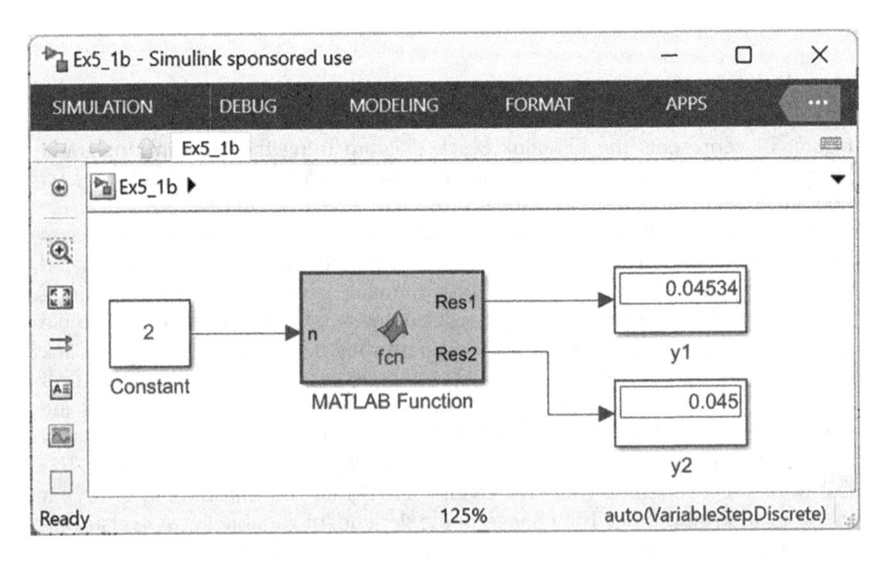

FIGURE 5.2 Simulink solution using the eigenvalue and eigenvector approach of ODEs defined in Example 5.1.

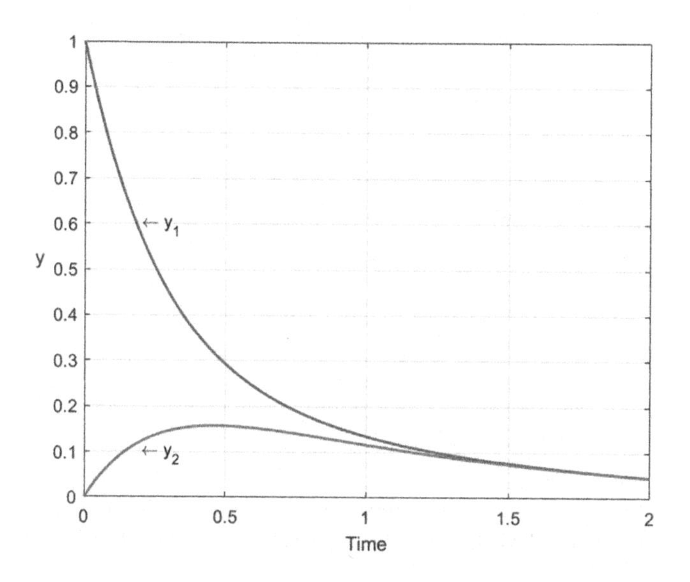

FIGURE 5.3 Simulink generates a plot of the two ODEs defined in Example 5.1.

```
% Example 5.1
function [Res1,Res2] = fcn(n)
% The coefficients of the Matrix
A = [-3 2;
    1 -2];
% Calculating the eigenvalue and eigenvectors
[V, D] = eig(A);
% initial conditions
Y0 = [1; 0];
% Calculating the constants C1 and C2
c = diag(V\Y0);
t = linspace(0,n);
T = [t;t];

y = V*c*exp(D*T);
plot(t, y, 'LineWidth', 1.5))
xlabel("Time")
ylabel("y ", rotation=0)
grid on
str = '\leftarrow y_1';
text(0.2,0.6,str)
str2 = '\leftarrow y_2';
text(0.2,0.1,str2)
Res1 = y(1,end);
Res2 = y(2,end);
```

Python Solution

The following program is a Python code that uses the eigenvalue and eigenvector approaches to solve the two ODEs equations given in Example 5.1. The plot of y_1 and y_2 versus time (t) is presented in Figure 5.4.

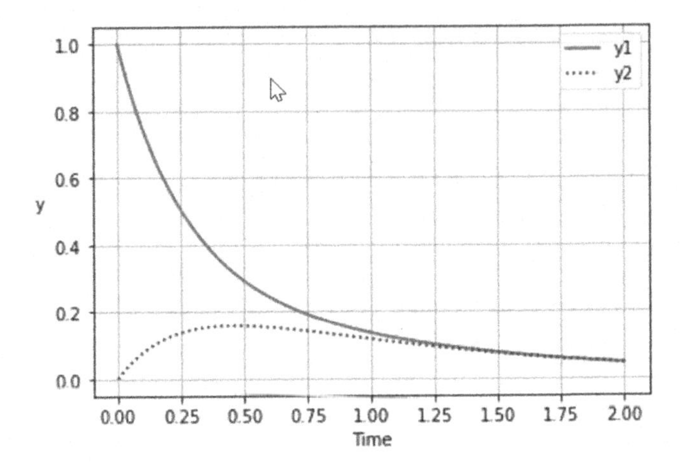

FIGURE 5.4 Plot of y_1 and y_2 versus time (t) generated by the Python program using eigenvalue and eigenvector represents the solution of two ODEs defined in Example 5.1.

```
# Example 5.1
import numpy as np
import matplotlib.pyplot as plt
A = np.array([[-3,2], [1,-2]])
# extract eigenvalues
[w, v]= np.linalg.eig(A)
y10 = 1
y20 = 0
Pinitial = np.array([y10, y20])
C1, C2 = np.linalg.solve(v, Pinitial)
# combine everything to get the solution to P1
#t = np.linspace(0,2)
fig = plt.figure()
t=np.linspace(0,2,100)
y1 = C1 * np.exp(w[0]*t)*v[0,0] + C2 * np.exp(w[1] * t) * v[0,1]
y2 = C1 * np.exp(w[0]*t)*v[1,0] + C2 * np.exp(w[1] * t) * v[1,1]

#plt.plot(t, y1, t, y2)
plt.plot(t, y1, '-r', label='y1')
plt.plot(t, y2, ':b', label='y2')
plt.grid(True)
plt.ylabel("y", rotation =0)
plt.xlabel("Time")
plt.legend()
```

Example 5.2 Determine the Eigenvalue and Eigenvector

Calculate the eigenvalue and eigenvector for the following two ODEs.

$$\frac{dy_1}{dt} = y_2, \quad y_1(0) = 1$$

$$\frac{dy_2}{dt} = -2y_1 - 3y_2, \quad y_2(0) = 0$$

Solve the two ODEs manually and support the manual solution using the Simulink/MATLAB programming and Python programming of the eigenvalue and eigenvector approaches.

Solution

Putting the two ODEs in terms of a 2×2 matrix

$$\boldsymbol{A}=\begin{bmatrix} 0 & 1 \\ -2 & -3 \end{bmatrix}$$

The characteristic equation is

$$|\boldsymbol{A}-\lambda I|=0$$

Substitute the values matrices $\boldsymbol{A}$ and $\boldsymbol{I}$

$$\begin{bmatrix} 0 & 1 \\ -2 & -3 \end{bmatrix}-\lambda\begin{bmatrix} 1 & 0 \\ 0 & 1 \end{bmatrix}=\begin{bmatrix} 0 & 1 \\ -2 & -3 \end{bmatrix}-\begin{bmatrix} \lambda & 0 \\ 0 & \lambda \end{bmatrix}=\begin{bmatrix} 0-\lambda & 1 \\ -2 & -3-\lambda \end{bmatrix}$$

The determinant of a matrix: A determinant is a particular number calculated from a matrix. The matrix must be square (the same number of rows and columns). This has two rows and two columns: The following matrix is a 2×2 :

$$|A|=\begin{bmatrix} a & b \\ c & d \end{bmatrix}=ad-bc$$

Accordingly,

$$\begin{bmatrix} 0-\lambda & 1 \\ -2 & -3-\lambda \end{bmatrix}=\left[(0-\lambda)(-3-\lambda)\right]-\left[(-2)(1)\right]=0$$

Simplify

$$\lambda^2+3\lambda+2=(\lambda+2)(\lambda+1)=0$$

And the two eigenvalues are

$$\lambda_1=-1,\ \lambda_2=-2$$

Find the two eigenvectors as follows:

To find the eigenvector, $\boldsymbol{X}_1$, associated with the eigenvalue, $\lambda_1=-1$

The eigenvectors for $\lambda_1 = -1$ are determined as follows:

$$A.X_1 = \lambda_1.X_1$$

$$(A - \lambda_1).X_1 = 0$$

$$\left(\begin{bmatrix} 0 & 1 \\ -2 & -3 \end{bmatrix} - \begin{bmatrix} \lambda_1 & 0 \\ 0 & \lambda_1 \end{bmatrix}\right).X_1 = \begin{bmatrix} -\lambda_1 & 1 \\ -2 & -3-\lambda_1 \end{bmatrix}.X_1 = 0$$

Substitute the value of $\lambda_1 = -1$

$$\begin{bmatrix} 1 & 1 \\ -2 & -2 \end{bmatrix}..x_1 = \begin{bmatrix} 1 & 1 \\ -2 & -2 \end{bmatrix}.\begin{bmatrix} x_{11} \\ x_{12} \end{bmatrix} = 0$$

So clearly, from the top row of the equation, we get

$$x_{11} + x_{12} = 0$$

Accordingly,

$$x_{11} = -x_{12}$$

If we use the second row, we will get the same

$$-2x_{11} - 2x_{12} = 0$$

Divide by 2 and rearrange, again

$$x_{11} = -x_{12}$$

The eigenvector

$$X_1 = \begin{bmatrix} x_{11} \\ x_{12} \end{bmatrix} = \begin{bmatrix} -1 \\ 1 \end{bmatrix}$$

We find that the first eigenvector is any vector composed of two components where the two components have the same magnitude and opposite signs. Follow the same procedure to determine the eigenvector for the second eigenvalue ($\lambda_2 = -2$):

$$A.X_2 = \lambda_2.X_2$$

Rearrange

$$(A - \lambda_2).X_2 = 0$$

$$\left(\begin{bmatrix} 0 & 1 \\ -2 & -3 \end{bmatrix} - \begin{bmatrix} \lambda_2 & 0 \\ 0 & \lambda_2 \end{bmatrix}\right).X_2 = 0$$

Subtract the two matrixes

$$\begin{bmatrix} 0-\lambda_2 & 1-0 \\ -2-0 & -3-\lambda_2 \end{bmatrix}.X_2 = 0$$

Substitute the value of $\lambda_2 = -2$

$$\begin{bmatrix} 2 & 1 \\ -2 & -1 \end{bmatrix}.X_1 = \left[\begin{bmatrix} 2 & 1 \\ -2 & -1 \end{bmatrix}\right].\begin{bmatrix} x_{21} \\ x_{22} \end{bmatrix} = 0$$

Multiply the first row in the first matrix with the first column of the second matrix

$$2x_{21} + x_{22} = 0$$

Rearrange

$$2\boldsymbol{x}_{21} = -\boldsymbol{x}_{22}$$

Accordingly,

$$X_2 = \begin{bmatrix} -1 \\ 2 \end{bmatrix}$$

Again, the eigenvector's choice of –1 and 2 was arbitrary; Only their proportion is essential. The general solution is

$$Y = c_1 e^{(\lambda_1 t)} X_1 + c_2 e^{\lambda_2 t} X_2$$

In another format,

$$\begin{Bmatrix} y_1 \\ y_2 \end{Bmatrix} = c_1 e^{(-1t)} \begin{Bmatrix} -1 \\ 1 \end{Bmatrix} + c_2 e^{-2t} \begin{Bmatrix} -1 \\ 2 \end{Bmatrix}$$

The solution is

$$y_1 = -c_1 e^{-t} - c_2 e^{-2t}$$

$$y_2 = c_1 e^{-t} + 2c_2 e^{-2t}$$

The values of c_1 and c_2 can be determined using the initial condition of y_1 and y_2

$$1 = c_1 e^{-0} - c_2 e^{-2(0)}$$

$$0 = -c_1 e^{-(0)} + 2c_2 e^{-2(0)}$$

Simplifying,

$$1 = -c_1 - c_2$$

$$0 = c_1 + 2c_2$$

Hence solving for c_1

$$c_1 = -2$$

$$c_2 = 1$$

Accordingly,

$$Y = -2\begin{Bmatrix} -1 \\ 1 \end{Bmatrix} e^{-t} + 1\begin{Bmatrix} -1 \\ 2 \end{Bmatrix} e^{-2t}$$

The final solution is

$$y_1 = -2e^{-t} - e^{-2t}$$

$$y_2 = 2e^{-t} + 2e^{-2t}$$

When $t = 2$

$$y_1 = 2e^{-2} - 1e^{-2(2)} = 0.25235$$

$$y_2 = 2e^{-(2)} + 2e^{-2(2)} = -0.23404$$

Simulink Solution

Figure 5.5 is the Simulink block diagram representing the solution of the two nonlinear ODEs equations defined in Example 5.2. The following program is the MATLAB code utilizing the eigenvalue and eigenvector approach associated with the Simulink MATLAB function block dragged from the User-Defined Functions Simulink library. The MATLAB function input port is connected with a constant (n), representing the simulation end time. The output port is connected with two Display blocks that yield the values of dependent variables at the simulation stop time.

The following program signifies the MATLAB code implanted in the MATLAB function present in Figure 5.5. The program uses eigenvalue and eigenvector methodology to solve the two ODEs defined in Example 5.2.

```
% Example 5.2
function [Res1,Res2] = fcn(n)
% The coefficients of the Matrix
A = [0 1;
     -2 -3];
% Calculating the eigenvalue and eigenvectors
```

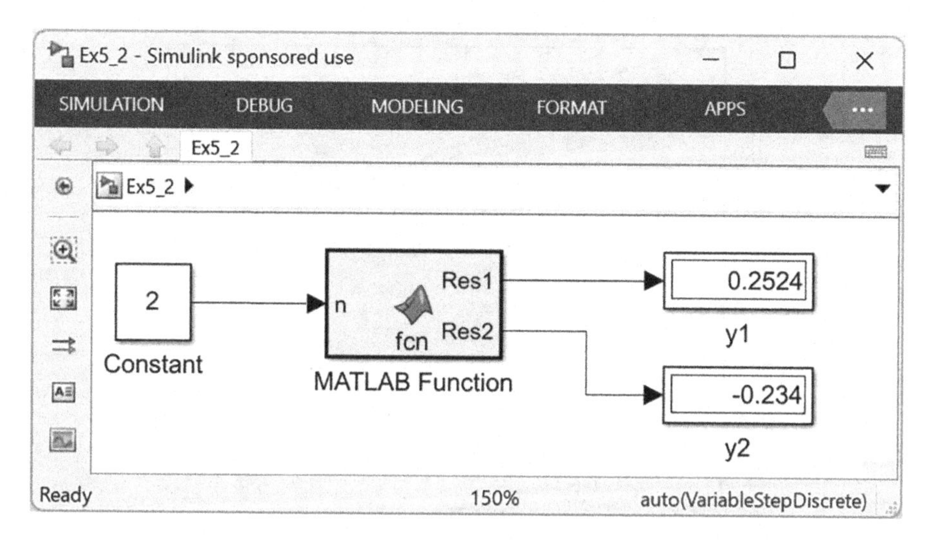

FIGURE 5.5 The Simulink block diagram using the eigenvalue and eigenvector approach of two ODEs defined Example 5.2.

```
[V, D] = eig(A);
% initial conditions
Y0 = [1; 0];
% Calculating the constants C1 and C2
c = diag(V\Y0);
t = linspace(0,n);
T = [t;t];

y = V*c*exp(D*T);

% Plot sextion
plot(t, y, 'LineWidth', 1.5)
xlabel("Time")
ylabel("y ", rotation=0)
grid on
str1 ='\leftarrow y_1';
str2 = '\leftarrow y_2';
text (0.65,0.8,str1,'Color', 'b', 'FontSize', 15);
text (0.4,-0.4,str2,'Color', 'r', 'FontSize', 15);
ylim([-0.6 1]);

Res1 = y(1,end);
Res2 = y(2,end);
```

Executing the described Simulink block diagram generates the plot shown in Figure 5.6.

Python Simulation

The following program is a Python code that simultaneously uses the eigenvalue and eigenvector approaches to solve the two nonlinear equations given in Example 5.2. Running the Python program generates the data plotted in Figure 5.7.

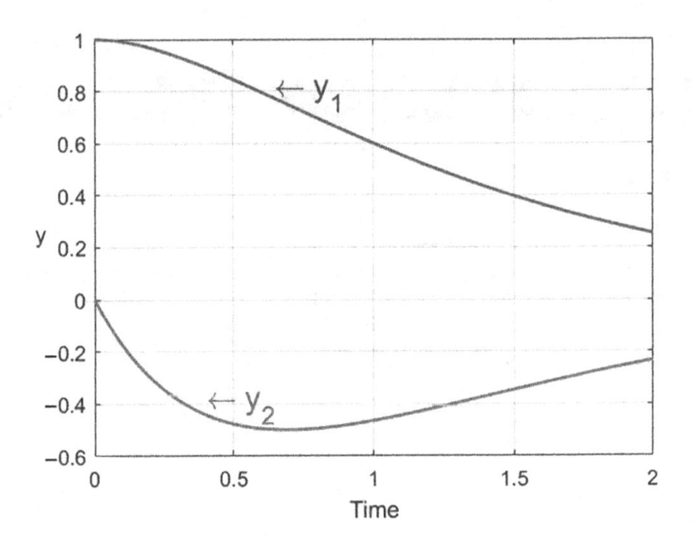

FIGURE 5.6 Plot of y_1 and y_2 versus time (t) using eigenvalue and eigenvector in MATLAB of the two equations defined in Example 5.2.

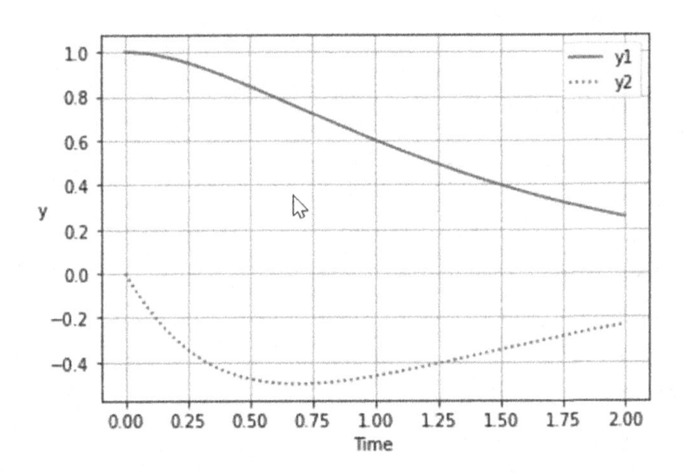

FIGURE 5.7 Plot of y_1 and y_2 versus time (t) utilizing the eigenvalue and eigenvector approach programmed in Python of the two equations described in Example 5.2.

```
# Example 5.2
import numpy as np
import matplotlib.pyplot as plt
A = np.array([[0,1], [-2,-3]])
# extract eigenvalues
[w, v]= np.linalg.eig(A)
y10 = 1
y20 = 0
Pinitial = np.array([y10, y20])
C1, C2 = np.linalg.solve(v, Pinitial)
```

```
# combine everything to get the solution to P1
#t = np.linspace(0,2)
fig = plt.figure()
t=np.linspace(0,2,100)
y1 = C1 * np.exp(w[0]*t)*v[0,0] + C2 * np.exp(w[1] * t) * v[0,1]
y2 = C1 * np.exp(w[0]*t)*v[1,0] + C2 * np.exp(w[1] * t) * v[1,1]

#plt.plot(t, y1, t, y2)
plt.plot(t, y1, '-g', label='y1')
plt.plot(t, y2, ':r', label='y2')
plt.grid(True)
plt.ylabel("y ", rotation=0)
plt.xlabel("Time")
plt.legend()
```

Example 5.3 Application of Eigenvalue and Eigenvector Technique

Solve the following two ODEs using the eigenvalue-eigenvector technique.

$$\frac{dy_1}{dx} = 2y_1 + y_2, \quad y_1(0) = 1$$

$$\frac{dy_2}{dt} = 3y_1, \quad y_2(0) = 0$$

Solution

Arrange the equations in matrix form

$$\frac{dY}{dt} = AY = \begin{bmatrix} 2 & 1 \\ 3 & 0 \end{bmatrix} Y, \quad Y_o = \begin{bmatrix} 1 \\ 0 \end{bmatrix}$$

Using the identity matrix (*I*) and the eigenvalue (λ), first, we will determine the eigenvalues of matrix *A*

$$AX = \lambda X$$

Rearrange

$$AX - \lambda X = (A - \lambda I)X = 0$$

The $(A - \lambda I)X = 0$ in matrix format

$$\left(\begin{bmatrix} 2 & 1 \\ 3 & 0 \end{bmatrix} - \begin{bmatrix} \lambda & 0 \\ 0 & \lambda \end{bmatrix} \right) X = 0$$

The result of subtraction matrices

$$\begin{bmatrix} 2-\lambda & 1 \\ 3 & 0-\lambda \end{bmatrix} = 0$$

The determinant equation

$$\det\begin{bmatrix} 2-\lambda & 1 \\ 3 & 0-\lambda \end{bmatrix} = 0$$

The resultant quadratic equation

$$(2-\lambda)(-\lambda)-(3)(1)=0$$

Simplify

$$\lambda^2-2\lambda-3=0$$

Factorize

$$(\lambda-3)(\lambda+1)=0$$

The eigenvalues are

$$\lambda_1=3,\ \lambda_2=-1$$

Each eigenvalue has an eigenvector.
Eigenvector X_1 for $\lambda_1=3$

$$\begin{bmatrix} 2-\lambda_1 & 1 \\ 3 & -\lambda_1 \end{bmatrix}\begin{bmatrix} x_{11} \\ x_{12} \end{bmatrix} = \begin{bmatrix} 0 \\ 0 \end{bmatrix}$$

Substitute $\lambda_1=3$

$$\begin{bmatrix} -1 & 1 \\ 3 & -3 \end{bmatrix}\begin{bmatrix} x_{11} \\ x_{12} \end{bmatrix} = \begin{bmatrix} 0 \\ 0 \end{bmatrix}$$

$$-x_{11}+x_{12}=0$$

If we assume that $x_{11}=1$, then $x_{12}=1$

$$X_1=\begin{bmatrix} x_{11} \\ x_{12} \end{bmatrix} = \begin{bmatrix} 1 \\ 1 \end{bmatrix}$$

The eigenvector of λ_1 is $\begin{bmatrix} 1 \\ 1 \end{bmatrix}$

Eigenvector X_2 of $\lambda_2=-1$

$$\begin{bmatrix} 2-\lambda_2 & 1 \\ 3 & -\lambda_2 \end{bmatrix}\begin{bmatrix} x_{21} \\ x_{22} \end{bmatrix} = \begin{bmatrix} 0 \\ 0 \end{bmatrix}$$

Substitute $\lambda_2 = -1$

$$\begin{bmatrix} 2-(-1) & 1 \\ 3 & -(-1) \end{bmatrix}\begin{bmatrix} x_{21} \\ x_{22} \end{bmatrix} = \begin{bmatrix} 0 \\ 0 \end{bmatrix}$$

$$\begin{bmatrix} 3 & 1 \\ 3 & 1 \end{bmatrix}\begin{bmatrix} x_{21} \\ x_{22} \end{bmatrix} = \begin{bmatrix} 0 \\ 0 \end{bmatrix}$$

$$3x_{21} + 1x_{22} = 0$$

If we assume $x_{21} = 1$, then $x_{22} = -3$

According to the eigenvector for λ_2 is $\begin{bmatrix} 1 \\ -3 \end{bmatrix}$

The eigenvector matrix is

$$X = \begin{bmatrix} 1 & 1 \\ 1 & -3 \end{bmatrix}$$

The solution

$$Y = c_1 e^{(\lambda_1 t)} X_1 + c_2 e^{\lambda_2 t} X_2$$

Substitute the eigenvalues and eigenvectors

$$\begin{Bmatrix} y_1 \\ y_2 \end{Bmatrix} = c_1 e^{(\lambda_1 t)} \begin{Bmatrix} 1 \\ 1 \end{Bmatrix} + c_2 e^{\lambda_2 t} \begin{Bmatrix} 1 \\ -3 \end{Bmatrix}$$

The solution of the two equations

$$y_1 = c_1 e^{3t} + c_2 e^{t}$$

$$y_2 = c_1 e^{3t} - 3c_2 e^{t}$$

Substitute the initial conditions

$$y_1(0) = 1,\ y_2(0) = 0$$

After substituting the initial conditions,

$$1 = c_1 + c_2 \tag{5.1}$$

$$0 = c_1 - 3c_2 \tag{5.2}$$

Subtract equation (5.2) from (5.1)

$$1 - 0 = c_1 - c_1 + c_2 - (-3c_2) = 4c_2$$

Accordingly,

$$c_2 = \frac{1}{4},\ c_1 = \frac{3}{4}$$

The final solution for the two ODEs:

$$y_1 = \frac{3}{4}e^{3t} + \frac{1}{4}e^{t}$$

$$y_2 = \frac{3}{4}e^{3t} - \frac{3}{4}e^{t}$$

Simulink Solution

Figure 5.8 shows the Simulink block diagram representing the solution of the two ODEs defined in Example 5.3 using the Integrator block dragged from the most Simulink commonly used library. Figure 5.9 is obtained by double-clicking the Scope. To change the color of the plot generated by the Scope, click on view/style and change the Figure color to white and the axis background color to white. The figure should be like that in Figure 5.9.

The Simulink block diagram describes the solution using the eigenvalue and eigenvector of the two ODEs defined in Example 5.3, shown in Figure 5.10. The input port of the MATLAB function (dragged from the User-Defined Functions in the Simulink library) is connected with a constant (n), representing the simulation end time. The two outlet ports are connected with two Display blocks to yield the dependent variable values (y_1, y_2) at the end period of the independent variable (t).

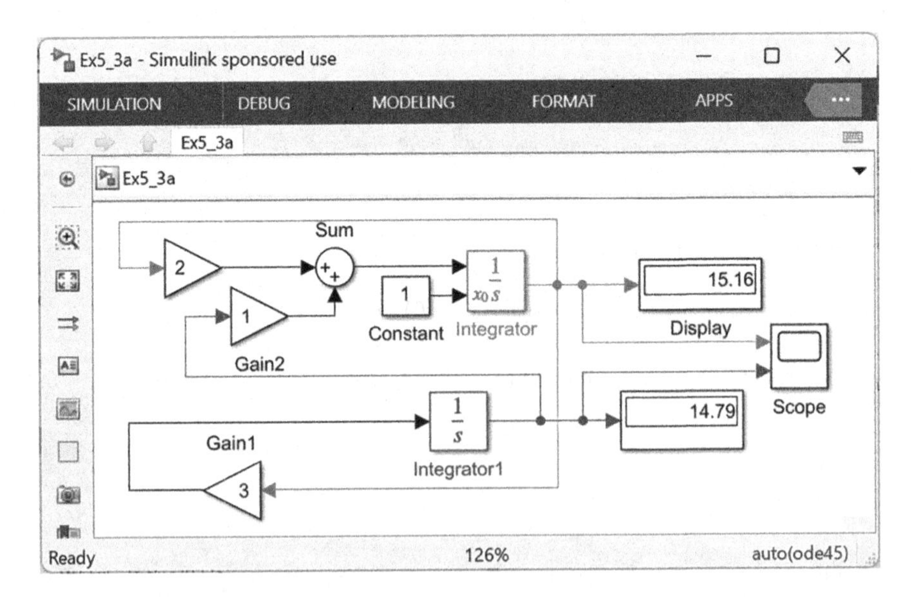

FIGURE 5.8 Simulink solution using the Integrator block to solve simultaneously the two ODEs defined in Example 5.3.

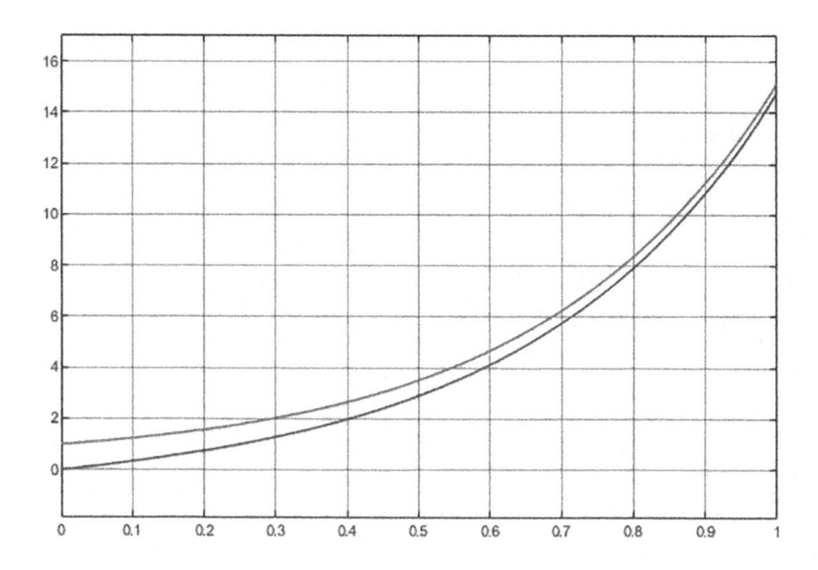

FIGURE 5.9 Plot generated by double-clicking the scope in the Simulink block diagram represents the solution of the two ODEs defined in Example 5.3.

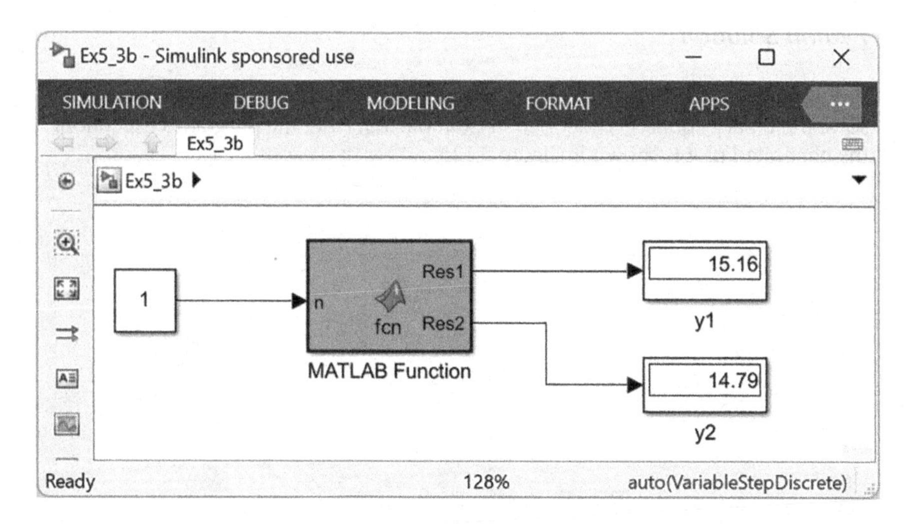

FIGURE 5.10 Simulink block diagram using eigenvalue and eigenvector of the two ODEs described in Example 5.3.

```
% Example 5.3
function [Res1,Res2] = fcn(n)
% The coefficients of the Matrix
A = [2 1;
     3 0];
% Calculating the eigenvalue and eigenvectors
[V, D] = eig(A);
```

```
% initial conditions
Y0 = [1; 0];
% Calculating the constants C1 and C2
c = diag(V\Y0);
t = linspace(0,n);
T = [t;t];

y = V*c*exp(D*T);
% plot section

plot(t, y, 'LineWidth', 1.2)
xlabel("Time", "FontSize", 15)
ylabel("y", 'rotation', 0,'FontSize', 15)
ylim([0 16]);
grid on;
str1 = '\leftarrow y_2';
str2 = 'y_1 \rightarrow';
text (0.6,4,str1,'color', 'b', 'FontSize', 15);
text (0.65,7.5,str2,'color', 'r', 'FontSize', 15);

% displayed results

Res1 = y(1,end);
Res2 = y(2,end);
```

The result of Example 5.3 generated by Simulink using the eigenvalue and eigenvector approach is shown in Figure 5.11.

Python Solution

The following is a Python program utilizing the eigenvalue and eigenvector to solve the binary equation described in Example 5.3. After the program's execution, the generated plot is shown in Figure 5.12.

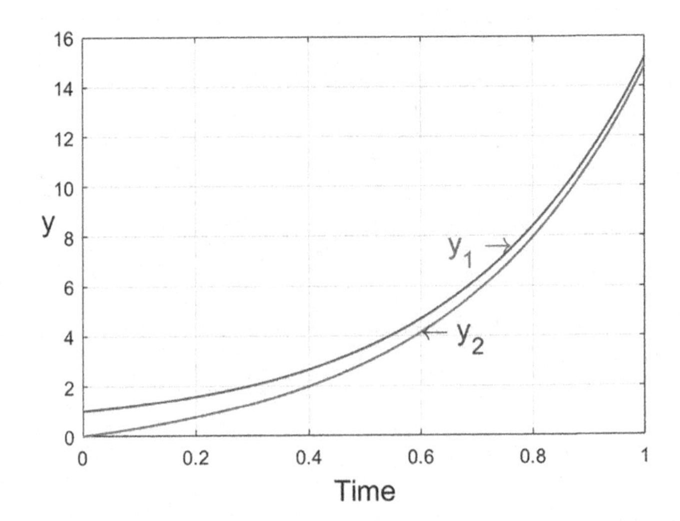

FIGURE 5.11 Simulink solution of the binary differential equations defined in Example 5.3 using the eigenvalue and eigenvector approach.

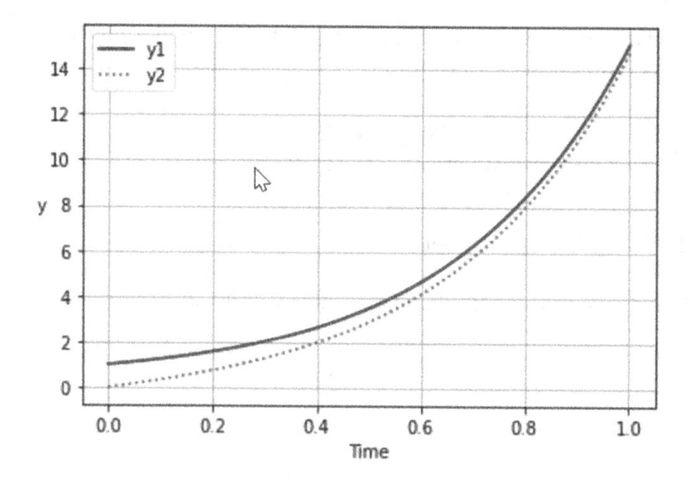

FIGURE 5.12 Python solution using eigenvalue and eigenvector technique of the two equations described in Example 5.3.

```
# Example 5.3
import numpy as np
import matplotlib.pyplot as plt
A = np.array([[2,1], [3,0]])
# extract eigenvalues
[w, v]= np.linalg.eig(A)
y10 = 1
y20 = 0
Pinitial = np.array([y10, y20])
C1, C2 = np.linalg.solve(v, Pinitial)
# combine everything to get the solution to P1
#t = np.linspace(0,2)
fig = plt.figure()
t=np.linspace(0,1,100)
y1 = C1 * np.exp(w[0]*t)*v[0,0] + C2 * np.exp(w[1] * t) * v[0,1]
y2 = C1 * np.exp(w[0]*t)*v[1,0] + C2 * np.exp(w[1] * t) * v[1,1]

#plt.plot(t, y1, t, y2)
plt.plot(t, y1, '-b', label='y1')
plt.plot(t, y2, ':r', label='y2')
plt.grid(True)
plt.ylabel("y ", rotation=0)
plt.xlabel("Time")
plt.legend()
```

5.3 ELIMINATION METHODS

Here, the procedure is the same as solving a system of linear equations, just using the operator "*D*." Solving a system of linear differential equations by applying the following steps to reach the solution:

1. Replace derivatives $\left(\dfrac{d}{d(\text{indep})}\right)$ with D.

2. Rearrange the equations such that they start with x and follows by y.
3. Solve by eliminating Dx or Dy.
4. Solve the differential equation.
5. Solve the other equation following the steps.
6. Eliminate constants using boundary conditions.

Example 5.4 Elimination Method

Solve the following two ODEs simultaneously using the elimination method.

$$\frac{dy}{dt} = -9x$$

$$\frac{dx}{dt} = -4y$$

Solution

Using the elimination method, follow the steps below:

1. Replace the derivative with D.

$$Dy = -9x$$

$$Dx = -4y$$

2. Organize the equation. Putting y first, as follows:

$$Dy + 9x = 0$$

$$4y + Dx = 0$$

3. Start the elimination process. Multiply the first equation by D and the second equation by 9.

$$D^2y + 9Dx = 0$$

$$36y + 9Dx = 0$$

Subtract the second from the first

$$D^2y - 36y = 0$$

4. Solve the differential equation. Let $y = ae^{at}$ and substitute the expression of y.

$$a^2e^{at} - 36e^{at} = 0.$$

Divide both sides by e^{at}

$$a^2 - 36 = 0$$

$$a = \mp 6$$

Accordingly,

$$y = c_1^{6t} + c_2 e^{-6t}$$

Repeat the same for x by eliminating y

$$Dy + 9x = 0$$

$$4y + Dx = 0$$

Multiply the first by 4

$$4Dy + 36x = 0$$

$$4Dy + D^2x = 0$$

Subtract the first from the second

$$D^2x - 36x = 0$$

Let $x = ae^{at}$

$$a^2e^{at} - 36e^{at} = 0$$

Divide by e^{at}

$$a^2 - 36 = 0$$

$$a = \mp 6$$

$$x = c_1e^{6t} + c_2e^{-6t}$$

Example 5.5 Applying the Elimination Method

Solve the following set of ODEs using the elimination method

$$\frac{dx}{dt} = 4x - 3y$$

$$\frac{dy}{dt} = 6x - 7y$$

Solution

Replace derivatives $\left(\frac{d}{d(\text{indep})}\right)$, with *D*. In other words, write in differential notation (*D*).

$$Dx = 4x - 3y$$

$$Dy = 6x - 7y$$

Move all terms with dependent variables to the left-hand side and arrange columns by variables.

$$(D-4)x + 3y = 0$$

$$-6x + (D+7)y = 0$$

Start the elimination process by eliminating x. Multiply the first equation by 6 and the second by $(D-4)$ as follows:

$$6x(D-4) + 18y = 0$$

$$-6x(D-4) + (D-4)(D+7)y = 0$$

Simplify and rearrange

$$18y + \left(D^2 + 3D - 28\right)y = 0$$

Divide both sides by *y* and rearrange

$$D^2 + 3D - 10 = 0$$

Find the roots

$$(D-2)(D+5) = 0$$

$$D = \frac{d}{dt} = 2 \text{ or } -5$$

Accordingly,

$$y(t) = c_1 e^{-5t} + c_2 e^{2t}$$

Solve for x by substituting *y* in one of the following equations:

$$\frac{dy}{dt} = 6x - 7y$$

Hence,

$$-5c_1 e^{-5t} + 2c_2 e^{2t} = 6x - 7(c_1 e^{-5t} + c_2 e^{2t})$$

Rearrange to obtain the solution of x

$$x(t) = \frac{1}{3}c_1 e^{-5t} + \frac{3}{2}c_2 e^{2t}$$

Example 5.6 Applying Eigenvalue and Eigenvector

Solve the following two IVPs by applying the eigenvalue and eigenvector method.

$$\frac{dy_1}{dt} = y_1 + 2y_2, \quad y_1(0) = 0$$

$$\frac{dy_2}{dt} = 3y_1 + 2y_2, \quad y_2(0) = -4$$

Solution

The first thing is to determine the eigenvalues for the matrix

$$\det(A - \lambda I) = \begin{bmatrix} 1-\lambda & 2 \\ 3 & 2-\lambda \end{bmatrix}$$

$$= \lambda^2 - 3\lambda - 4$$

$$= (\lambda + 1)(\lambda - 4) \rightarrow \lambda_1 = -1, \lambda_2 = 4$$

Now let us find the eigenvector for each of these

$$\lambda_1 = -1$$

You need to solve

$$\begin{bmatrix} 1 & 2 \\ 3 & 3 \end{bmatrix} \begin{bmatrix} x_{11} \\ x_{12} \end{bmatrix} = \begin{bmatrix} 0 \\ 0 \end{bmatrix}$$

$$2x_{11} + 2x_{12} = 0 \rightarrow x_{11} = -x_{12}$$

The eigenvector, in this case, is assumed, $x_{11} = -1$

$$X = \begin{bmatrix} -1 \\ 1 \end{bmatrix}$$

$$\lambda_2 = 4$$

We will need to solve

$$\begin{bmatrix} -3 & 2 \\ 3 & -2 \end{bmatrix} \begin{bmatrix} x_{21} \\ x_{22} \end{bmatrix} = \begin{bmatrix} 0 \\ 0 \end{bmatrix}$$

$$-3x_{21} + 2x_{22} = 0 \rightarrow x_{21} = \left(\frac{2}{3}\right)x_{22}$$

assuming, $x_{22} = 1, \quad x_{21} = 2/3$

The eigenvector in this case is $x = \begin{bmatrix} 2/3 \\ 1 \end{bmatrix}$

The general solution is

$$X(t) = c_1 e^{-t}\begin{pmatrix} -1 \\ 1 \end{pmatrix} + c_2 e^{4t}\begin{pmatrix} 2 \\ 3 \end{pmatrix}$$

Apply the initial conditions to find C_1 and C_2

$$\begin{pmatrix} 0 \\ -4 \end{pmatrix} = c_1(1)\begin{pmatrix} -1 \\ 1 \end{pmatrix} + c_2(1)\begin{pmatrix} 2 \\ 3 \end{pmatrix}$$

Multiply and solve for c_1 and c_2

$$0 = -c_1 + 2c_2$$

$$-4 = c_1 + 3c_2$$

Rearranging,

$$c_1 = -\frac{8}{5}, \; c_2 = -\frac{4}{5}$$

The equations in vector form are

$$X(t) = -\frac{8}{5}e^{-t}\begin{pmatrix} -1 \\ 1 \end{pmatrix} - \frac{4}{5}e^{4t}\begin{pmatrix} 2 \\ 3 \end{pmatrix}$$

The final solution of the individual equations is

$$y_1 = \frac{8}{5}e^{-t} - \frac{8}{5}e^{4t}$$

$$y_2 = -\frac{8}{5}e^{-t} - \frac{12}{5}e^{4t}$$

Simulink Solution

The Simulink integrator, dragged from the Simulink/Commonly used Blocks library, can be used to solve the two ODEs simultaneously, as shown in Figure 5.13.

Figure 5.14 is the plot generated by double-clicking on the Scope block of Figure 5.13. The generated figure is the solution of the two equations (y_1 and y_2) versus time (t).

An alternative Simulink solution uses the MATLAB function associated with the MATLAB function.

Figure 5.15 is the Simulink block diagram using the eigenvalue and eigenvector to solve Example 5.6.

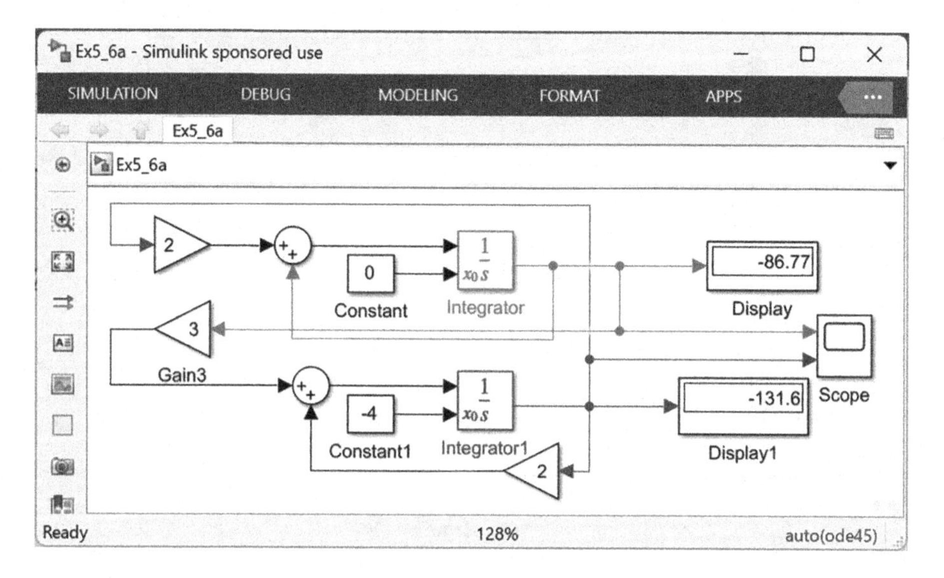

FIGURE 5.13 Solution of the equations in Example 5.6 using the Simulink integrator.

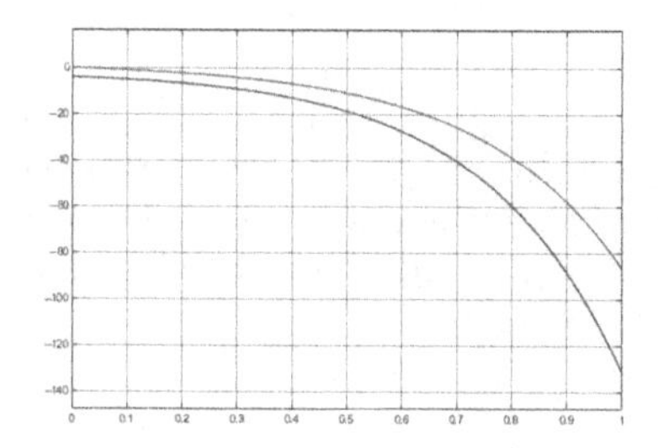

FIGURE 5.14 Plot generated by double-clicking the scope of Figure 5.13, generating the solution of y_1 and y_2 versus time.

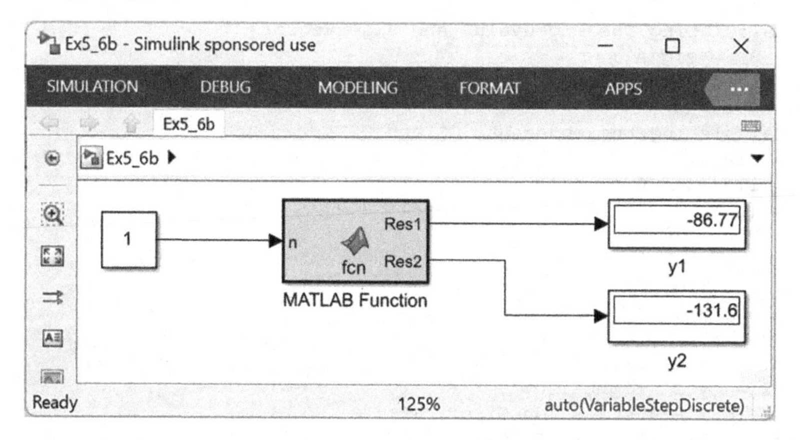

FIGURE 5.15 Simulink MATLAB function and the eigenvalue and eigenvector technique to solve the two ODEs defined in Example 5.6.

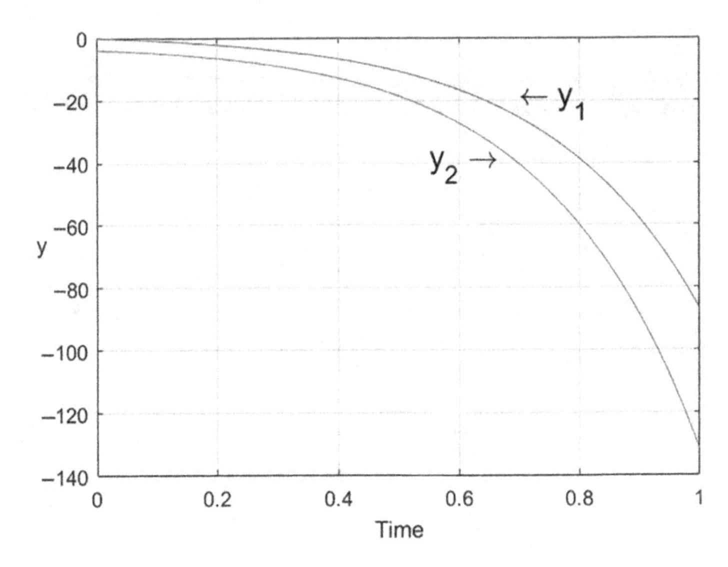

FIGURE 5.16 Simulink generated plot signifies the solution of the two ODEs in Example 5.6 using eigenvalue and eigenvector.

The following is the MATLAB program implanted in the MATLAB function graphically programmed in Figure 5.15. The MATLAB code simultaneously programmed the eigenvalue and eigenvector for solving the two ODE equations described in Example 5.6. Figure 5.16 shows the plotted result of the solution of the two ODEs defined in Example 5.6.

```
% Example 5.6
function [Res1,Res2] = fcn(n)
% The coefficients of the Matrix
A = [1 2;
3 2];
% Calculating the eigenvalue and eigenvectors
[V, D] = eig(A);
% initial conditions
Y0 = [0; -4];
% Calculating the constants C1 and C2
c = diag(V\Y0);
t = linspace(0,n);
T = [t;t];
y = V*c*exp(D*T);
plot(t, y)
xlabel("Time")
ylabel("y ", rotation =0 )
grid on;
str1 = '\leftarrow y_1';
str2 = 'y_2 \rightarrow';
text (0.7,-20,str1,'FontSize', 15);
text (0.55,-40,str2,'FontSize', 15);
Res1 = y(1,end);
Res2 = y(2,end);
```

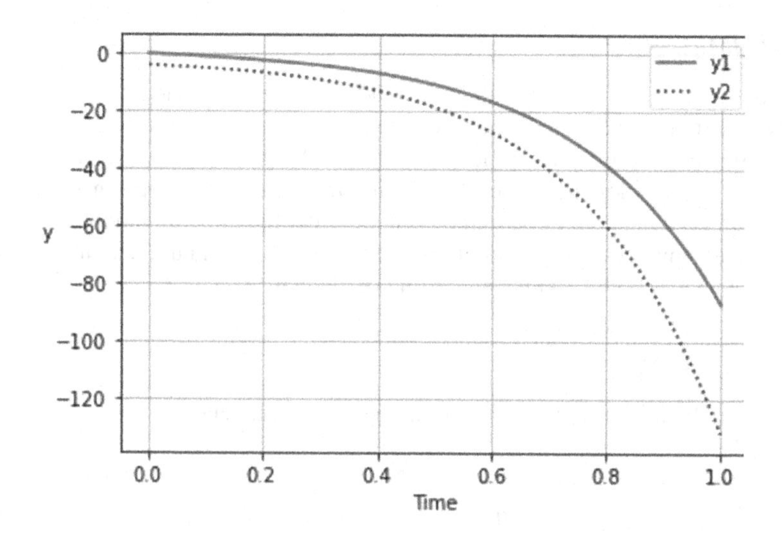

FIGURE 5.17 Plot of the y_1 and y_2 versus time (t) generated by the Python program uses the eigenvalue and eigenvector for solving the two equations defined in Example 5.6.

Python Solution

The following Python code utilizes the eigenvalue and eigenvector approaches to solve the two equations described in Example 5.6. Running the Python program generates the plot shown in Figure 5.17.

```
# Example 5.6
import numpy as np
import matplotlib.pyplot as plt
A = np.array([[1,2], [3,2]])
# extract eigenvalues
[w, v]= np.linalg.eig(A)
y10 = 0
y20 = -4
Pinitial = np.array([y10, y20])
C1, C2 = np.linalg.solve(v, Pinitial)
# combine everything to get the solution to P1
#t = np.linspace(0,2)
fig = plt.figure()
t=np.linspace(0,1,100)
y1 = C1 * np.exp(w[0]*t)*v[0,0] + C2 * np.exp(w[1] * t) * v[0,1]
y2 = C1 * np.exp(w[0]*t)*v[1,0] + C2 * np.exp(w[1] * t) * v[1,1]
#plt.plot(t, y1, t, y2)
plt.plot(t, y1, '-b', label='y1')
plt.plot(t, y2, ':r', label='y2')
plt.grid(True)
plt.ylabel("y ", rotation = 0)
plt.xlabel("Time")
plt.legend()
```

5.4 SUMMARY

The aim of this chapter is to acquaint the reader with the numerical method required in solving initial value ODEs that arise in chemical engineering topics such as fluid flow, diffusion, reaction, mass, and heat transfer. IVP is an ODE with an initial condition that determines the value of the unknown function at a given point in the domain. In this chapter, we solved systems of two linear differential equations in which the eigenvalues are real numbers. The eigenvalue is a number, and the eigenvector is a vector, both of which are hiding in the matrix.

5.5 PROBLEMS

1. Solve the following two ODE using eigenvalue and eigenvector techniques.

$$\frac{dy}{dt} = 3x + y, \quad y(0) = -1$$

$$\frac{dx}{dt} = 5x - y, \quad x(0) = 2$$

Answer:

$$x(t) = -\frac{3}{2}e^{2t} + \frac{7}{2}e^{4t}$$

$$y(t) = -\frac{9}{2}e^{2t} + \frac{7}{2}e^{4t}$$

2. Solve the following IVPs using the elimination method.

$$\frac{dx}{dt} = -(y + 5x), \quad x(1) = 0$$

$$\frac{dy}{dt} = 4x - y, \quad y(1) = 1$$

Answer:

$$x(t) = c_1e^{-3t} + c_2te^{-3t}$$

$$y(t) = -(2c_1 + c_2)e^{-3t} - 2c_2te^{-3t}$$

3. Solve the following system of IVPs using the elimination method.

$$\frac{dx}{dt} = 6y$$

$$\frac{dy}{dt} = x - y$$

Answer:

$$x(t) = c_1 e^{-3t} + c_2 e^{2t}$$

$$y(t) = -\frac{1}{2} c_1 e^{-3t} + \frac{1}{3} c_2 e^{2t}$$

4. Solve the following system of IVPs using the elimination method.

$$\frac{dx}{dt} = x - 2y$$

$$\frac{dy}{dt} = 2x - 3y$$

Answer:

$$x(t) = c_1 e^{-t} + c_2 t e^{-t}$$

$$y(t) = c_1 e^{-t} - \frac{c_2}{2} e^{-t} + c_2 t e^{-t}$$

5. Solve the following ODEs using eigenvalue and eigenvector.

$$\frac{dx}{dt} = 3x - 13y, \;\; x(0) = 3$$

$$\frac{dy}{dt} = 5x + y, \;\; y(0) = -10$$

Answer:

$$x(t) = 3\cos(8t)e^{2t} + \frac{133}{8}\sin(8t)e^{2t}$$

$$y(t) = -10\cos(8t)e^{2t} + \frac{25}{8}\sin(8t)e^{2t}$$

6. Solve the following two ODEs using the eigenvalue and eigenvector technique.

$$\frac{dx}{dt} = 2x + 3y$$

$$\frac{dy}{dt} = 4x + y$$

Answer:

$$x(t) = c_1e^{5t} + 3c_2e^{-2t}$$

$$y(t) = c_1e^{5t} - 4c_1e^{-2t}$$

7. Solve the following two ODEs using the elimination method.

$$\frac{dy}{dt} = 3x$$

$$\frac{dx}{dt} = y$$

Answer:

$$x(t) = c_1e^{\sqrt{3}t} + c_2e^{-\sqrt{3}t}$$

$$x(t) = \sqrt{3}c_1e^{\sqrt{3}t} - \sqrt{3}c_2e^{-\sqrt{3}t}$$

8. Solve the following two ODEs using the elimination method.

$$\frac{dy}{dt} = 2x$$

$$\frac{dx}{dt} = 3y$$

Answer:

$$x(t) = c_1 e^{\sqrt{6t}} + c_2 e^{-\sqrt{6}} t$$

$$y(t) = \frac{\sqrt{6}}{3} c_1 e^{\sqrt{6t}} - \frac{\sqrt{6}}{3} c_1 e^{-\sqrt{6t}}$$

9. Solve the following two ODEs using the elimination method.

$$\frac{dy}{dt} = x$$

$$\frac{dx}{dt} = 2x - y$$

Answer:

$$x(t) = c_1 e^t + c_2 t e^t$$

10. Solve the following two ODEs using eigenvalue and eigenvector techniques.

$$\frac{dx}{dt} = 2y$$

$$\frac{dy}{dt} = 7y - 6x$$

Answer:

$$x(t) = 2c_1 e^{3t} + c_2 e^{4t}$$

$$y(t) = 3c_1 e^{3t} + 2c_2 e^{4t}$$

REFERENCE

1. Griffiths, D.F. and Higham, D.J., 2011. *Numerical Methods for Ordinary Differential Equations*. New York, NY: Springer.

6 Numerical Integration of Definite Functions

Numerical integration estimates the values of definite integrals when a closed integral form is challenging to find. The integral is evaluated at a finite set of points called the points of integration, and a weighted sum of these values is used to approximate the integral. This chapter aims to calculate definite integrals using various numerical integration methods and validate the manual calculations with those programmed using Python and the Simulink graphical programming of MATLAB. The integration methods include Simpson, Trapezoidal, Midpoint, Boole, and Romberg rules.

LEARNING OBJECTIVES

1. Calculate a definite integral using Simpson's rule.
2. Estimate a definite integral using Trapezoidal rule.
3. Assess a definite integral using Midpoint rule.
4. Evaluate a definite integral using Boole's rule.
5. Use Romberg's rule to estimate the value of a definite integral.
6. Apply Simulink and Python to compute definite integrals.

6.1 INTRODUCTION

There are cases where it is difficult to find an analytical solution to a definite integral. However, it is possible to approximate the integral by dividing the function into small subintervals and approximating the area. Numerical integration includes a wide range of algorithms for calculating the numerical value of a definite integral. Integration is often used to find the area under the graph of a function, and the area can be found by adding small slices close to zero width. The definite integral is a formal computation of the area under a function using tiny fragments or lines from the area. A definite integral has start and end values: in other words, there is an interval $[a, b]$, where a and b are called limits, bounds, or boundaries.

Numerical integration methods combine evaluations of the integral to obtain an approximation of the definite integral. Integration is evaluated at a finite set of points called integration points, and a weighted sum of these values is used to approximate the integral. In numerical analysis, numerical integration constitutes a wide range of algorithms for calculating the numerical value of a definite integral, and thus, the term is sometimes also used to describe the numerical solution of differential equations. The fundamental problem that is considered by numerical integration is calculating an approximate solution to a definite integral. They differ from analytic integration in two ways: The first is approximate and will not yield an exact answer; error analysis is an essential aspect of numerical integration. Second, it does not

DOI: 10.1201/9781003360544-6

result in an initial function from which the area can be determined due to arbitrary bounds; it only produces an approximate numerical value for the area [1]. A definite integral is an area under a curve between two fixed limits. The following is a definite integral of $f(x)$ concerning dx from a to b:

$$\int_a^b f(x)dx \tag{6.1}$$

where a and b are lower and upper integral limits, $f(x)$ is the integrand, and dx is the integrating agent. The definite integral of the real-valued function:

$$\int_a^b f(x)dx = F(b) - F(a) \tag{6.2}$$

When having a continuous function $f(x)$ of the interval $[a, b]$, the interval is divided into n subdivisions of equal width (h), and from each interval, choose a point, x_i. Then the definite integral of the mentioned function is

$$\int_a^b f(x)dx = \lim_{n\to\infty}\sum_{i=1}^{n} f(x_i)\Delta x \tag{6.3}$$

The number at the bottom of the integral sign (a) is the integral lower limit, and (b) at the top of the integral sign is the integral upper limit. Numerical integration estimates the values of definite integrals when an approximate value only of the definite integral is needed or when a closed-form integral is difficult to find. The most used numerical integration techniques are the Simpson, Trapezoidal, and Midpoint rules. Simpson's rule approximates the definite integral using quadratic functions; the Trapezoidal rule approximates the definite integral using trapezoidal regions; the Midpoint rule approximates the definite integral using rectangular approximation; Boole's rule represents the area under the curve $y = f(x)$ and between the coordinates $x = a$ and $x = b$. Romberg's method evaluates the integrand at equally spaced points, and the integrand must have continuous derivatives. The Romberg integration (I) is approximated using the composite Trapezoidal rule.

6.2 SIMPSON'S RULE

Simpson's rule is a numerical technique approximating a definite integral's value using a quadratic function. The method is named after the English mathematician Thomas Simpson (1710–1761). In Simpson's rule, we estimate the areas of regions under curves by using rectangles. Simpson's rule is a method for approximating a definite integral using a multidefined quadratic function; the method uses parabolas rather than a straight line. In numerical analysis, Simpson's 1/3 rule is a method for the numerical approximation of definite integrals. Specifically, it is the following approximation:

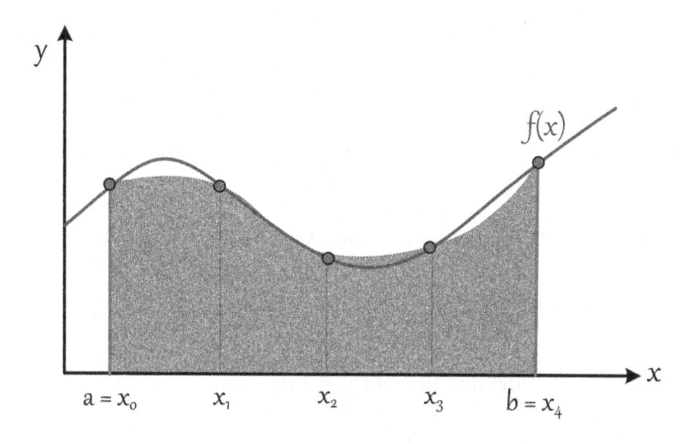

FIGURE 6.1 Graphical representation of Simpson's rule.

$$\int_a^b f(x)dx = \frac{h}{3}\{f(x_o)+4f(x_1)+2f(x_2)+4f(x_3)+\ldots+f(x_n)\} \tag{6.4}$$

To obtain an approximate value of the definite integral, split the main interval into subintervals with an even number of n subintervals (Figure 6.1). Each step size (width), h, can be defined as follows:

$$h = \frac{b-a}{n} \tag{6.5}$$

Simpson's rule works best when the function has a similar shape to the approximation device within a narrow interval.

Example 6.1 Application of Simpson's Rule

Use Simpson's rule to integrate the following definite integral, dividing it into two subintervals.

$$\int_0^1 x^3 dx$$

Evaluate the definite integral manually and confirm the manual calculations with the following approaches:

a. Simulink/MATLAB programming of Simpson's rule with n subinterval.
b. Write a Python program to evaluate a definite integral using Simpson's rule with n subdivisions.

Solution

Divide the main interval [0,1] into two subintervals with a width, h,

$$h = \frac{1-0}{2} = \frac{1}{2}$$

Substitute in Simpson's rule formula as follows:

$$S_n = \frac{h}{3}\left(f(0) + 4f\left(\frac{1}{2}\right) + f(1) \right)$$

Substitute the values of the function at the subdivisions

$$S_n = \frac{\left(\frac{1}{2}\right)}{3}\left(0 + 4\frac{1}{8} + 1\right)$$

The approximate numerical value of the definite integral is

$$S_n = \frac{1}{4} = 0.25$$

The exact value of the definite integral is

$$\int_0^1 x^3 dx = \frac{x^4}{4} = \frac{1}{4} - 0 = \frac{1}{4} = 0.25$$

The approximate numerical value agreed with the exact value of the definite integral.

Simulink Solution

Using the Simpson rule, the Simulink block diagram for solving Example 6.1 is described in Figure 6.2. The figure is followed by the MATLAB program associated with the Simulink MATLAB function. The MATLAB function is connected to three input ports: the lower integral limit (a), the upper integral limit (b), and the number of subintervals (n). There is one output port connected to the Display block to release the output value of the integral. The following program is the MATLAB code associated with the Simulink MATLAB function [2].

```
function res = simp(a, b, n)
f=@(x)x^3; %the desired function
h=(b-a)/n;
s0=f(a)+f(b);
s1=0;
s2=0;
for i=1:(n-1)
x=a+i*h;
if rem(i, 2)==0
```

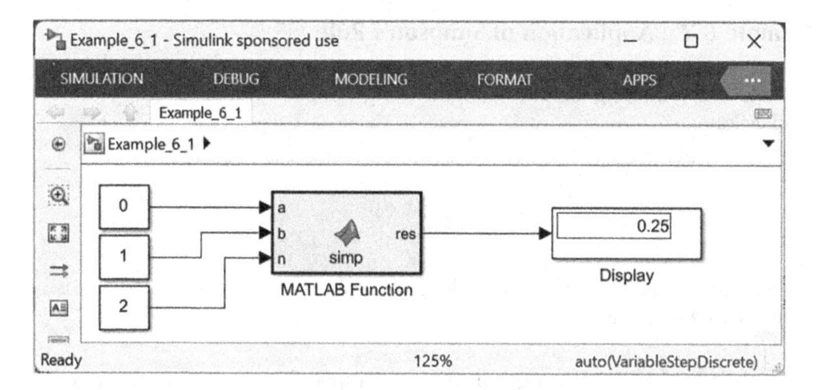

FIGURE 6.2 A Simulink solution using the Simpson rule of the definite integral defined in Example 6.1.

```
s2=s2+f(x);
else
s1=s1+f(x);
end
end
res=(h/3)*(s0+2*s2+4*s1);
end
```

Python Solution

This Python program uses Simpson's 1/3 rule to determine the approximate solution of the definite integral required in Example 6.1. In the implementation, the function input parameters *a*, *b*, and *n* return the approximation results into the Display block [3].

```
# Example 6.1
# Solution of a definite integral using Simpson's rule with n
# subdivisions
def f(x):
return x**3
# the lower limit, upper limit, subinterval number
a=0
b=1
n=2
h= (b-a)/n
sum = f(a)+f(b)
d=4
for k in range(1,n):
x= a+k*h
sum= sum+ d*f(x)
d= 6-d
sum= h/3*sum
print("value of the integral= ", sum)
```

The executed result is

```
value of the integral= 0.25
```

Example 6.2 Application of Simpson's Rule

Estimate the following definite integral using Simpson's rule, dividing it into six subdivisions.

$$\int_1^4 \frac{1}{2}x^2 dx$$

Evaluate the definite integral manually and confirm the manual calculations with the following approaches:

a. Simulink/MATLAB programming of Simpson's rule with n subinterval.
b. Write a Python program to evaluate a definite integral using Simpson's rule with n subdivisions.

Solution

The step size (h) for six subintervals

$$h = \frac{4-1}{6} = \frac{1}{2}$$

Starting with Simpson's rule formula,

$$S_n = \frac{h}{3}\left(f(x_0) + 4f(x_1) + 2f(x_2) + 4f(x_3) + 2f(x_4) + 4f(x_5) + f(x_6)\right)$$

As a function of x,

$$S_n = \frac{h}{3}\left(f(1) + 4f\left(\frac{3}{2}\right) + 2f(2) + 4f\left(\frac{5}{2}\right) + 2f(3) + 4f\left(\frac{7}{2}\right) + f(4)\right)$$

Substitute the values obtained from function $f(x)$ at each node

$$S_n = \frac{\left(\frac{1}{2}\right)}{3}\left(0.5 + 4\,(1.125) + 2(2.0) + 4(3.125) + 2(4.5) + 4(6.125) + 8\right) = 10.5$$

The correct solution is

$$\int_1^4 \frac{1}{2}x^2 dx = \frac{x^3}{(2)(3)} = \frac{4^3 - 1^3}{6} = 10.5$$

The approximate numerical value of the definite integral is identical to the exact value.

Simulink Solution

Figure 6.3 is the Simulink graphical programming using Simpson's rule, representing the solution of the definite integral defined in Example 6.2. The figure is followed by the MATLAB code associated with the MATLAB function, where *a* and *b* are the lower and upper interval limits, respectively, and *n* is the number of subdivisions. The output port is connected to the Display block and outputs the integration value between the two limits (*a* and *b*).

The following program is the MATLAB code using Simpson's rule and implanted in the Simulink MATLAB function.

```
function res = simp(a, b, n)
f=@(x)0.5*x^2; %the desired function
h=(b-a)/n;
s0=f(a)+f(b);
s1=0;
s2=0;
for i=1:(n-1)
x=a+i*h;
if rem(i, 2)==0
s2=s2+f(x);
else
s1=s1+f(x);
end
end
res=(h/3)*(s0+2*s2+4*s1);
end
```

Python Solution

The "SciPy. Integration" subpackage includes many functions for approximating integrals numerically and solving differential equations. Using Simpson's rule available in Scipy, the following program is the Python code, which resulted in an integral execution value of 10.5.

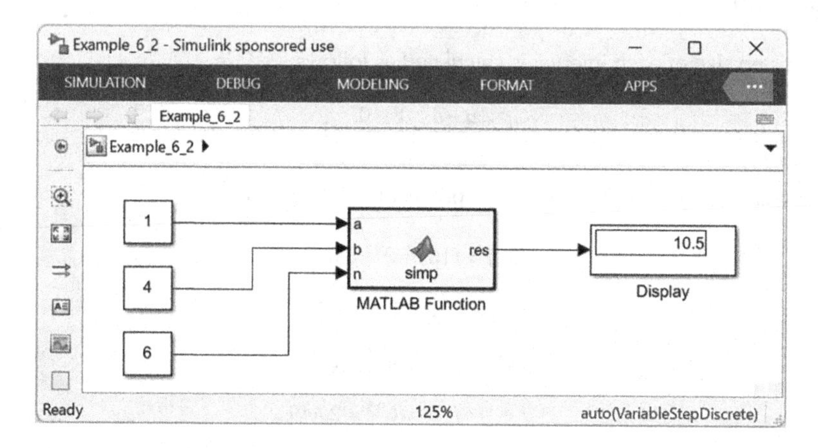

FIGURE 6.3 A Simulink block diagram using the Simpson rule demonstrating the solution of the definite integral specified in Example 6.2.

```
# Example 6.2
# Simpson's rule is available in Scipy.
import numpy as np
import scipy.integrate as spi
a=1
b=4
N=6
x=np.linspace(a, b, N-1)
y=0.5*x*x
approx=spi.simps(y, x)
print("value of the integral= ", approx)
```

The program execution result is

```
value of the integral= 10.5
```

Example 6.3 Application Simpson's Rule

Use Simpson's rule with four subdivisions ($n = 4$) to approximate the following definite integral.

$$f(x) = \int_0^8 x^{0.5} dx$$

Evaluate the definite integral manually and check the manual calculations with the following approaches:

1. Simulink/MATLAB programming of Simpson's rule with n subintervals.
2. Write a Python program to evaluate a definite integral using Simpson's rule with n subdivisions.

Solution

The step size of each interval is calculated as follows:

$$h = \frac{b-a}{n} = \frac{8-0}{4} = 2$$

Calculate the function values at the points of x_i : [0, 2, 4, 6, 8]

$$f(x_o) = f(0) = 0^{0.5} = 0$$

$$f(x_1) = f(2) = 2^{0.5} = 1.414$$

$$f(x_2) = f(4) = 4^{0.5} = 2$$

$$f(x_3) = f(6) = 6^{0.5} = 2.449$$

$$f(x_4) = f(8) = 8^{0.5} = 2.828$$

Substitute these into Simpson's rule formula as follows:

$$\int_0^8 x^{0.5}dx \cong \frac{h}{3}\left[f(x_o)+4f(x_1)+2f(x_2)+4f(x_3)+f(x_4)\right]$$

Substitute the values of the function $f(x)$ for each subinterval

$$\int_0^8 x^{0.5}dx \cong \frac{2}{3}\left[0+4(1.414)+2(2)+4(2.449)+(2.828)\right] \approx 14.85$$

The correct (analytical) solution

$$\int_0^8 x^{0.5}dx = \left[\frac{x^{0.5+1}}{1.5}\right]_0^8 = (8^{1.5}-0^{1.5})/1.5 \approx 15.09$$

The percent error in approximating the integral compared with the exact analytical solution

$$|E| = \left|\frac{15.09-14.86}{15.09}\right| \times 100\% = 1.5\%$$

Simulink Solution

Figure 6.4 represents Simpson's rule programmed in MATLAB and linked to the Simulink MATLAB function. The Simulink block diagram can be used to solve other functions by changing the values of integral limits (*a*, *b*) and the number of subintervals (*n*). Other functions can replace the desired integral function in the MATLAB code to solve other definite integrals. The following program is the MATLAB code utilizing the Simpson rule and embedded in the Simulink MATLAB function located at the center of Figure 6.4.

```
function res = simp(a, b, n)
f=@(x)x^0.5; %the desired function
h=(b-a)/n;
s0=f(a)+f(b);
s1=0;
s2=0;
for i=1:(n-1)
x=a+i*h;
if rem(i, 2)==0
s2=s2+f(x);
else
s1=s1+f(x);
end
end
res=(h/3)*(s0+2*s2+4*s1);
end
```

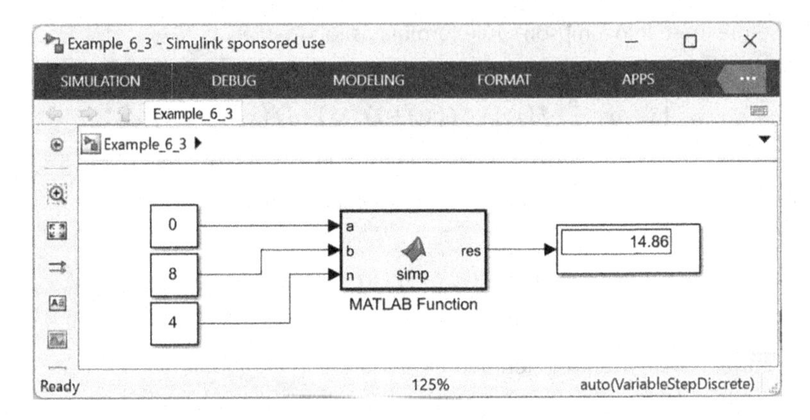

FIGURE 6.4 Simulink solution employing Simpson's rule for solving the definite integral associated with Example 6.3.

Python Solution

In this example, the Simpson rule function is included, and there is no need to import the "SciPy. integrate" library. The Simpson function keeps track of the running sum and the multiplicands associated with each evaluation of the function, but the function to be evaluated is an actual Python program that takes in a single argument, x, and returns its value. The Trapezoidal rule yields comparable results, and the following is the Python program written to evaluate the desired definite integral using Simpson's rule with n subdivisions.

```
# Example 6.3
# The program evaluates a definite integral using Simpson's
# rule with n subdivisions
# Implementing Simpson's 1/3 rule
def f(x):
return x**0.5 # The desired function
def simpson(x0,xn, n):
h= (xn-x0)/n # The step size
Sum=f(x0)+f(xn)
for i in range(1,n):
j = x0 + i*h
if i%2 == 0:
Sum = Sum + 2 * f(j)
else:
Sum= Sum + 4 * f(j)
#Finding value
Sum = Sum * h/3
return Sum
result = simpson(0,8, 4)
print("The value is: %0.5f" % (result))
```

The execution result is

```
The value is: 14.85549
```

6.3 TRAPEZOIDAL RULE

The Trapezoidal rule is a rule that evaluates the area under the curve. The area under the curve is calculated by dividing the total area into small trapezoids instead of the rectangles that divide the interval [a, b] into n equal subintervals, each of width, h, as shown in Figure 6.5.

$$h = \frac{b-a}{n}$$

Such that,

$$a = x_o < x_1 < x_2 \ldots < x_n = b$$

The Trapezoidal rule approximates the definite integrals and uses the linear approximation of the functions. When the underlying function is smooth, the Trapezoidal rule does not give values as accurate as Simpson's rule because Simpson's rule uses the quadratic approximation rather than the linear approximation used by the Trapezoidal rule. Both the Trapezoidal rule and Simpson's rule give the approximation value. By contrast, Simpson's rule gives more accurate values of the definite integral.

The function, $f(x)$, is a continuous function over the interval [a, b]. The approximate solution of such a function is given by

$$\int_a^b f(x)dx = \frac{h}{2}\left(f(x_0) + 2f(x_1) + 2f(x_2) + \ldots + 2f(x_{n-1}) + f(x_n)\right) \tag{6.5}$$

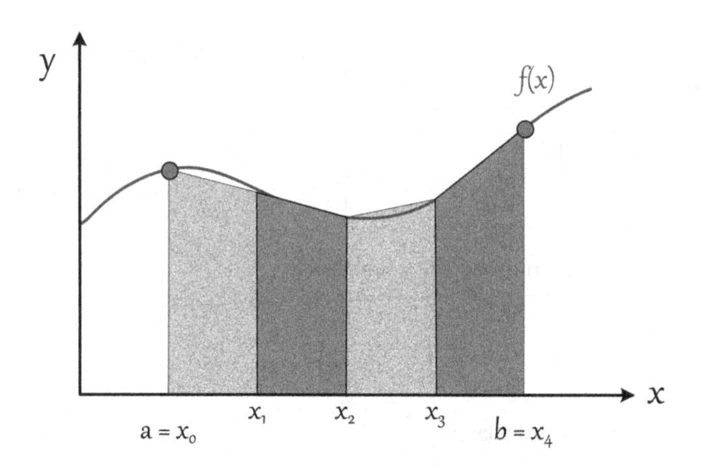

FIGURE 6.5 Graphical interpretation of Trapezoidal rule.

Example 6.4 Application of Trapezoidal Rule

Estimate the following definite integral using the Trapezoidal rule utilizing four subintervals, estimating the absolute and relative error.

$$\int_0^1 x^2 dx$$

Evaluate the definite integral manually and support the manual calculations with the following approaches:

1. Simulink/MATLAB programming of the Trapezoidal rule.
2. Write a Python program to evaluate a definite integral using the Trapezoidal rule with n subdivisions.

Solution

First, calculate the step size, h, using the integral upper limit (b) and lower limit (a) and the number of the subintervals (n):

$$h = \frac{b-a}{n} = \frac{1-0}{4} = \frac{1}{4}$$

The Trapezoidal rule

$$T_n = \int_0^1 x^2 dx = \frac{\left(\frac{1}{4}\right)}{2}\left(f(0) + 2f\left(\frac{1}{4}\right) + 2f\left(\frac{1}{2}\right) + \ldots + 2f\left(\frac{3}{4}\right) + f(1)\right)$$

Find the functions at different intervals and substitute them into the Trapezoidal technique formula

$$T_n = \int_0^1 x^2 dx = \frac{\left(\frac{1}{4}\right)}{2}\left(0 + 2\left(\frac{1}{16}\right) + 2\left(\frac{1}{4}\right) + \ldots + 2\left(\frac{9}{16}\right) + (1)\right) = \frac{11}{32} = 0.344$$

The calculated value from the direct integration

$$\int_0^1 x^2 dx = \left.\frac{x^3}{3}\right]_0^1 = \frac{1}{3}$$

The absolute error is calculated

$$\left|\frac{1}{3} - \frac{11}{32}\right| = \frac{1}{192}$$

The percent relative error is

$$\frac{\left|\frac{1}{3}-\frac{11}{32}\right|}{\frac{1}{3}}=\frac{\frac{1}{192}}{\frac{1}{3}}\times 100\% = 3.1\%$$

Simulink Solution

Using the MATLAB function that utilizes the Trapezoidal rule implanted in the Simulink MATLAB function (Figure 6.6), the embedded MATLAB code follows Figure 6.6. The Simulink block diagram can be used to solve other functions by changing the values of integral limits (*a*, *b*), the number of subintervals (*n*), and the desired function in MATLAB code.

The following program is the MATLAB code embedded in the Simulink MATLAB function and centered in Figure 6.6. The function can be used to solve other definite integrals by changing the desired function.

```
function result = Trapz(a, b, n)
f=@(x)x^2; %the desired function
h=(b-a)/n;
sum = 0;
for i=1:(n-1)
sum = sum +f(a+i*h);
end
result = (h/2)*(f(a)+f(b)+2*sum);
end
```

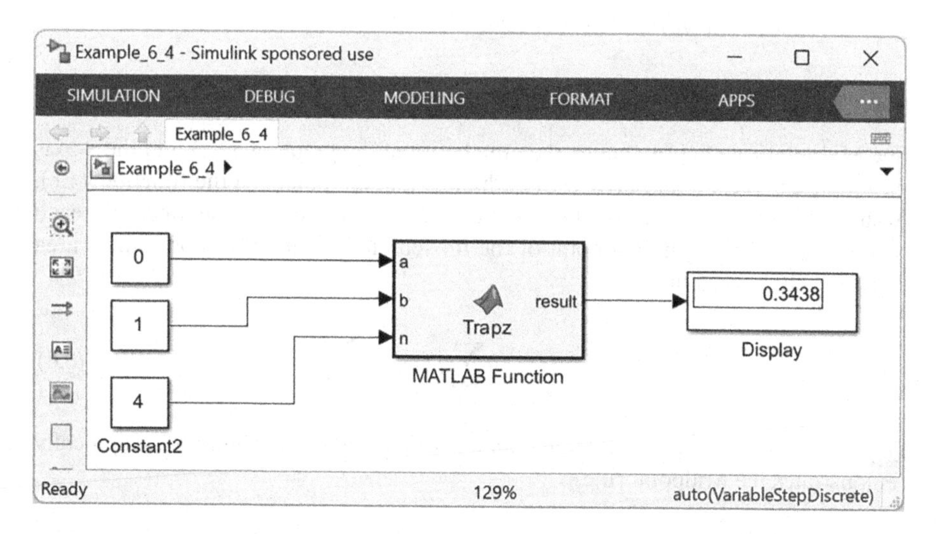

FIGURE 6.6 Trapezoidal rule programmed in Simulink/MATLAB function for solving the definite integral specified in Example 6.4.

Python Solution

The Trapezoidal algorithm is used without importing the function from the Python library in the solution of this example using Python programming. The function (f) can be changed to any other desired function. The Trapezoidal rule yields comparable results to Simpson's rule.

```
# Example 6.4
# Python program to evaluate a definite integral using
Trapezoidal's # rule with n subdivisions
# Implementing Trapezoidal's rule
def trapezoid(a, b, n):
f= lambda x: x**2 # the desired function
sum = 0
h= (b - a ) / n
for k in range(n + 1):
x = a + (k * h)
summand = f(x)
if (k != 0) and (k != n):
summand *= 2
sum += summand
return sum * h / 2
# Example of use
result =trapezoid(0, 1, 4)
print("The calculated value is: %0.5f" % (result))
```

The execution result is

```
The calculated value is: 0.34375
```

6.4 RECTANGLE RULE (MIDPOINT RULE)

The Midpoint rule approximates the area between the graph of $f(x)$ and the x-axis by adding the rectangular area with midpoints that are points on the function $f(x)$. Assume $f(x)$ is continuous on the interval $[a, b]$. If divided into n subintervals, each of length h and m_i is the midpoint of the ith subinterval, and the approximate area under the curve is estimated as

$$M_n = h\sum_{i=1}^{n} f(m_i) \tag{6.6}$$

The Midpoint rule is more accurate than the Trapezoidal rule. Figure 6.7 graphically demonstrates the Midpoint rule.

Example 6.5 Applying the Midpoint Rule

Use the Midpoint rule with four subdivisions to evaluate the following definite integral.

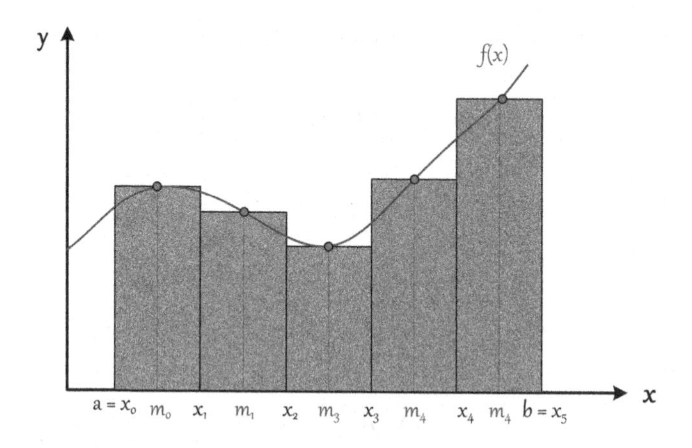

FIGURE 6.7 Graphical clarification of Midpoint rule.

$$\int_0^1 x^2 dx$$

Compare the approximate numerical integration result using the Midpoint rule with actual analytical integration value and Python and Simulink/MATLAB programming of the Midpoint rule.

Solution

The actual value of this integral is as follows:

$$\int_0^1 x^2 dx = \left.\frac{x^3}{3}\right]_0^1 = \frac{1}{3} - \frac{0}{3} = \frac{1}{3}$$

The interval length

$$h = \frac{b-a}{n} = \frac{1-0}{4} = \frac{1}{4}$$

There are four subintervals

$$\left[0 - - \frac{1}{4} - - \frac{1}{2} - - \frac{3}{4} - - 1\right]$$

The middle point of these subintervals

$$\left[0 - \frac{1}{8} - \frac{1}{4} - \frac{3}{8} - \frac{1}{2} - \frac{5}{8} - \frac{3}{4} - \frac{7}{8} - 1\right]$$

The integral is calculated using the Midpoint rule as follows:

$$\int_0^1 x^2 dx = \frac{1}{4}\left\{f\left(\frac{1}{8}\right)+f\left(\frac{3}{8}\right)+f\left(\frac{5}{8}\right)+f\left(\frac{7}{8}\right)\right\}$$

Find the function and substitute

$$\int_0^1 x^2 dx = \frac{1}{4}\left\{\frac{1}{64}+\frac{9}{64}+\frac{25}{64}+\frac{21}{64}\right\} = \frac{21}{64} = 0.328$$

The absolute error

$$|A - B|$$

$$\left|\frac{1}{3} - \frac{21}{64}\right| = \frac{1}{192}$$

The relative error is the error as a percentage of the actual value and is given by

$$\frac{|A - B|}{A} \times 100\%$$

where *A* is the actual value and *B* is our estimated value. Figure 6.8 shows the Simulink block diagram representing the Midpoint rule function. The Simulink block diagram can be utilized to solve other functions by changing the values of integral limits (*a*, *b*), the number of subintervals (*n*), and the desired function in MATLAB code.

The following MATLAB program of the Midpoint method is rooted in the MATLAB function (Figure 6.8).

```
function Result = Midp(a, b, n)
f=@(x)x^2; %the desired function
h=(b-a)/n;
sum = 0;
```

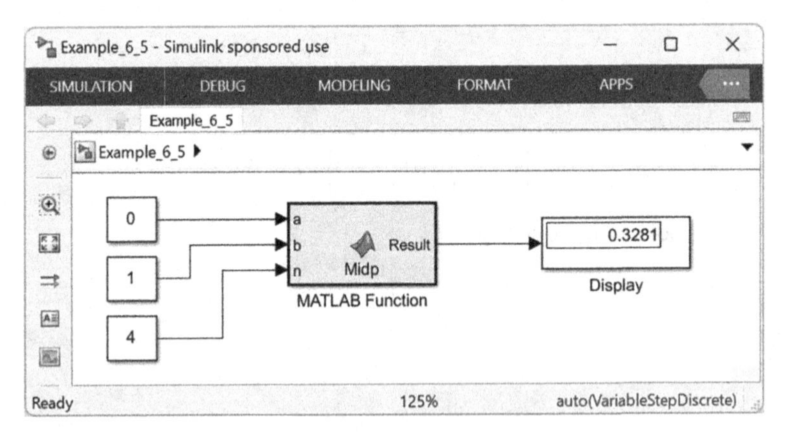

FIGURE 6.8 Simulink midpoint block diagram represents the solution of the definite integral required in Example 6.5.

```
for i=1:n
sum = sum +h*(f((a+(a+h))/2));
a=a+h;
end
Result = sum;
End
```

Python Solution

The Midpoint rule evaluates the desired function between each point in the required interval. Using the functions from the Python library is certainly much faster to implement. By contrast, the following Midpoint method in Python was interesting to learn how the function works.

```
# Example 6.5
# Python program to evaluate a definite integral using Midpoint's
# rule with n subdivisions
# implementing Midpoint's Rule
def midpoint(a, b, n):
f=lambda x:x**2 # The desired function
sum = 0
x_in=((2*n+1)*a-b)/(2*n)
h=(b-a)/n
for k in range (1,n+1):
x=x_in+(k*h)
sum= sum + f(x)
return sum*h
# Example to use
Res = midpoint (0,1,4)
print("The calculated value is: %0.5f" % (Res))
```

The result after the program execution is

```
The calculated value is: 0.32812
```

6.5 BOOLE'S RULE

The method is named after a famous English mathematician George Boole (1815–1864). Boole's rule is derived by putting $n=4$ in the general quadrature formula, which means $f(x)$ can be approximated by a polynomial of the fourth degree. It is used to evaluate the definite integral, and it can only be used for problems with sub-interval, i.e., $n=4$ (Figure 6.9).

$$\int_a^b f(x)dx = \frac{2h}{45}\left(7y_o + 32y_1 + 12y_2 + 32y_3 + 14y_4 + 32y_5 + 12y_6 + 32y_7 + 14y_8 + ..\right)$$

Example 6.6 Applying Boole's Rule to Estimate a Definite Integral

Estimate for following definite integral using Boole's rule, assuming the number of intervals equals 4.

$$\int_0^1 \frac{dx}{1+x^2}$$

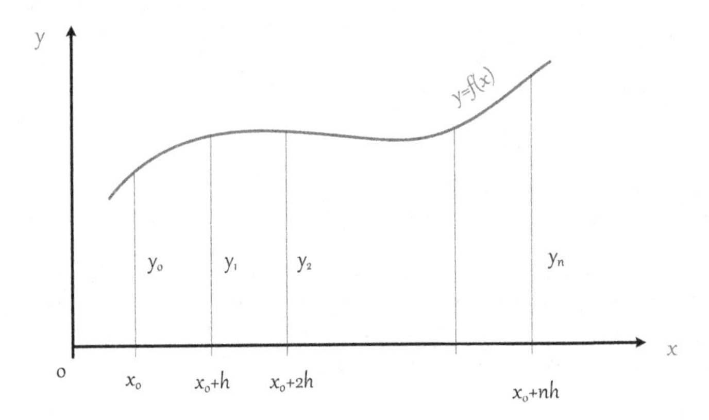

FIGURE 6.9 Graphical representation of Boole's rule.

Compare the result of the approximate numerical integration using Boole's rule with actual analytical integration value and employ Python and Simulink/MATLAB programming of Boole's rule.

Solution

From the given data, assuming $n = 4$, then

$$h = \frac{b-a}{4} = \frac{1-0}{4} = \frac{1}{4}$$

Applying Boole's formula,

$$\int_0^1 \frac{dx}{1+x^2} = \frac{2h}{45}\left[7y_0 + 32y_1 + 12y_2 + 32y_3 + 7y_4\right]$$

Then substitute the obtained values at each subinterval

$$\int_0^1 \frac{dx}{1+x^2} = \frac{2h}{45}\left[7f(0) + 32f\left(\frac{1}{4}\right) + 12f\left(\frac{1}{2}\right) + 32f\left(\frac{3}{4}\right) + 7f(1)\right]$$

Substitute

$$\int_0^1 \frac{dx}{1+x^2} = \frac{2h}{45}\left[7(1) + 32(0.9412) + 12(0.8) + 32(0.64) + 7(0.5)\right] = 0.7855$$

Accordingly, the obtained approximate solution of the defined integral is 0.7855.

Simulink Solution

The Simulink block diagram represents the solution of the definite function specified in Example 6.6, shown in Figure 6.10. Boole's method programmed in the MATLAB code is associated with the Simulink MATLAB function. The Simulink block diagram can be utilized to solve other integrals by changing the values of

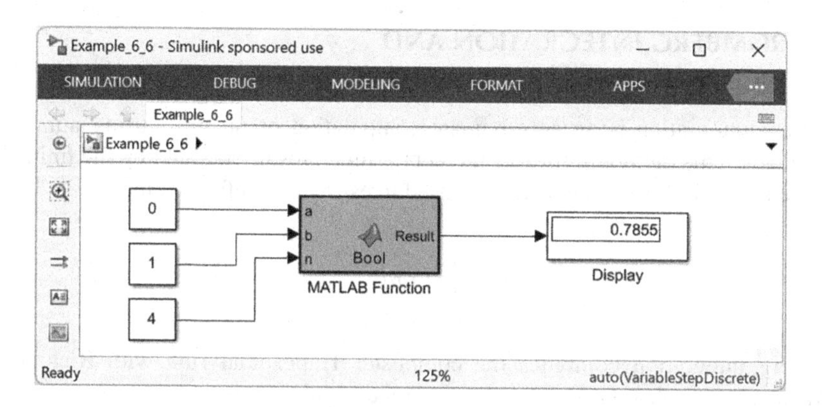

FIGURE 6.10 Simulink block diagram utilizing Boole method to estimate the definite integral needed in Example 6.6.

integral limits (a, b), the number of subintervals (n), and the desired function in MATLAB code.

The following is the MATLAB code implanted in the Simulink MATLAB function to solve the equation required in Example 6.6.

```
function Result = Bool(a, b, n)
f=@(x)1/(1+x^2); %the desired function
h=(b-a)/n;
sum = 7*(f(a)+f(b));
sum = sum +32*(f(a+h)+f(b-h));
sum = sum +12*(f(a+2*h));
Result = 2*sum*h/45;
```

Python Solution

Here, Example 6.6 is solved by programming Boole's rule in Python. The program evaluates the required integral within the interval [0, 1] and with n subdivisions.

```
# Example 6.6
# Python program to evaluate a definite integral using Boole
method
# implementing the Boole method
def Boole(a, b, n):
f=lambda x:1/(1+x**2 )# The desired function
h = (b-a)/n
sum = 7*(f(a)+f(b))
sum = sum +32*(f(a+h)+f(b-h))
sum = sum+ 12*(f(a+2*h))
res = 2*sum*h/45
return res
# Example to use
res = Boole (0,1,4)
print("The calculated value is: %0.5f" % (res))
```

The result after running the program is

```
The calculated value is: 0.78553
```

6.6 ROMBERG INTEGRATION AND RICHARDSON EXTRAPOLATION

Romberg integration is an approximate computation of integrals using numerical techniques. This method improves the approximate result obtained by the finite difference method. Romberg's method is used to estimate a definite integral

$$\int_a^b f(x)dx$$

Romberg integration combines the composite Trapezoidal rule with Richardson extrapolation (Table 6.1). The composite Trapezoidal rule

$$I = \int_a^b f(x)dx = \frac{h}{2}\left[f(a) + 2\sum_{i=1}^{n-1} f(x_i) + f(b)\right]$$

where

$$h = \frac{b-a}{n}$$

$$x_i = x_{i-1} + h$$

This procedure ends when two successive values are remarkably close (Figure 6.11).

TABLE 6.1
The Romberg Integration Table

h_i	I_i'	I_i'	I_i''	I_i'''
h	I_1	$I_1' = I_2 + \frac{1}{3}(I_2 - I_1)$		
			$I_1'' = I_1' + \frac{1}{3}(I_2' - I_1)$	
$\frac{h}{2}$	I_2	$I_2' = I_3 + \frac{1}{3}(I_3 - I_2)$		$I_1''' = I_2'' + \frac{1}{3}(I_2'' - I_1'')$
			$I_2'' = I_3' + \frac{1}{3}(I_3' - I_2)$	
$\frac{h}{4}$	I_3	$I_3' = I_4 + \frac{1}{3}(I_4 - I_3)$		
$\frac{h}{8}$	I_4			

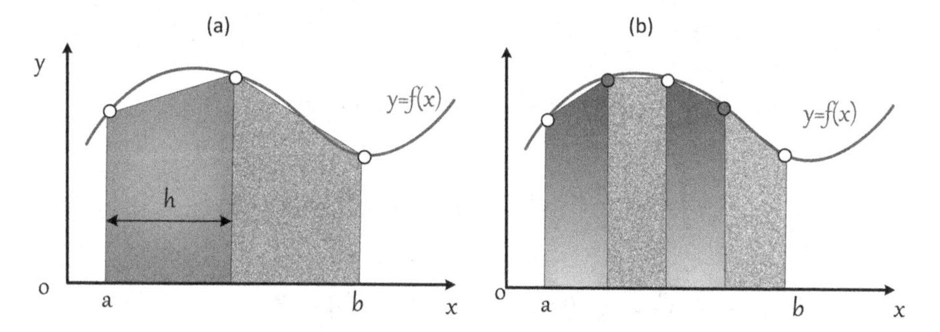

FIGURE 6.11 Graphical representation of the (a) Romberg integration and (b) Richardson extrapolation.

Example 6.7 Applying Romberg's Rule

Evaluate for following definite integral by Boole's rule in four subintervals.

$$\int_0^8 x^2 dx$$

Compare the approximate numerical integration result of manual calculation with actual analytical integration value and the result from utilizing Python and Simulink/MATLAB programming of Romberg's rule.

Solution

Find the step size

$$h = \frac{b-a}{n} = \frac{8-0}{4} = 2$$

$$f(x) = x^2$$

The interval will be

$$h = 4$$

$$\int_0^8 x^2 dx = \frac{h}{2}\left\{f(x_o) + 2\ f(x_1) + f(x_4)\right\}$$

$$I_1 = \int_0^8 x^2 dx = \frac{4}{2}\left\{f(0) + 2\left(f(4)\right) + f(x_4)\right\}$$

$$I_1 = \int_0^8 x^2 dx = \frac{4}{2}\left\{\left(0^2\right) + 2\left(4^2\right) + \left(8^2\right)\right\} = 192$$

$$h = 2$$

$$\int_0^8 x^2 dx = \frac{h}{2}\left\{f(x_o) + 2\left(f(x_1) + f(x_2) + f(x_3)\right) + f(x_4)\right\}$$

$$I_2 = \int_0^8 x^2 dx = \frac{h}{2}\left\{f(0) + 2\left(f(2) + f(4) + f(6)\right) + f(8)\right\}$$

$$I_2 = \int_0^8 x^2 dx = \frac{2}{2}\left\{0^2 + 2\left(2^2 + 4^2 + 6^2\right) + 8^2\right)\right\} = 176$$

$$h = 1$$

$$I_3 = \frac{1}{2}\left\{0^2 + 2\left(1^2 + 2^2 + 3^2 + 4^2 + 5^2 + 6^2 + 7^2\right) + 8^2\right\} = 172$$

$$h = 0.5$$

$$I_3 = \frac{0.5}{2}\left\{0^2 + 2\left(0.5^2 + 1^2 + 1.5^2 + .. + 7.5^2\right) + 8^2\right\} = 171$$

The rest of the calculation is listed in Table 6.2.

Finally,

$$I_1''' = I_2 + \frac{1}{3}(I_2'' - I_1'') = 170.66 + \left(\frac{1}{3}\right)(170.66 - 170.66) = 170.66$$

$$\int_0^8 x^2 dx = 170.66$$

Simulink Solution

The Romberg method programmed in the MATLAB code is associated with the Simulink MATLAB function. The Simulink block diagram can be utilized to solve other integrals by changing the values of integral limits (a,b), the number of subintervals

TABLE 6.2
This Arithmetic Procedure Ends When Two Consecutive Values Are Very Close (170.66)

h_i	I_i	I_i'	I_i''
h	192	176 + 1/3 (176 − 192) = 170.66	170.66 + 1/3 (170.66 − 170.66) = 170.66
$\frac{h}{2}$	176	172 + 1/3 (172 − 176) = 170.66	170.66 + 1/3 (170.66 − 170.66) = 170.66
$\frac{h}{4}$	172	171 + 1/3 (171 − 172) = 170.66	
$\frac{h}{8}$	171		

(n), and the desired function in MATLAB code. The Simulink block diagram for solving the definite integral required in Example 6.7 is shown in Figure 6.12. To reach $h=0.5$, the program's n value equals 16.

The following is the MATLAB program refers to the Romberg integration method associated with the Simulink MATLAB function.

```
function Result = Romberg(a, b, n)
f=@(x)x) x^2; %the desired function
h=(b-a);
r=zeros(2,n+1);
r(1,1)=(f(a)+f(b))/(2*h);
for i = 2:n
rv=0;
for k = 1:2^(i-2)
rv=rv+f(a+(k-0.5)*h);
end
r(2,1)=(r(1,1) + h*rv)/2;
for j=2:i
l=2^(2*(j-1));
r(2,j)=r(2,j-1)+(r(2,j-1)-r(1,j-1))/(l-1);
end
for k=1:i
fprintf('%7.5f', r(2,k));
end
fprintf('\n\n');
h=h/2;
for j=1:i
r(1,j)=r(2,j);
end
end
Result=r(1,n);
End
```

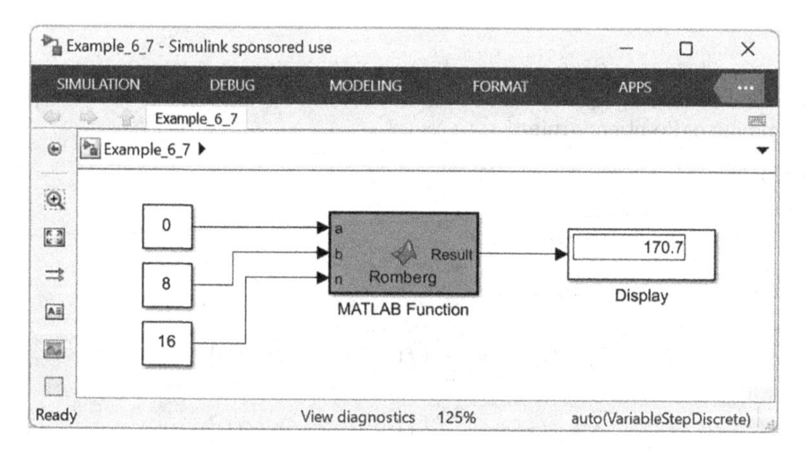

FIGURE 6.12 Simulink solution applying Romberg's rule programmed to solve the definite integral identified in Example 6.7.

Python Solution

Romberg integration method of a definite integral returns the integral of a function of one variable over the interval [*a*, *b*]. The printed solution represents the triangular array of the intermediate results. The following Python program uses Romberg's rule available in Scipy to evaluate the definite integral specified in Example 6.7.

```
# Example 6.7
# The program evaluates a definite integral using Romberg's rule
# Implementing the Romberg method
import numpy as np
from scipy import integrate
f = lambda x: x**2 # The desired function
# using scipy.integrate.romberg()
res = integrate.romberg(f, 0, 8, show = True)
print("The calculated value is: %0.5f" % (res))
```

The result of the definite integral of the function defined in Example 6.7 is as follows:

```
Steps StepSize Results
1 8.000000 256.000000
2 4.000000 192.000000 170.666667
4 2.000000 176.000000 170.666667 170.666667
The final result is 170.66666666666666 after 5 function
evaluations.
The calculated value is: 170.66667
```

Example 6.8 Applying the Romberg Integration Technique

Solve the following integral utilizing the Romberg integration method

$$\int_0^1 \frac{1}{x+1}dx$$

Compare the approximate numerical integration result of manual calculation with actual analytical integration value and utilize Python and Simulink/MATLAB programming of Romberg's rule.

Solution

First, start with the Trapezoidal rule

$$h = 0.5$$

$$I_1 = f(x)dx = \frac{0.5}{2}\{f(0) + 2f(0.5) + f(1)\}$$

$$I_1 = f(x)dx = \frac{0.5}{2}\{f(0) + 2f(0.5) + f(1)\}$$

$$h = \frac{0.5}{2} = 0.25$$

$$I_2 = f(x)dx = \frac{0.25}{2}\{f(0) + 2(f(0.25) + f(0.5) + f(0.75)) + f(1)\}$$

$$I_2 = f(x)dx = \frac{0.25}{2}\{1 + 2(0.8 + 0.667 + 0.5714) + 0.5\} = 0.697$$

$$h = \frac{0.25}{2} = 0.125$$

$$I_3 = \frac{0.125}{2}\left\{f(0) + 2\begin{pmatrix} f(0.125) + f(0.25) + f(0.375) + f(0.5) + f(0.625 + f(0.75) \\ +f(0.875 \end{pmatrix} + f(1)\right\}$$

$$I_3 = \frac{0.125}{2}\{(1) + 2(0.89 + 0.8 + 0.727 + 0.667 + 0.615 + 0.57 + 53) + 0.5\} = 0.694$$

Table 6.3 lists the rest of the calculation.

The Simulink graphical programming utilizing the Romberg rule is demonstrated in Figure 6.13, followed by the MATLAB code associated with the MATLAB function.

By double-clicking the Simulink MATLAB function block, the following code will appear.

```
function Result = Romberg(a, b, n)
f=@(x)1/(1+x); %the desired function
h=(b-a);
r=zeros(2,n+1);
r(1,1)=(f(a)+f(b))/(2*h);
for i = 2:n
rv=0;
for k = 1:2^(i-2)
rv=rv+f(a+(k-0.5)*h);
end
```

TABLE 6.3
Romberg Integration Method of the Definite Integral Specified in Example 6.8

h_i	I_i	I_i'	I_i''
h	0.708	0.697 + 1/3 (0.697 – 0.708)=0.6933	0.6931 + 1/3 (0.693 – 0.6933)=0.6929
$\frac{h}{2}$	0.697	0.694 + 1/3 (0.694 – 0.697)=0.693	
$\frac{h}{4}$	0.694		

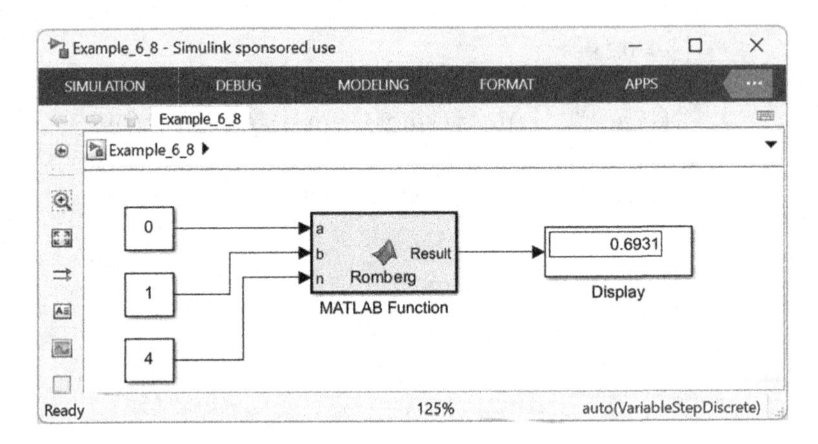

FIGURE 6.13 Simulink program utilizing Romberg integration method for solving the definite integral identified in Example 6.8.

```
r(2,1)=(r(1,1) + h*rv)/2;
for j=2:i
l=2^(2*(j-1));
r(2,j)=r(2,j-1)+(r(2,j-1)-r(1,j-1))/(l-1);
end

for k=1:i
fprintf('%7.5f', r(2,k));
end

fprintf('\n\n');
h=h/2;
for j=1:i
r(1,j)=r(2,j);
end
end

Result=r(1,n);
end
```

Python Solution

In this example, the Romberg function is imported from the built-in Python library "SciPy. integrate". We used Romberg integration of a callable function from limit (*a* to *b*) by using "SciPy. integrate". Romberg's rule is available in SciPy.

```
# Example 6.8
# The program evaluates the definite integral using Romberg's
rule
# import scipy.integrate
from scipy import integrate
f = lambda x: 1/(1+x) # desired function
# using scipy.integrate.romberg()
res = integrate.romberg(f, 0, 1, show = True)
print(res)
```

The results of the program execution are

```
Steps StepSize Results
1 1.000000 0.750000
2 0.500000 0.708333 0.694444
4 0.250000 0.697024 0.693254 0.693175
8 0.125000 0.694122 0.693155 0.693148 0.693147
16 0.062500 0.693391 0.693148 0.693147 0.693147 0.693147
32 0.031250 0.693208 0.693147 0.693147 0.693147 0.693147 0.693147
The final result is 0.6931471805622968 after 33 function
evaluations.
0.6931471805622968
```

6.7 SUMMARY

Numerical integration evaluates a definite integral from a set of integrand $f(x)$ numerical values. The most widely used methods of numerical integration are the Midpoint rule, Trapezoidal rule, and Simpson's rule. The Midpoint rule approximates the definite integral using rectangular areas, while the Trapezoidal rule approximates the definite integral using a trapezoidal approximation. We can use numerical integration to estimate the values of definite integrals when it is difficult to find a closed form or when only an approximate value of the definite integral is needed. The most widely used methods of numerical integration are the Midpoint rule, Trapezoidal rule, and Simpson's rule.

6.8 PROBLEM

1. Solve the integral by the Trapezoidal rule with $h = 0.5$, 25, and 0.125. Then obtain a better estimate using the Romberg integration method. Compared with Simulink and the analytical solution.

$$f(x) = \int_0^1 \frac{1}{1+x} dx$$

2. Using Simpson's rule, estimate the approximate value with $n = 4$ (n = Even). Compared with the Simulink and the analytical solution.

$$\int_2^{10} x^3 dx$$

Answer: 2496

3. Calculate the approximate value of the following definite integral using the Trapezoidal rule with $n = 10$. Compared with the Simulink and the analytical solution ($\ln(2) = 0.69314$).

$$f(x) = \int_{1}^{2} \frac{1}{x} dx$$

Answer: 0.694

4. Calculate the approximate value of the following definite integral using Simpson's rule with $n = 6$ (note that n should be an even number). Compared with the Simulink and the analytical solution.

$$\int_{0}^{2} f(x) dx = \int_{0}^{2} \frac{1}{(x+1)^{0.5}} dx$$

Answer: 1.4642

5. Use the Midpoint rule to get the approximate value ($n = 4$) of the following definite integral. Compared with the Simulink and the analytical solution.

$$\int_{-0.5}^{3.5} f(x) dx = \int_{-0.5}^{3.5} 0.25\, x^3 dx$$

Answer: 9

6. Use the Midpoint rule to get the approximate value ($n = 3$) of the following definite integral. Compared with the Simulink and analytical solution.

$$\int_{1}^{5} x(2-x) dx$$

Answer: −16.7407

7. Use Boole's rule to get the approximate value ($n = 4$) of the following definite integral. Compared with Simulink and the analytical solution.

$$\int_{0.2}^{0.6} \left(x^2+1\right)^{-1} dx$$

Answer: 0.343

8. Use Boole's rule to get the approximate value (h = 0.5) of the following definite integral. Compared with Simulink and the analytical solution.

$$\int_0^4 \left(x^2+1\right)^{-1} dx$$

Answer: 1.3624

9. Use the Romberg integration method and Trapezoidal rule to get the approximate value ($h = 0.5$, 0.25, and 0.125). Compare the approximate value with Simulink and the analytical solution.

$$\int_0^1 (x+1)^{-1} dx$$

Answer: 0.6931

10. Use the Romberg integration method and the Trapezoidal rule to get the approximate value (taking h = 0.5) of the following definite integral. Compare the approximate value with Simulink and the analytical solution.

$$\int_0^1 \left(x^2+1\right)^{-1} dx$$

Answer: 0.7853

REFERENCES

1. Davis, P.J. and Rabinowitz, P., 2014, *Methods of Numerical Integration*. Orlando, FL: Academic Press Inc.
2. Woodford, C. and Phillips, C., 2012, *Numerical Methods with Worked Examples*. Cham: Springer International Publishing,.
3. Kong, Q., Siauw, T. and Bayen, A., 2020, *Python Programming and Numerical Methods: A Guide for Engineers and Scientists*. Amsterdam: Elsevier Science.

7 Numerical Solution of Ordinary Differential Equations

An ordinary differential equation (ODE) is a mathematical equation relating one or more functions of the independent variable with their derivatives. Differential equations are important to chemical engineers as they are essential tools for the mathematical modeling of any problem involving rate changes. The chapter presents Euler, Midpoint, Heun, Runge-Kutta, and Picard's iteration methods to solve different ODEs. The methods produce the integral value of the desired ODE in different approaches and numerical precision.

LEARNING OBJECTIVES

1. Solve ODEs using the Euler method.
2. Utilize the Midpoint method to solve ODE.
3. Use Heun's method to solve ODE.
4. ApplyRunge-Kutta fourth-order (RK4) method.
5. Apply Picard's iterative method.
6. Employ Python and Simulink to solve ODE.

7.1 INTRODUCTION

Solving differential equations is an essential skill for mathematicians and engineers. However, many differential equations cannot be solved using traditional symbolic calculations (analysis). Therefore, a more advanced solution method is required to solve more complex or lengthy differential equations. Accordingly, a numerical approximation of the solution is used as a method of solution and is often sufficient for practical purposes.

A differential equation is an equation that contains one or more functions and their derivatives; it contains derivatives that are either ordinary derivatives or partial derivatives. An ODE contains one or more functions of one independent variable. Different numerical methods are used to solve the numerical approximation for ODEs; depending on the desired value, the methods are used to find the numerical approximations to the solution of ODEs. The integral solution for ODEs is desired in several cases, and this use is known as 'numerical integration'. The reason behind numerical integration is that many differential equations cannot be solved using symbolic computation as they require long computation and are complex equations. In engineering and for practical purposes, an ODE is extracted from a practical application, and the integral is required, so the numerical approximation to the solution is

DOI: 10.1201/9781003360544-7

sufficient to provide accurate results. Various algorithms studied in this chapter can be used to compute the integral value for ODEs, each with a different approach and accuracy level. A manual numerical solution solves each example in this class, and the manual solution is confirmed using Simulink/MATLAB and Python. The objective of explaining different solution methods is to educate the reader on software usage, keep up with the advancement of technology, and utilize available resources. The Euler solution has two main methods, namely explicit and implicit, each presented in this chapter. Furthermore, other methods, such as Midpoint, Heun, Runge-Kutta, and Picard's iteration, are explained in the solution method.

7.2 EULER METHOD (EXPLICIT METHOD)

In finding the integral value for an ODE, a solution method called the Euler method is used. Two leading solutions depend on the case study, either explicit or implicit. The parameters are calculated based on previous levels in the definitive solution, so it uses forward or central difference methods. The Euler method is a numerical procedure for solving ODEs with a given initial value. Although this method does not output a highly accurate value, it is easily solvable and helps understand the numerical methods approach for ODEs. The Euler method can be used to solve ODEs as long as the initial value is stated. Although the method is neither accurate nor efficient as other numerical solution methods, it helps understand the general ODE numerical method solution. Figure 7.1 shows the Euler method solution approach, where the initial coordinate point is taken and the following coordinate is solved depending on the step size and the trend the values are showing. The Euler method is a first-order method that uses the following first-order ODE.

$$\frac{dy}{dx} = f(x, y), \quad y(x_o) = y_o$$

The Euler method solution is based on the following equation:

$$y_{i+1} = y_i + hf(x_i, y_i)$$

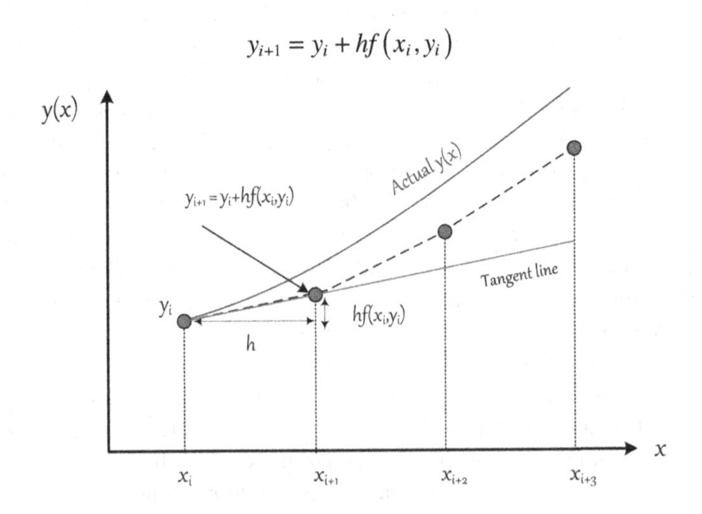

FIGURE 7.1 Euler's interpretation numerical method.

where h is the step size.

Example 7.1 Applying the Forward Euler Method

Solve the following initial value ODE using the explicit Euler method and step size, $h = 0.1$, calculate the value of y at $x = 1.4$.

$$\frac{dy}{dx} = 3x^2y$$

Consider the following initial conditions

$$y(x_o) = y_o \rightarrow y(1) = 2$$

Compare the approximate results obtained from the Euler method, Simulink, and Python to the actual analytical solution values

$$y(x) = 2 * e^{x^2 - 1}$$

Solution

In the Euler equation, the $f(x_i, y_i)$ is the derivative equation $\frac{dy}{dx}$, and is substituted as such. Knowing the initial condition, at $x_o = 1$, $y(1) = 2$, we start by adding the step size to the initial value of x, $x = x + h$. At each step, the values of x and y are taken from the previous step solved, and this is why an initial value is required for this method to work.

$$x_1 = 1.1$$

$$y_1 = y_o + hf(x_o, y_o)$$

$$y_1 = 2 + 0.1f(1,2)$$

$$y_1 = 2 + 0.1\left[3(1)^2 2\right] = 2.6$$

$$x_2 = 1.2$$

$$y_2 = y_1 + hf(x_1, y_1)$$

$$y_2 = 2.6 + 0.1f(1.1, 2.6)$$

$$y_2 = 2.6 + 0.1\left[3(1.1)^2 2.6\right] = 3.544$$

$$x_3 = 1.3$$

$$y_3 = y_2 + hf(x_2, y_2)$$

$$y_3 = 3.544 + 0.1f(1.2, 3.544)$$

$$y_3 = 3.544 + 0.1\left[3(1.2)^2 3.544\right] = 5.075$$

$$x_4 = 1.4$$

$$y_4 = y_3 + hf(x_3, y_3)$$

$$y_4 = 5.0747 + 0.1f(1.3, 5.0747)$$

$$y_4 = 5.0747 + 0.1\left[3(1.3)^2 5.0747\right] = 7.648$$

Simulink and MATLAB Solution

Solving ODEs in Simulink/MATLAB will help understand the diagram of operation used in the ODE solution and equip the reader with more solution options. Therefore, a detailed step-by-step solution is presented. First, start MATLAB, then Simulink, while in Simulink and from modeling/setting, on the solver page, specify the start time, stop time, the fixed-step size, and the solver (Figure 7.2).

Figures 7.3 and 7.4 show the Simulink block diagram that describes the solution of Example 7.1 without writing any MATLAB code. The first integrator generates the variable *x*, while the second integrator produces the value of *y*. The results shown are for the required value at $x_4 = 1.4$, which is 7.648, a similar value as obtained in the Euler method solution by hand correct to three decimal points. An alternative method is using the 'MATLAB function' available in the Simulink library under the 'user-defined functions', where a code is extracted from the Simulink-created diagram. The MATLAB function required for solving the ODE integral and outputting the value at a specific point at the normal function is presented in Example 7.1. The presented MATLAB code can be used for solving similar ODEs by changing the equation at the right-hand side of the function '*f*' (line 2). Furthermore, the integration interval between [*a*, *b*] needs to be specified along with step size, *h*, and the initial value y_o. The code is written in the area displayed after double-clicking the 'MATLAB function' (Figure 7.4).

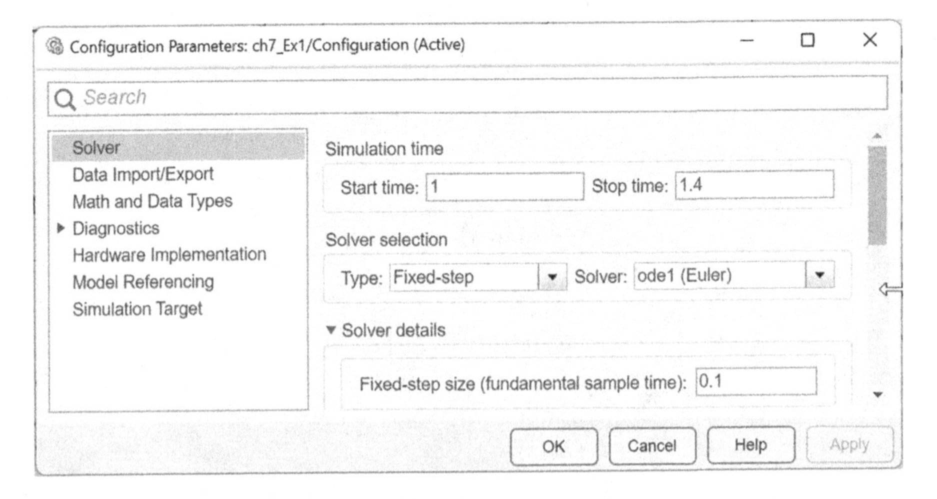

FIGURE 7.2 Solver setup of a Simulink integral to use the forward Euler method in the solution case in Example 7.1.

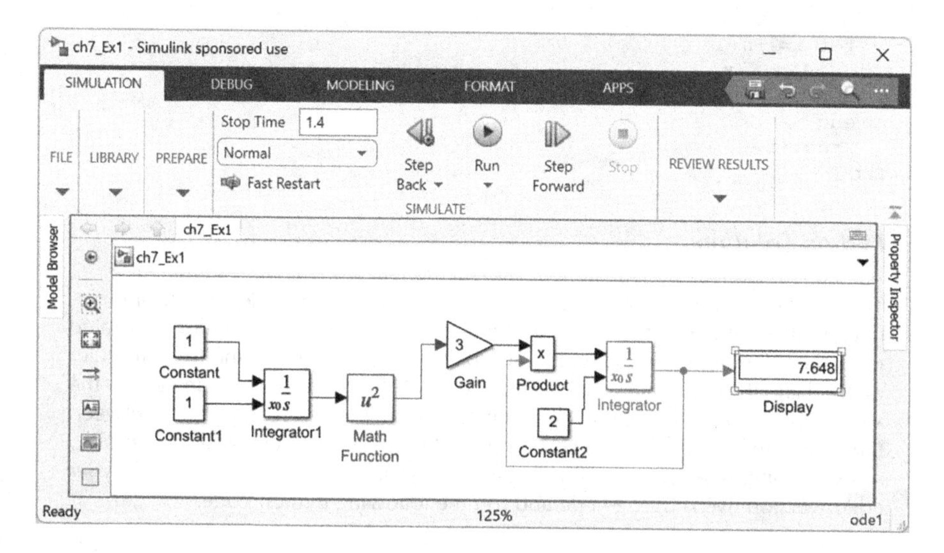

FIGURE 7.3 Simulink block diagram using the Euler method of the equation defined in Example 7.1.

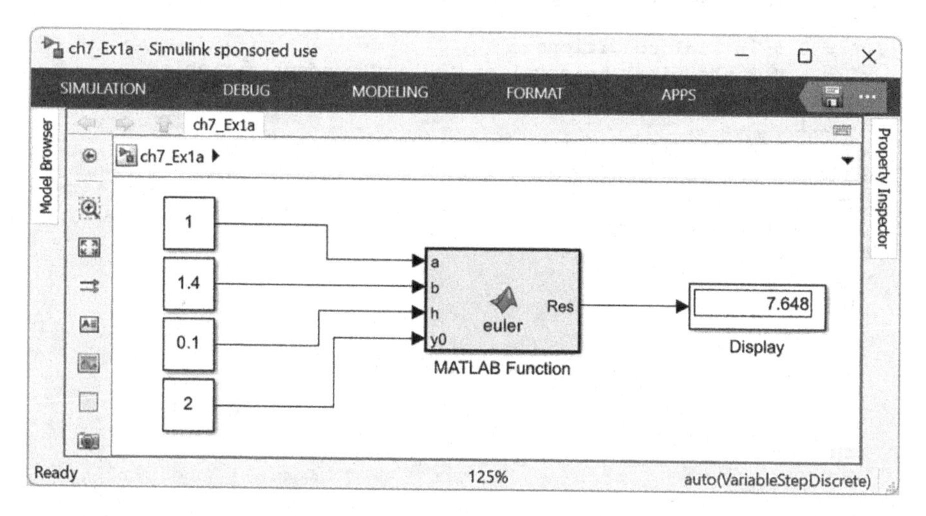

FIGURE 7.4 Simulink block diagram using the Euler method of equation specified in Example 7.1, the stop time is four.

The following MATLAB code represents forward Euler's method that combines the Simulink MATLAB function to solve the following initial value problem (IVP) defined in Example 7.1.

```
function Res = euler(a, b, h, y0)
f = @(x, y)(3*x^2*y);% ODE
x = a:h:b;
n = (b-a)/h;
y=zeros(size(x));
y(1)=y0;
```

```
    for i=1:n+1
        k1 = f(x(i), y(i));
        y(i+1) =y(i) + h*(k1);
    end
    Res=y(i+1);
End
```

Python Solution

The Python program is utilized to solve the IVP stated in Example 7.1. The approximate numerical integration value and the exact analytical solution are illustrated in Figure 7.5. The figure shows the approximate solution using the forward Euler method and the exact analytical solution. Figure 7.5 shows an increasing gap as the x value is increased due to an accumulated error. The discrepancy gap between the approximated and the exact solution can be shrunk by decreasing the step size, as decreasing the step size will increase the accuracy of the solution. The Spyder editor version five is used to edit and run the following Python code.

```
# Euler method numerical approximation
import numpy as np
from matplotlib import pyplot as plt
x0 = 1
y0 = 2 # initial conditions
xf = 1.4 # the final interval of the independent variable
n = 5 # number of steps
h= 0.1
x = np.linspace(x0,xf, n)
# define the function dy/dx = x+2y
def f(x, y):
     return 3*(x**2)*y
# set the intial condtios
y = np.zeros([n])
y[0] = y0
```

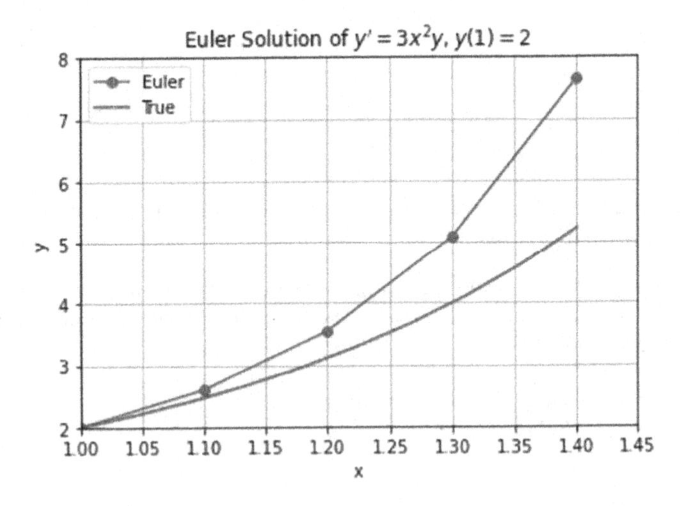

FIGURE 7.5 Comparison of the Euler's method programmed in Python and the accurate analytical solution of the ODE defined in Example 7.1.

```
for i in range(1,n):
    y[i] = y[i-1] + h*f(x[i-1], y[i-1])
print("x_n\t y_n")
for i in range(n):
    print(x[i], "\t", format(y[i], '6f'))
# plot section
x_true = np.linspace(1,1.4,10)
y_true = 2*np.exp(x_true**2-1)
plt.plot(x, y, 'o-', x_true, y_true, 'r')
plt.axis([1,1.45,2,8])
plt.legend(['Euler', 'True'])
plt.xlabel("x")
plt.ylabel("y")
plt.grid(True)
plt.title("Euler Solution of $y'=3x^2y, y(1)=2$")
plt.show()
```

Example 7.2 Solving ODE with the Euler Explicit Method

The forward Euler (explicit) method is a first-order numerical procedure for solving ODEs with a given initial value. Consider the following first-order differential equation with an initial condition $y(0)=1$, and a step size, $h=0.025$, find $y(0.1)$.

$$\frac{dy}{dx}=x+y+xy$$

Manual Solution

Specify the function as the right-hand side of the ODE.

$$f(x,y)=x+y+xy$$

With the initial conditions and the following step size,

$$x_o=0,\ y_o=1,\ h=0.025$$

Utilizing the Euler method formula,

$$y_i=y_{i-1}+hf(x_{i-1},y_{i-1})$$

Substitute the initial value in the Euler format to calculate y_1,

$$y_1=y_o+hf(x_o,y_o)=1+0.025f(0,1)$$

$$y_1=1+0.025\times(0+1+0\times1)=1.025$$

Accordingly,

$$y(0.025)=1.025$$

Similarly, we can calculate, $y(0.05)$ to $y(0.1)$ as tabulated in Table 7.1.

The exact analytical solution at $x=0.1$ is 1.11589, so the value obtained has an error % of 0.38% when compared with the explicit Euler method.

Simulink/MATLAB Solution

Simulink is a graphical programming environment in MATLAB. The Euler method settings are shown in Figure 7.6, where the Euler method is selected by following, modeling/model settings. While on the configuration parameters page, change the default values to those shown in Figure 7.6 (start time, stop time, type, solve, and the fixed-step size). The entire Simulink program utilizing the Euler method is described in Figure 7.7. Executing the Simulink program outputs the value y at $x=0.1$ as 1.112. The value is identical to that found in manual calculations.

An alternative way is to program the Euler method in MATLAB and embed the MATLAB code into the MATLAB function (Figure 7.8). The MATLAB function got four inputs: the lower limit interval (a), higher limit interval (b), step size (h), and the initial condition of $y(y_o)$. Running the Simulink program leads to the same previous results of y (1.112).

The following program is the MATLAB codes implanted in the MATLAB function of Figure 7.8. The program represents Euler's method combined with the Simulink MATLAB function for solving the ODE described in Example 7.2. The MATLAB code can be used to solve other ODEs by changing the equation after the function:

TABLE 7.1
Explicit Euler Method

i	x	y
0	0	1
1	0.025	1.025
2	0.05	1.051891
3	0.075	1.080753
4	0.1	1.111673

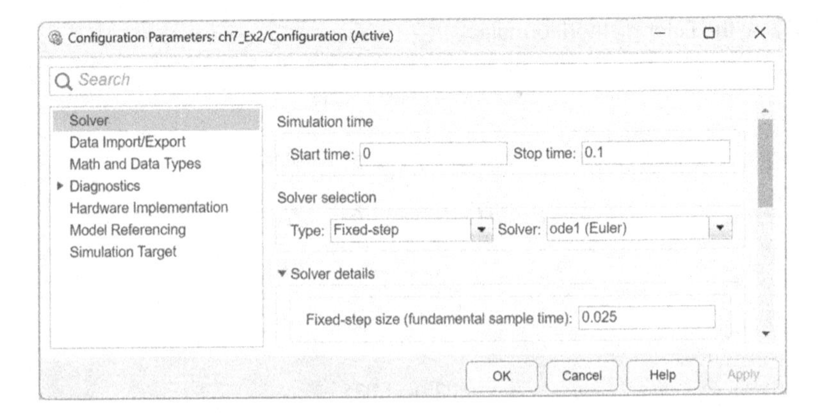

FIGURE 7.6 Setting the solver to the Euler forward method of Example 7.2.

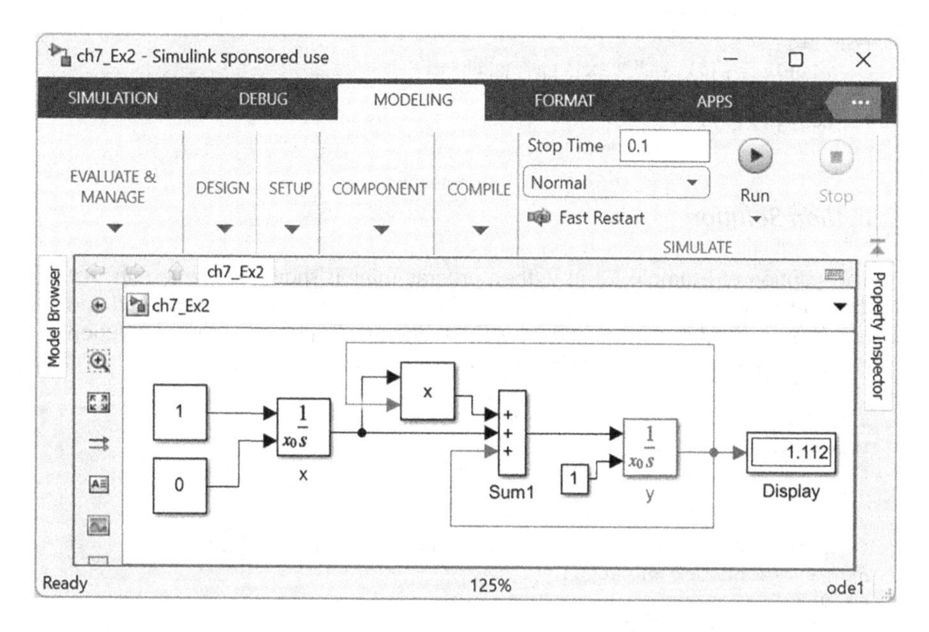

FIGURE 7.7 Simulink solution using fixed-point iteration Euler method of Example 7.2.

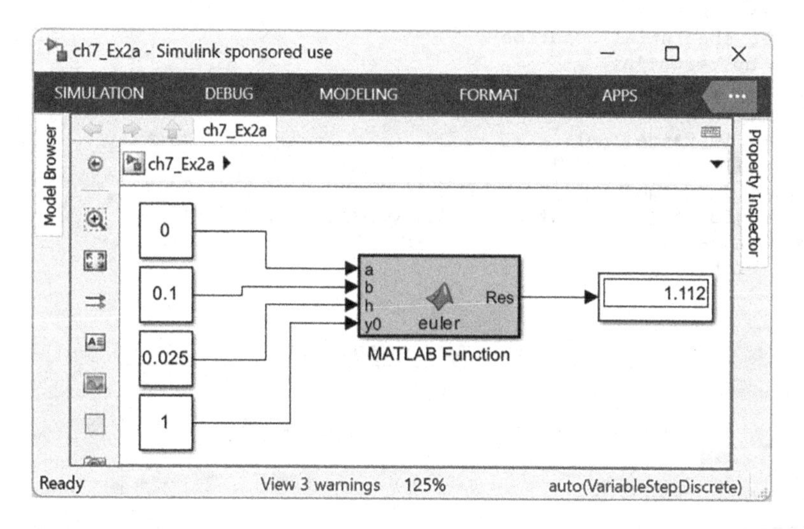

FIGURE 7.8 Simulink MATLAB function using the Euler method of the ODE defined in Example 7.2.

```
f=@(x, y).
function Res = euler(a, b, h, y0)
f = @(x, y)(x+y+x*y);% ODE
x = a:h:b;
n = (b-a)/h;
y=zeros(size(x));
y(1)=y0;
```

```
for i=1:n
   y(i+1) =y(i) + h*f(x(i),  y(i));
end
   Res=y(i+1);
end
```

Python Solution

The solution of Example 7.2 in Python programming is shown below to enhance the reader's programming skills. In this solution, the function Euler is created and called at the code's end. Figure 7.9 depicts the approximate numerical solution generated after running the Python program.

```
# Euler method numerical approximation
import numpy as np
from matplotlib import pyplot as plt
x0 = 0
y0 = 1 # initial conditions
xf = 0.1  # the final interval of the independent variable
n = 5   # number of steps
h= 0.025
x = np.linspace(x0,xf, n)
def f(x, y):
   return x+y+x*y
# set the intial condtios
y = np.zeros([n])
y[0] = y0
print("x_n\t                y_n")
for i in range(1,n):
   y[i] =  y[i-1] + h*f(x[i-1], y[i-1])
for i in range(n):
print(format(x[i], '4f'), "\t", format(y[i], '4f'))
# plot section
plt.plot(x, y, '-o')
```

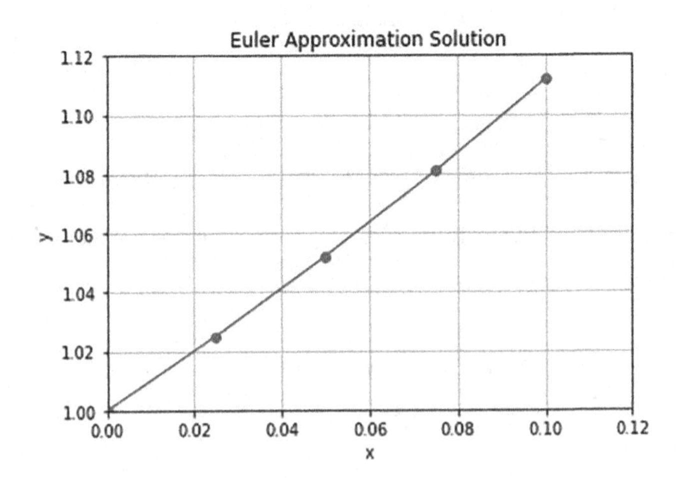

FIGURE 7.9 Python numerical approximate solution of the ODE specified in Example 7.2.

```
plt.xlabel("x")
plt.ylabel("y")
plt.grid(True)
plt.axis([0,0.12,1,1.12])
plt.title("Euler Approximation Solution")
plt.grid(True)
plt.show()
```

The results

```
x_n          y_n
0.000000     1.000000
0.025000     1.025000
0.050000     1.051891
0.075000     1.080753
0.100000     1.111673
```

7.3 BACKWARD EULER METHOD (IMPLICIT METHOD)

In numerical analysis and scientific computing, the backward Euler method (or implicit Euler method) is one of the most basic numerical methods for solving ODEs. Implicit methods find a solution by solving an equation involving both the current state of the system and the latter one. In this method, we must solve an equation to find y_{i+1}. Solving this method outcomes in an algebraic equation requires, in some cases, using fixed-point iteration. It requires more time to solve this equation than the Euler explicit methods but results in higher accuracy of the solution. Considering the following first-order initial value ODE,

$$\frac{dy}{dx} = f(x,y), \quad y(x_o) = y_o$$

The backward Euler (Implicit) method is

$$y_{i+1} = y_i + hf(x_{i+1}, y_{i+1})$$

The new approximation y_{i+1} appears on both sides of the equation, and thus the method needs to solve an algebraic equation for unknown y_{i+1}. Figure 7.10 shows the presentation of the backward Euler method.

Example 7.3 Solving ODE with the Backward Euler Method (Implicit Method)

The backward Euler method is a first-order numerical procedure for solving ODEs with a given initial value. Consider the following first-order differential equation with an initial condition $y(0)=1$, and a step size, $h=0.025$, find $y(0.1)$.

$$\frac{dy}{dx} = x + y + xy$$

Confirm the manual solution with Python and Simulink/MATLAB programming.

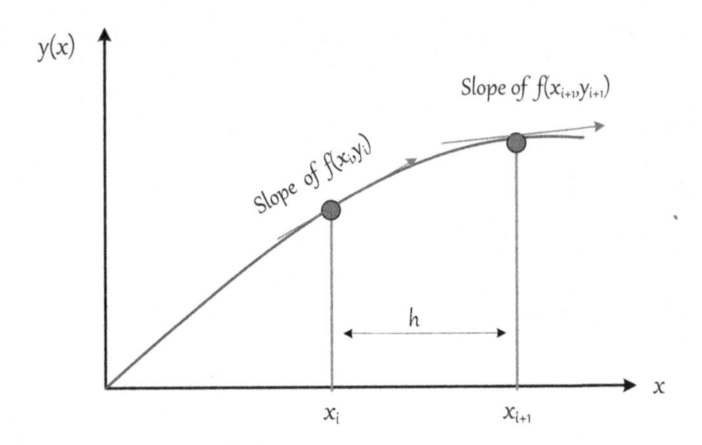

FIGURE 7.10 Graphical explanation of the backward Euler method.

Manual Solution

The formula of the backward Euler method

$$y_i = y_{i-1} + hf(x_i, y_i)$$

Place the function equal to the right-hand side of the equation

$$f(x,y) = x + y + xy$$

Substituting,

$$i = 1,\ x = 0.025$$

$$y_1 = y_o + hf(x_1, y_1) = y_o + h(x_1 + y_1 + x_1 y_1)$$

$$i = 2,\ x = 0.05$$

$$y_2 = y_1 + hf(x_2, y_2) = y_1 + h(x_2 + y_2 + x_2 y_2)$$

$$i = 3,\ x = 0.075$$

$$y_3 = y_2 + hf(x_3, y_3) = y_2 + h(x_3 + y_3 + x_3 y_3)$$

$$i = 4,\ x = 0.1$$

$$y_4 = y_3 + hf(x_4, y_4) = y_3 + h(x_4 + y_4 + x_4 y_4)$$

Rearranging,

$$(h + x_1 h - 1) y_1 \ + \ 0 + \ 0 + \ 0 \ = -hx_1 - y_o$$

$$y_1 + (h + x_3 h - 1) y_2 + 0 + 0 = -hx_2$$

$$0 + y_2 + (h + x_3 h - 1) y_3 + 0 = - \ hx_3$$

$$0 + 0 + y_3 + (h + x_4 h - 1) y_4 = -hx_4$$

The below Python program solves the above set of linear algebraic equations. The result for x_1 to x_4 are shown below the Python program.

```
# Solving a set of linear algebraic equations,
import numpy as np
x0=0
y0=1
h=0.025
a1=h+h*0.025-1
a2=h+h*0.05-1
a3=h+h*0.075-1
a4=h+h*0.1-1
A1=[a1, 0,0,0]
A2=[1, a2,0,0]
A3=[0, 1,a3,0]
A4=[0,0,1,a4]
A = np.array([A1,A2 , A3 , A4 ])
B = np.array([-h*0.025-y0, -h*0.05, -h*0.075, -h*0.1])
X = np.linalg.inv(A).dot(B)
print(" ")
print(X)
```

The answer is shown in a 1×4 matrix, each column representing a step in the solution, and the last column is at $x=0.1$.

```
[1.02694035 1.05590793 1.08699595 1.12030431]
```

Simulink and MATLAB Solution

Simulink is a graphical programming language of MATLAB that helps understand the solution's connections and the bigger picture. While on the Simulink page, click on modeling in the toolbar, then 'model setting' to change the configuration parameters. Select the type of solver, and specify the start time, stop time, and step size (Figure 7.11).

Figure 7.12 demonstrates the Simulink block diagram. The displayed value of y is 1.12 at the stop time of 0.1. An alternative way to solve Example 7.3 in Simulink is to use the MATLAB function (Figure 7.13). The Simulink MATLAB function requires writing a MATLAB code describing the backward Euler method. The MATLAB function contains four input ports: the lower limit interval (a), upper limit interval (b), step size (h), and the initial condition (y_o), and one output port connected to the Display. The Display outputs the value of y at the upper limit interval.

The MATLAB code shown below represents the implicit Euler method combined with the Simulink MATLAB function for the solution of the following ODE defined in Example 7.3.

```
function Res = Im_euler(a, b, h, y0)
f = @(x, y) (x+y+x*y);% ODE
   %f - this is your y prime
   xinit = a;
   yinit = y0;
   xfinal = b;
%h - step size

n = (xfinal-xinit)/h; %Calculate steps
```

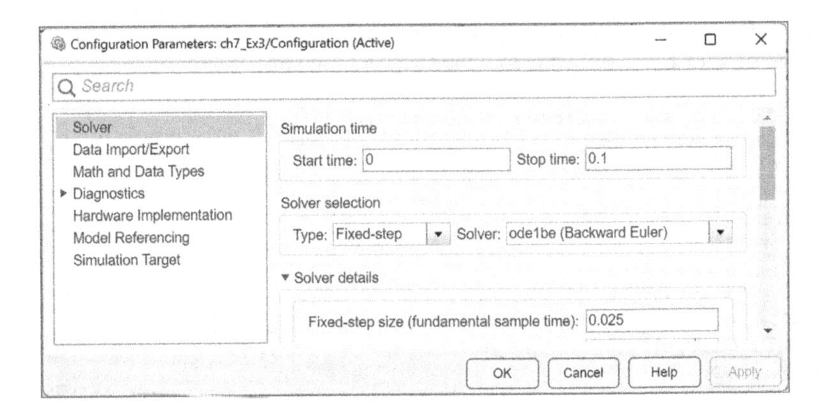

FIGURE 7.11 Simulink modeling configuration menu for setting the simulation time, solver selection, and solver details of the backward Euler method for solving the equation defined in Example 7.3.

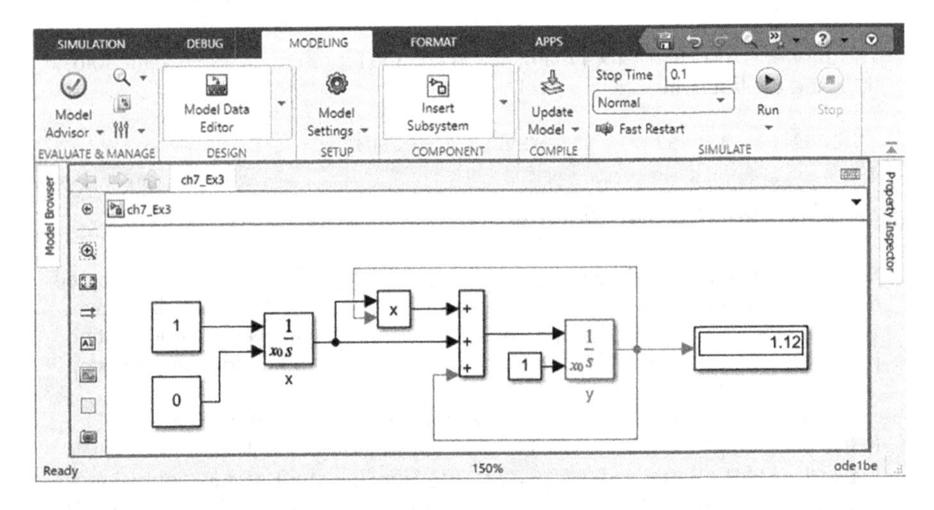

FIGURE 7.12 Backward Euler method solution of the equation specified in Example 7.3.

```
    % Initialize arrays...
    % The first elements take xinit
    % yinit correspondingly, the rest fill with 0s.
    x = [xinit zeros(1,n)];
    y = [yinit zeros(1,n)];

%Numeric routine
for i = 1:n
    x(i+1) = x(i)+h;
    ynew = y(i)+h*(f(x(i), y(i)));
    y(i+1) = y(i)+h*f(x(i+1), ynew);
end
Res = y(i+1);
End
```

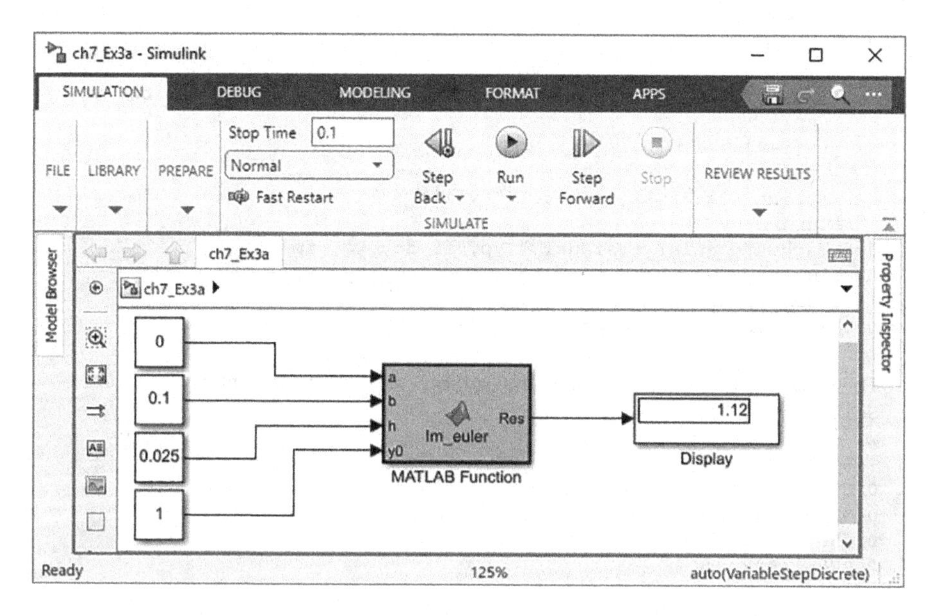

FIGURE 7.13 Simulink block diagram using the backward Euler (implicit) method for solving the equation required in Example 7.3.

Python Solution

The Python program below uses the backward Euler method to solve the equation given in Example 7.3. Running the Python program produces the following numerical approximation results containing the values of y at different values of x with a step size of h (0.025).

```
#Backward Euler method (Implicit method)
from scipy.optimize import fsolve
import numpy as np
x0=0
tspan =[0,0.1]
y0= 1
n=5
x1=0
def f( x, y ):
    return (x + y + x * y)
# Backward Euler method (Implicit)
def backward_euler ( f, tspan, y0, n ):
  x = np.zeros ( n + 1 )
  y = np.zeros ( n + 1 )
  h = 0.025
  x[0] = 0
  y[0] = 1
  print (" "), print (" Solution ")
  print ("x y"), print ("--- -----")
  for i in range ( 0, n ):
     to = x[i]
     yo = y[i]
```

```
        tp = x[i] + h
        yp = yo + h * f ( to, yo )
        yp = fsolve (backward_euler_residual, yp, args = (f, to, yo,
        tp))
        x[i+1] = tp
        y[i+1] = yp
        print ("%.3f"%x[i], "%.3f"%y[i])
    return x, y
def backward_euler_residual ( yp, f, to, yo, tp ):
    value = yp - yo - ( tp - to ) * f ( tp, yp );
    return value
backward_euler ( f, tspan, y0, n )
```

The results obtained from running the Python code are shown below:

```
Solution
x   y
--- -----
0.000 1.000
0.025 1.027
0.050 1.056
0.075 1.087
0.100 1.120
```

7.4 MIDPOINT METHOD

The Midpoint method solves for the integral of the first-order ODEs and finds the function's value at a specific point. The midpoint numerical method uses the derivative at the midpoint $\left(x_{i+\frac{1}{2}}, y_{i+\frac{1}{2}}\right)$ to compute y_{i+1} with an IVP. The Midpoint method is a point where the desired solution is solved by studying the trend of the values in the midpoint between the initial value and the desired point. Finding the slope and value of the midpoint increases the accuracy of the solution and outputs a lower error percentage. Figure 7.14 is a representation of the Midpoint numerical method. It first estimates $y_{i+\frac{1}{2}}$, then using the derivative at the midpoint to compute y_{i+1}, for the following ODE

$$\frac{dy}{dx} = f(x,y), \quad y(0) = y_o$$

Using the Midpoint numerical method,

$$y_{i+1} = y_i + k_2 h$$

here k_1

$$k_1 = f(x_i, y_i)$$

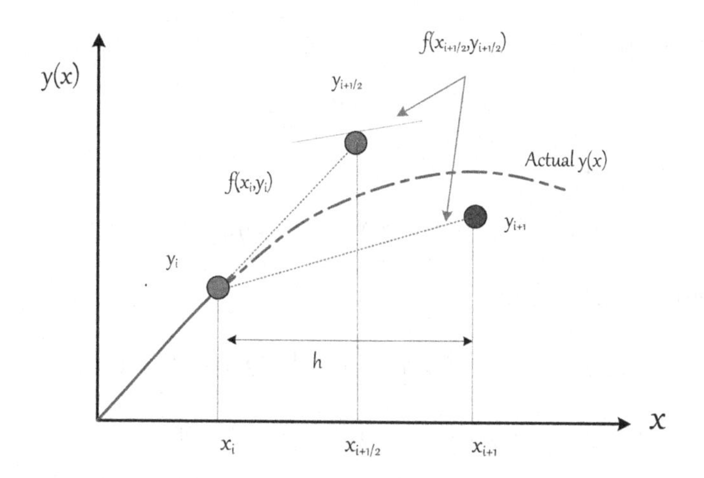

FIGURE 7.14 Graphical representation of the Midpoint numerical method.

and k_2

$$k_2 = f\left(x + \frac{h}{2}, y_i + \frac{h}{2} k_1\right)$$

Example 7.4 Applying Midpoint Numerical Method

Solve the initial value ODE using the Midpoint method and a step size of $1.5 (h = 1.5)$ to find $y(3)$.

$$\frac{dy}{dx} = 3e^{-x} - 0.4y, \quad y(0) = 5$$

Confirm the manual calculations with Python and Simulink/MATLAB programming.

Solution

The initial condition to start with is $x_0 = 0, \; y_o = 5, \; h = 1.5$

Accordingly, at $x_1 = 0 + h = 0 + 1.5 = 1.5$

Find

$$k_1 = f(x_i, y_i) = f(0, 5) = 3 * e^0 - 0.4(5) = 1$$

Calculate

$$k_2 = f\left(x + \frac{h}{2}, y_i + \frac{h}{2} k_1\right) = f\left(0 + \frac{1.5}{2}, 5 + \frac{1.5}{2}(1)\right)$$

$$= f(0.75, 5.75) = 3 * e^{-0.75} - 0.4(5.75) - 0.8829$$

The value of y_1 at $x_1 = 1.5$

$$y_1 = y_0 + k_2 h = 5 - 0.8829(1.5) = 3.676$$

At $x_2 = 3$, the value of k_1

$$k_1 = f(x_1, y_1) = f(1.5,\ 3.676) = 3 * e^{-1.5} - 0.4(3.676) = -0.8$$

The calculated value of k_2

$$k_2 = f\left(x + \frac{h}{2},\ y_i + \frac{h}{2}\ k_1\right) = f\left(1.5 + \frac{1.5}{2},\ 3.676 + \frac{1.5}{2}(-0.8)\right)$$

$$= f(2.25,\ 3.075) = 3 * e^{2.25} - 0.4(3.075) = -0.9138$$

The calculated value of y_2 at $x_2 = 3$

$$y_2 = y_1 + k_2 h = 3.676 - 0.9138(1.5) = 2.305$$

Simulink and MATLAB Solution

For this example, a Simulink diagram is created, and the MATLAB function is dragged from the Simulink library to the work environment to solve the IVP described in Example 7.4. Figure 7.15 shows the Simulink block diagram centered by the MATLAB function and connected between four input ports and one output Display.

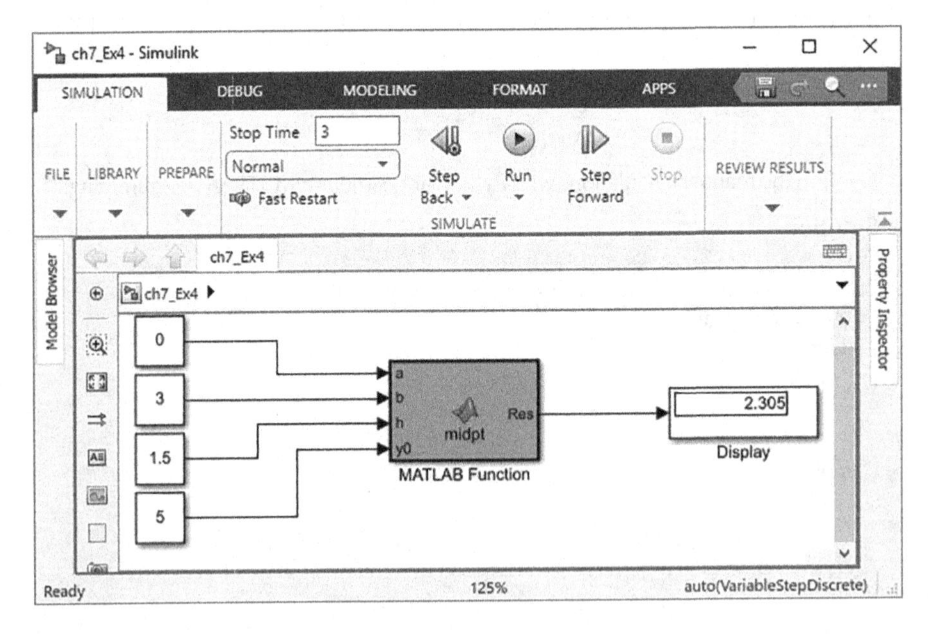

FIGURE 7.15 Simulink solution using MATLAB function and the Midpoint method of the ODE defined in Example 7.4.

The four input ports are the lower limit interval (*a*), the upper limit interval (*b*), the step size (*h*), and the initial value of $y(y_0)$. The Display icon yields the value of the y at the upper limit interval or the simulation stop time (i.e., 3). The figure is followed with embedded MATLAB functions. The following MATLAB codes programmed the Midpoint method combined with the Simulink MATLAB function for solving the ODE described in Example 7.4.

```
function Res = midpt(a, b, h, y0)
f = @(x, y)(3*exp(-x)-0.4*y);% ODE
x = a:h:b;
n = (b-a)/h;
y=zeros(n);
y(1)=y0;
  for i=1:n
      k1 = f(x(i), y(i));
      k2 = f(x(i)+h/2, y(i)+k1*h/2);
      y(i+1) =y(i) + h*(k2);
  end
  Res=y(i+1);
end
```

Python Solution

The following Python code solves the ODE specified in Example 7.4 using the midpoint method. Unless the right-hand side of the ODE is linear in the dependent variable, each of the midpoint steps requires solving an implicit nonlinear equation. Using the midpoint formula, the user can calculate the curve representing the solution to the normal differentiation of an equation with an initial value.

```
# Midpoint method python program
import numpy as np
# initial conditions
x0 = 0
y0 = 5
# The calculating point
xn = 3
n = 2
h = (xn-x0)/n
step= h
print('\n--------SOLUTION--------')
# function to be solved
def f(x, y):
    return 3*np.exp(-x)-0.4*y
# using the midpoint formula
for i in range(n):
    k1 = h * (f(x0, y0))
    k2 = h * (f((x0+h/2), (y0+k1/2)))
    yn = y0 + k2
    x0 = x0+h
    print("at x = ", x0,", yn =", yn)
    y0 = yn
```

The results obtained from running Python code with midpoint in solving the given equation in Example 7.4 are shown as follows:

```
--------SOLUTION--------
at x = 1.5 , yn = 3.6756494873345655
at x = 3.0 , yn = 2.304947496982057
```

7.5 HEUN'S METHOD

Heun's method considers the tangent lines of the solution curve at both ends of the interval, one overestimates and the other underestimates the ideal vertical coordinates (Figure 7.16). Heun's method involves the determination of two derivatives at the interval at the initial and end points. This process relies on predicting the new value of y, then correcting it based on the slope calculated at that new value. The two derivatives are then averaged to obtain an improved slope estimate for the entire interval. The prediction line should be constructed based on the slope of the right endpoint tangent alone, approximated using the Euler method.

$$\frac{dy}{dx} = f(x,y), \quad y(0) = y_o$$

Heun's method involves the determination of two derivatives for the interval at the initial point and the endpoint.

$$k_1 = f(x_i, y_i)$$

$$k_2 = f(x + h, \; y_i + k_1 h)$$

The calculated y_{i+1}

$$y_{i+1} = y_i + h\left(\frac{k_1 + k_2}{2}\right)$$

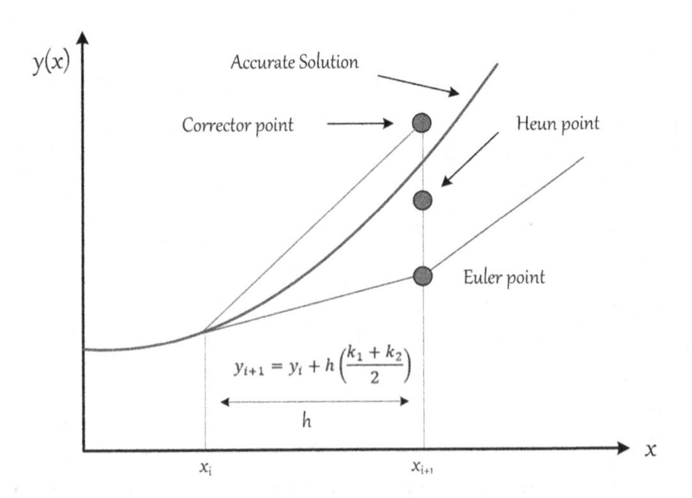

FIGURE 7.16 Graphical description of Heun's numerical method.

Example 7.5 Applying the Heun Method

Solve the following initial value nonlinear ODE using Heun's method to find y at $x=5$ using a step size, $h=0.5$

$$\frac{dy}{dx}=\frac{100+y}{5x}, \quad y(0.5)=20$$

Confirm the manual answer with Python and Simulink/MATLAB programming of the Heun method.

Solution

The initial condition, at $x_o=0.5$, $y_o=20$, or $y(0.5)=20$

Set the function equal to the right-hand side of the equation.

$$f(x,y)=\frac{100+y}{5x}$$

At

$$x_1=0.5+h=0.5+0.5=1.0$$

$$k_1=f(x_0,y_o)=f(0.5,\ 20)=\frac{100+20}{5(0.5)}=48$$

$$k_2=f(x_o+h,\ y_o+k_1h)=f(0.5+0.5,\ 20+48*0.5)=f(1,\ 44)=\frac{100+44}{0.5(1)}=28.8$$

The calculated y_1

$$y_1=y_0+h\left(\frac{k_1+k_2}{2}\right)=20+0.5\left(\frac{48+28.8}{2}\right)=39.2$$

The exact analytical solution of the ODE using separable integration

$$\frac{dy}{dx}=\frac{100+y}{5x}$$

Rearrange to the separable integration format

$$\int_{y_o}^{y}\frac{dy}{100+y}=\frac{1}{5}\int_{x_o}^{x}\frac{dx}{x}$$

Integrate

$$\ln\left(\frac{100+y}{100+y_o}\right)=\frac{1}{5}\ln\left(\frac{x}{x_o}\right)$$

Simplify

$$y = (100 + y_o)\left(\frac{x}{x_o}\right)^{\frac{1}{5}} - 100 = 120\left(\frac{1}{0.5}\right)^{\frac{1}{5}}\left(x^{\frac{1}{5}}\right) - 100 = 137.844x^{\frac{1}{5}} - 100$$

Accordingly, the accurate analytical solution is

$$y = 137.844x^{\frac{1}{5}} - 100$$

Table 7.2 lists the approximate numerical solution using Heun's method and the exact analytical solution of the ODE specified in Example 7.5.

Simulink and MATLAB Solution

The configuration parameters need to be changed to solve the ODE in Simulink using the Heun's method, as shown in Figure 7.17. Click on modeling in the Simulink toolbar, then click on the gear shape model/setting. Fix the solver type, the start time, the stop time, and the fixed-step size. The entire Simulink block diagram is shown in Figure 7.18. The first integrator is to generate x with an initial value of 0.5 since the integration starts from 0.5 and ends at 5.

An alternative Simulink solution is to use the MATLAB function from the Simulink library (Figure 7.19). The MATLAB function requires writing a program describing the solution of the required ODE using the Heun method and implanting it in the Simulink MATLAB function.

The following is a MATLAB code implanted in the MATLAB function pulled from the Simulink library corresponding to the Heun method used for the ODE solution presented in Example 7.5.

```
function Res = Heun(x0,xf, h, y0)
     f= @(x, y)((100+y)/(5*x)); % ODE
     n= (xf-x0)/h;
     yi=y0;
```

TABLE 7.2
Heun's Method of Numerical Approximation and the Exact Analytical Solution of the Equation Defined in Example 7.5

x	k_1	k_2	y_{Heun}	y_{exact}
0.5			20.000	20.000
1	48.000	28.800	39.200	37.844
1.5	27.840	20.416	51.264	49.488
2	20.169	16.135	60.340	58.341
2.5	16.034	13.469	67.715	65.568
3	13.417	11.628	73.977	71.717
3.5	11.598	10.273	79.445	77.093
4	10.254	9.229	84.315	81.886
4.5	9.216	8.397	88.718	86.222
5	8.387	7.716	92.744	90.187

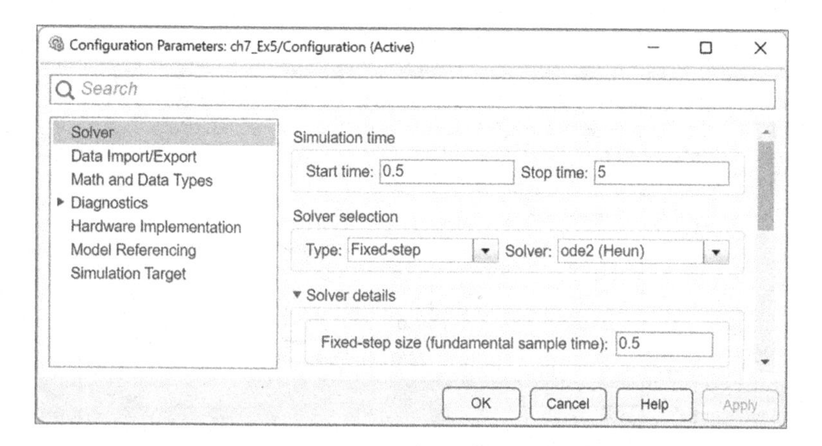

FIGURE 7.17 Simulink configuration parameters utilizing the Heun method to solve the ODE specified in Example 7.5.

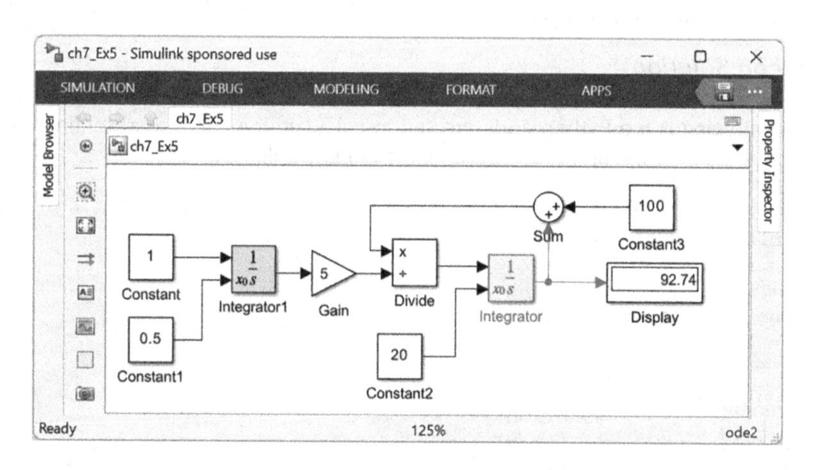

FIGURE 7.18 Simulink block diagram using fixed-step size Heun's method for the equation solution specified in Example 7.5, the stop time equals five.

```
      xi=x0;
for i =1:n
      k1= f(xi, yi);
      k2= f(xi+h, yi+(h*k1));
      yf =yi+(h/2)*(k1+k2);
      yi=yf;
      xi = xi +h;
end
   Res=yi;
End
```

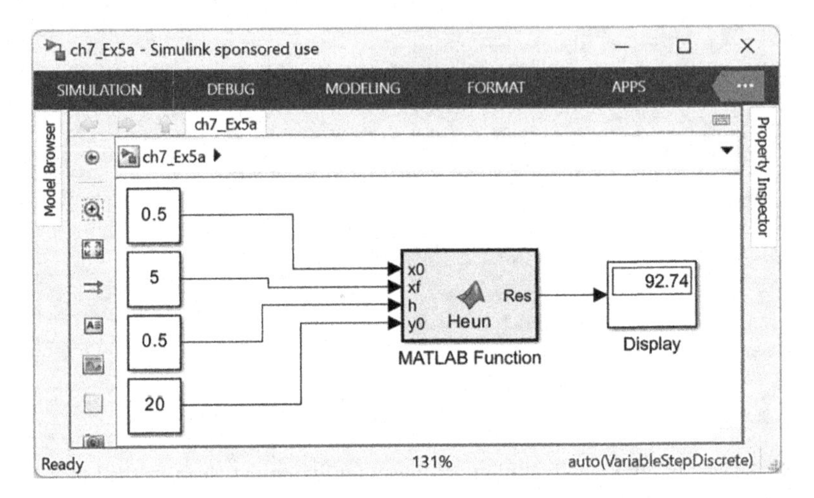

FIGURE 7.19 Simulink block diagram using the Heun method to solve the equation defined in Example 7.5, stop time equals five.

Python Solution

The following is the Python code created to calculate the curve representing the Heun method of the first-order ODE defined in Example 7.5.

```
# Heun method python program
# initial conditions
x0 = 0.5
y0 = 20
# The calculating point
xn = 5
n = 9
h = (xn-x0)/n
print('\n--------SOLUTION--------')
print("at x= ", x0," yn =", y0)
# function to be solved
def f(x, y):
     return (100+y)/(5*x)
# using Heun formula
for i in range(n):
     k1 = (f(x0, y0))
     k2 = (f((x0+h), (y0+k1*h)))
     yn = y0 + h*(k1+k2)/2
     x0 = x0+h
     print("at x= ", x0," yn =", yn)
     y0 = yn
```

The results obtained after running the Python program that solves the ODE specified in Example 7.5 are as follows:

```
--------SOLUTION--------
at x= 0.5 yn = 20
at x= 1.0 yn = 39.2
at x= 1.5 yn = 51.264
at x= 2.0 yn = 60.33984
```

```
at x= 2.5 yn = 67.71547264
at x= 3.0 yn = 73.97685028522667
at x= 3.5 yn = 79.4446941513338
at x= 4.0 yn = 84.31533584972715
at x= 4.5 yn = 88.71842442835953
at x= 5.0 yn = 92.7444174828312
```

7.6 RUNGE-KUTTA METHOD

Various types of Runge-Kutta methods can be derived by employing different terms in the increment function specified by the order term. The fourth-order Runge-Kutta is a method of numerically integrating ODEs using a trial step at the midpoint of an interval. The method is an effective and widely used method for solving the IVPs of differential equations, and the method is graphically illustrated in Figure 7.20. The formulas of the RK4 method are as follows:

$$k_1 = f(x_i, y_i)$$

$$k_2 = f\left(x_i + \frac{h}{2},\ y_i + \frac{h}{2}k_1\right)$$

$$k_3 = f\left(x_i + \frac{h}{2},\ y_i + \frac{h}{2}k_2\right)$$

$$k_4 = f(x_i + h,\ y_i + hk_3)$$

The solution of the ODE, y_{i+1}

$$y_{i+1} = y_i + \frac{h}{6}(k_1 + 2k_2 + 2k_3 + k_4)$$

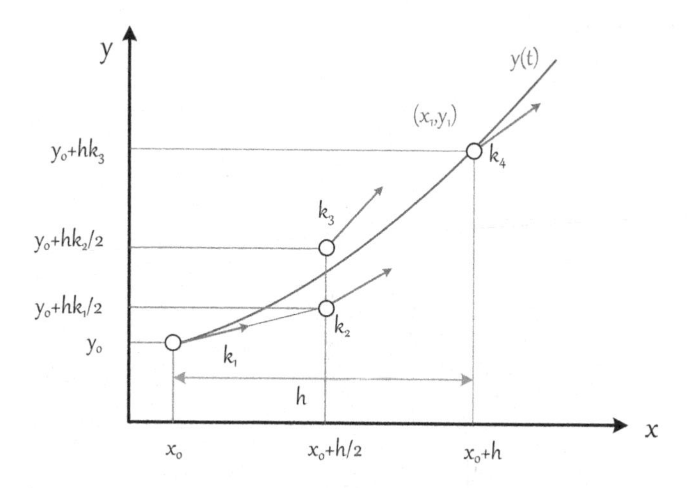

FIGURE 7.20 Graphical explanation of the RK4 method.

Example 7.6 Applying RK4 Method

Solve the following initial value ODE using the RK4 method to find the value of y at $x = 1.5$. Use a step size of $0.1 (h = 0.1)$.

$$\frac{dy}{dx} = 3x^2 y, \quad x(1) = 2$$

Confirm the manual calculations with Python and Simulink/MATLAB programming of the RK4.

Solution

The initial value, $x_0 = 1,\ y_0 = 2,\ h = 0.1$

At $x_1 = 1.1,\ y_1 = ?$

$$k_1 = f(x_0, y_0) = f(1,2) = 3(1)^2\, 2 = 6$$

$$k_2 = f\left(x_0 + \frac{h}{2},\ y_0 + \frac{h}{2}k_1\right) = f\left(1 + \frac{0.1}{2},\ 2 + \frac{0.1}{2}6\right) = 3(1.05)^2 (2.3) = 7.607$$

$$k_3 = f\left(x_0 + \frac{h}{2},\ y_0 + \frac{h}{2}k_2\right) = f\left(1 + \frac{0.1}{2},\ 2 + \frac{0.1}{2}7.607\right) = 7.873$$

$$k_4 = f(x_o + h,\ y_o + hk_3) = f(1 + 0.1,\ 2 + 0.1(7.873)) = f(1.1,\ 2.7873) = 2.785$$

The solution of the ODE, y_{i+1}

$$y_{i+1} = y_i + \frac{h}{6}(k_1 + 2k_2 + 2k_3 + k_4)$$

$$y_1 = y_0 + \frac{0.1}{6}(6 + 2(7.607) + 2(7.873) + 10.118) = 2.785$$

At $x = 1.1,\ y_1 = 2.785$

The analytical solution is defined as follows:

$$\frac{dy}{dx} = 3x^2 y$$

Using the separable integration,

$$\int_2^y \frac{dy}{y} = 3\int_{x_o}^x x^2 dx$$

The solution of the integral

$$\ln\left(\frac{y}{y_o}\right) = \frac{3(x^3 - x_o^3)}{3} = x^3 - x_o^3$$

Rearrange

$$y = y_o e^{x^3 - x_o^3} = 2e^{x^3 - 1}$$

Table 7.3 lists the solution of the ODE in Example 7.4 using RK4 (step size, $h = 0.1$) and the analytical solution. The RK4 calculations are in good agreement with the analytical solution.

Simulink and MATLAB Solution

Change the setting to the fixed-value RK4 method as follows: after launching Simulink, click on modeling in the toolbar, and then from the pulldown menu, select Model/Setting. While on the configuration parameters page, change the default values of the start time, stop time, type and solver, and fixed-step size to those shown in Figure 7.21. The Simulink block diagram that graphically describes the solution of the equation defined in Example 7.6 is shown in Figure 7.22. Running the model outputs the value of y (21.48) in the Display at x =1.5.

The alternative way to use Simulink is embedding a MATLAB code into the Simulink function block (Figure 7.23). The MATLAB function is attached to four

TABLE 7.3
Comparison of the Runge-Kutta Method and the Exact Solution of the ODE Defined in Example 7.6

x	k_1	k_2	k_3	k_4	y_{RK4}	y_{exact}
1.000					2.000	2.000
1.100	6.000	7.607	7.873	10.118	2.785	2.785
1.200	10.108	13.053	13.638	17.921	4.141	4.142
1.300	17.891	23.606	24.946	33.645	6.619	6.620
1.400	33.558	45.362	48.589	67.489	11.435	11.440
1.500	67.236	93.329	101.558	145.736	21.480	21.502

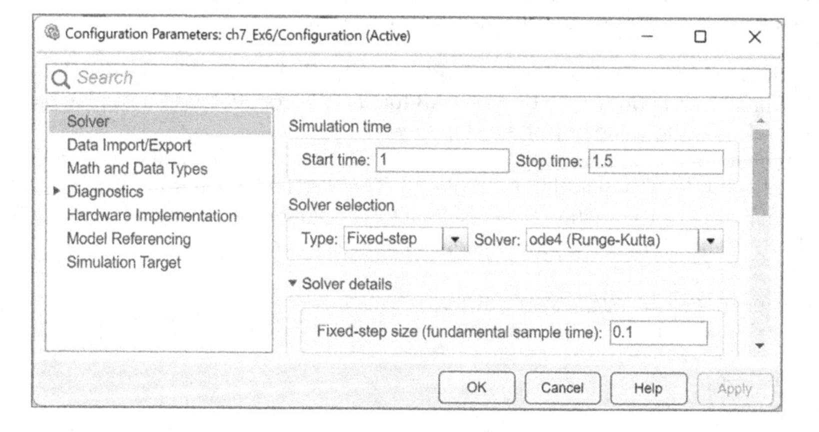

FIGURE 7.21 Model setting using fixed-step Runge-Kutta method, $h = 0.1$ (Example 7.6).

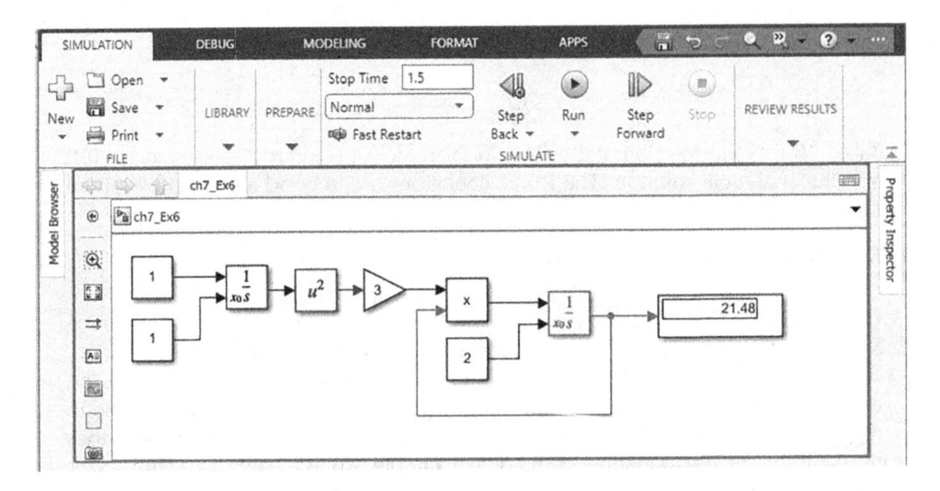

FIGURE 7.22 Simulink solution using fixed-step Runge-Kutta method of the ODE required in Example 7.6.

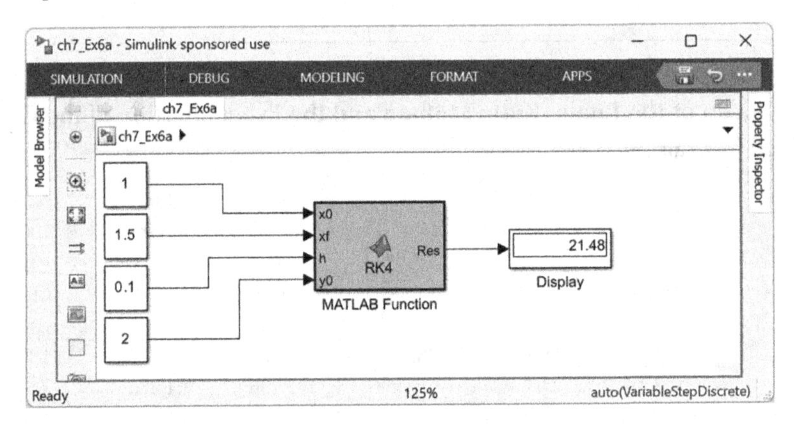

FIGURE 7.23 Simulink block diagram symbolizes the solution using the RK4 method with a stop time of five to the equation defined in Example 7.6.

input ports, lower limit interval (x_o), upper limit interval (x_f), the step size (*h*), and the initial conditions (y_0). The MATLAB function is connected to a display block that releases the value of *y* at the stop time *x*.

The following program represents the MATLAB code utilizing the RK4 method to solve the ODE required in Example 7.6. The MATLAB code is implanted in the Simulink MATLAB function. The resultant ODE solution found in the Simulink display block is 21.48 at the independent variable stop time (i.e., $x = 0.2$).

```
function Res = RK4(x0,xf, h, y0)
   f= @(x, y)(3*x^2*y); % ODE
   n= (xf-x0)/h;
   yi=y0;
   xi=x0;
for i =1:n
```

```
   k1 = f(xi, yi);
   k2 = f(xi+h/2, yi+ h*k1/2);
   k3 = f(xi+h/2, yi+k2*h/2);
   k4 = f(xi+h, yi+k3*h);
   k = k1+2*k2+2*k3+k4;
   yf = yi + k*h/6;
   yi=yf;
   xi = xi +h;
end
   Res=yi;
end
```

Python Solution

A Python code program that utilizes the RK4 method to solve an ODE in the Python programming language is presented. In this Python program, x_0 and y_o represents the initial condition. x_n is the calculation point on which the value of y_n corresponding to x_n is to be calculated using the RK4 method. The step size is *h*, which represents the number of finite steps before reaching to x_n, desired value.

```
# RK4 method method python program
# initial conditions
x0 = 1
y0 = 2
# The calculating point
xn = 1.5
n = 5
#h = (xn-x0)/n
h =0.1
step= h
print('\n--------SOLUTION--------')
print("at x= ", x0," yn =", y0)
# function to be solved
def f(x, y):
   return 3*(x**2)*y
# using midpint formula
for i in range(n):
   k1 = (f(x0, y0))
   k2 = (f((x0+h/2), (y0+k1*h/2)))
   k3 = (f((x0+h/2), (y0+k2*h/2)))
   k4 = (f((x0+h), (y0+k3*h)))
   k = (k1+2*k2+2*k3+k4)
   yn = y0 + k*h/6
   x0 = x0+h
   print("at x= ", x0," yn =", yn)
   y0 = yn
```

Running the Python code releases the following results:

```
 --------SOLUTION--------
at x= 1 yn = 2
at x= 1.1 yn = 2.7846419118859376
at x= 1.2000000000000002 yn = 4.141490537335979
at x= 1.3000000000000003 yn = 6.618844434974083
at x= 1.4000000000000004 yn = 11.434686303979241
at x= 1.5000000000000004 yn = 21.480436540827917
```

Example 7.7 Applying RK4 Method

Use the RK4 method to find the approximate value of y at $x = 0.2$

$$\frac{dy}{dx} = x + y^2, \ y(0) = 1$$

Given that $y = 1$ when $x = 0$, in a step size, $h = 0.1$

Verify the manual calculations with Python and Simulink/MATLAB of the utilization of RK4 method in solving the required ODE.

Solution

The following is an initial value nonlinear ODE

$$\frac{dy}{dx} = x + y^2$$

The right-hand side of the ODE is

$$f(x, y) = x + y^2, \ y(0) = 1$$

At $x_1 = x_0 + h = 0 + 0.1 = 0.1$

$$k_1 = f(x_o, y_o) = x_o + y_o^2 = 0 + 1^2 = 1$$

$$k_1 = 1$$

Calculate k_2

$$k_2 = f\left(x_o + \frac{h}{2}, y_o + \frac{h}{2}k_1\right) = f\left(0 + \frac{0.1}{2}, 1 + \frac{0.1}{2}(1)\right) = f(0.05, 1.1)$$

$$k_2 = 0.05 + 1.05^2 = 1.1525$$

$$k_2 = 1.1525$$

Calculate k_3

$$k_3 = f\left(x_o + \frac{h}{2}, y_o + \frac{h}{2}k_2\right) = f\left(0 + \frac{0.1}{2}, 1 + \frac{0.1}{2}(1.1525)\right) = f(0.05, 1.0576)$$

$$k_3 = 0.05 + 1.0576^2 = 1.169$$

$$k_3 = 1.169$$

Calculate k_4

$$k_4 = f(x_o + h, y_o + hk_2) = f(0 + 0.1, 1 + 0.1(1.169)) = f(0.1, 1.1169)$$

$$k_4 = 0.1 + 1.1169^2 = 1.3474$$

$$k_4 = 1.374$$

Calculate y_1 at x_1

$$y_1 = y_0 + \frac{h}{6}(k_1 + 2k_2 + 3k_3 + k_4)$$

$$y_1 = 1 + \frac{0.1}{6}(1 + 2(1.1525) + 2(1.169) + 1.374) = 1.1165$$

$$y_1(0.1) = 1.1165$$

Hence

$$x_1 = 0.1,\ y_1 = 1.1165$$

Calculate y_2 at $x_2 = 0.2$

Calculate k_1:

$$k_1 = f(x_1, y_1) = f(0.1, 1.1165) = 0.1 + 1.117^2 = 1.3466$$

Calculate k_2:

$$k_2 = f\left(x_o + \frac{h}{2},\ y_o + \frac{k_1}{2}\right) = f\left(0.1 + \frac{0.1}{2},\ 1.1165 + \frac{0.1}{2}1.3466\right)$$

Simplify

$$k_2 = f(0.15,\ 1.18383) = 0.15 + 1.18383^2 = 1.5514$$

Calculate k_3:

$$k_3 = f\left(x_o + \frac{h}{2},\ y_o + \frac{h}{2}k_2\right) = f\left(0.1 + \frac{0.1}{2},\ 1.1165 + \frac{0.1}{2}1.5514\right)$$

Simplify

$$k_3 = f(0.15,\ 1.1947) = 1.5758$$

Calculate k_4

$$k_4 = f(x_o + h,\ y_o + hk_3) = f(0.1 + 0.1,\ 1.1165 + 0.1(1.5758))$$

Simplify

$$k_4 = f(0.2,\ 1.274) = 0.2 + 1.274^2 = 1.8233$$

Calculate y_2

$$y_2 = y_1 + \frac{h}{6}\{k_1 + 2(k_2) + 2(k_3) + k_4\}$$

$$y_2 = y_1 + \frac{h}{6}\{1.3466 + 2(1.5514) + 2(1.5758) + 1.8233\} = 1.2736$$

Hence, $x_2 = 0.2,\ y_2 = 1.12736$.

Simulink and MATLAB Solution

While on the Simulink page, click on modeling, then model setting, and fill in the following information, start time and stop time values of 0 and 0.2, respectively. The solver type is set as a fixed type with a fixed-step size value of 0.1, as shown in Figure 7.24.

The Simulink block diagram solution of Example 7.7 is shown in Figure 7.25, and the obtained results agree with the manual calculations.

A diagram is created using Simulink's built-in function (Figure 7.26). Example 7.7 can be solved in the following alternative method, which extracts the MATLAB code from the Simulink diagram.

The following MATLAB code corresponds to the RK4 method planted in the Simulink function for the solution of the following ODE solution (Example 7.7).

```
function Res = RK4(x0,xf, h, y0)
   f= @(x, y)(x+y^2); % ODE
   n= (xf-x0)/h;
   yi=y0;
   xi=x0;
for i =1:n
   k1 = f(xi, yi);
   k2 = f(xi+h/2, yi+ h*k1/2);
   k3 = f(xi+h/2, yi+k2*h/2);
   k4 = f(xi+h, yi+k3*h);
   k = k1+2*k2+2*k3+k4;
   yf = yi + k*h/6;
```

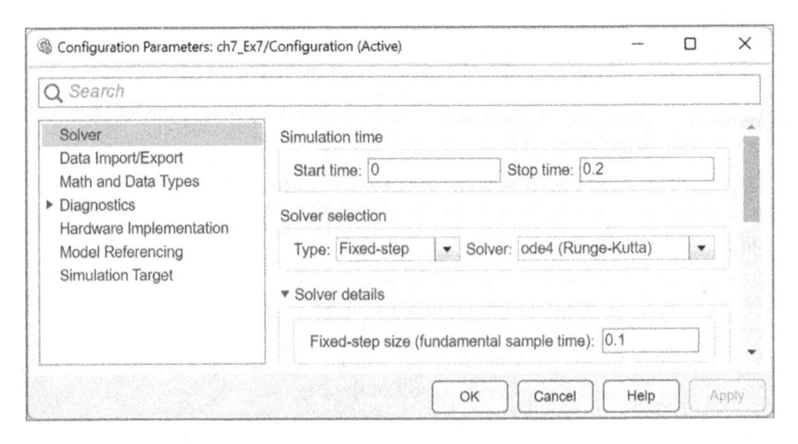

FIGURE 7.24 Model setting using the fixed step of 0.1 of the Runge-Kutta method for solving the ODE defined Example 7.7.

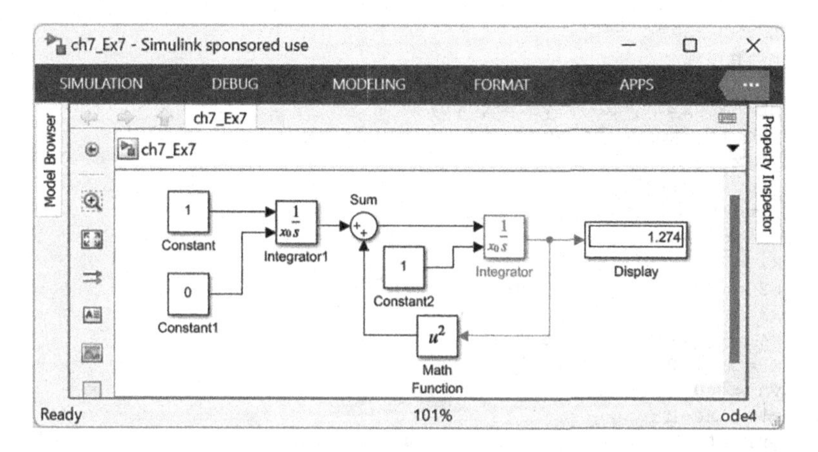

FIGURE 7.25 Simulink solution with a stop time of 0.2 of the equation defined in Example 7.7.

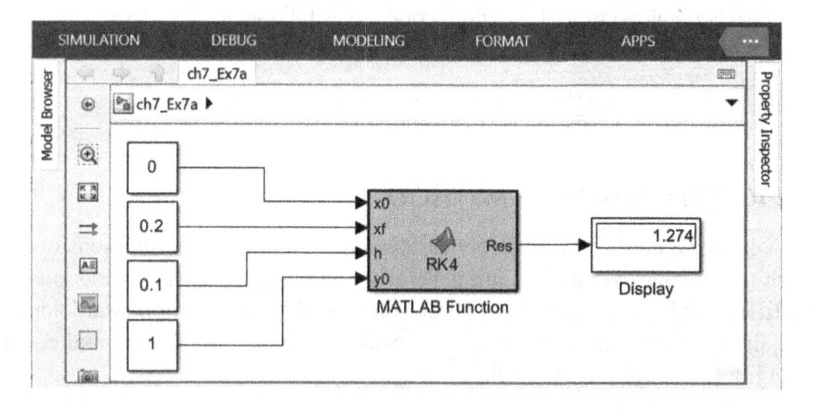

FIGURE 7.26 Simulink solution using the RK4 method, a stop time equal to 5, of the equation defined in Example 7.7.

```
    yi=yf;
    xi = xi +h;
end
    Res=yi;
end
```

Python Solution

The below Python program, followed by the execution results, corresponds to the approximate numerical solution of Example 7.7 using the RK4 method.

```
# RK4 method method python program
# initial conditions
x0 = 0
y0 = 1
# The calculating point
xn = 0.2
n = 2
#h = (xn-x0)/n
```

```
h =0.1
step= h
print('\n--------SOLUTION--------')
print("at x= ", x0," yn =", y0)
# function to be solved
def f(x, y):
    return x+y**2
# using RK4 formula
for i in range(n):
    k1 = (f(x0, y0))
    k2 = (f((x0+h/2), (y0+k1*h/2)))
    k3 = (f((x0+h/2), (y0+k2*h/2)))
    k4 = (f((x0+h), (y0+k3*h)))
    k = (k1+2*k2+2*k3+k4)
    yn = y0 + k*h/6
    x0 = x0+h
    print("at x= ", x0," yn =", yn)
    y0 = yn
```

The results obtained from the Python code are as follows:

```
--------SOLUTION--------
at x= 0 yn = 1
at x= 0.1 yn = 1.1164918497132719
at x= 0.2 yn = 1.2735625426752228
```

7.7 PICARD'S ITERATIVE METHOD

Picard's iterative method is relatively easy to implement, and the solutions obtained through this numerical analysis are generally power series. This method is used to solve differential equations with accurate solutions. It is a straight forward approach but requires lengthy calculations. If we consider the first-order differential equation with an initial condition, $y = y_0$ when $x = x_o$.

$$\frac{dy}{dx} = f(x,y)$$

Integrate between limits, we get

$$\int_{y_0}^{y} dy = \int_{x_o}^{x} f(x,y)\,dx$$

Integrating,

$$y = y_o + \int_{x_0}^{x} f(x,\, y)\,dx$$

Now we replace y by y_o as the first approximation

$$y_1 = y_o + \int_{x_0}^{x} f(x,\, y_o)\,dx$$

The second approximation

$$y_2 = y_o + \int_{x_0}^{x} f(x,\ y_1)\,dx$$

For the nth approximation,

$$y_n = y_o + \int_{x_0}^{x} f(x,\ y_{n-1})\,dx$$

The Picard's method generates a sequence of approximations as y_1, y_2, and y_3 that converges to the exact solution $y(x)$.

Example 7.8 Application of Picard's Iterative Method

Determine the value of y at $x = 0.3$ of the following ODE using Picard's iterative approximation method.

$$\frac{dy}{dx} = x + y^2 \ ,\ y(0) = 0$$

Solve manually and with Python and Simulink/MATLAB programming.

Solution

Using Picard's method,

1. First iteration when $y = 0$

$$y(x) = y_o + \int_{x_0}^{x} f(x + y^2)\,dx = y_o + \int_{x_0}^{x} f(x + 0)\,dx = \frac{x^2}{2}$$

For $x = 0.3$, integration and substitute $x = 0.3$

$$y_1(0.3) = y_o + \int_{x_0}^{x} f(x + 0)\,dx = 0 + \int_{x_0}^{x} x\,dx = \frac{x^2}{2} = \frac{0.3^2}{2} = 0.045$$

2. Second iteration

$$\frac{dy}{dx} = x + y_1^2 = x + \left(\frac{x^2}{2}\right)^2 = x + \frac{x^4}{4}$$

Therefore,

$$y_2(x) = y_o + \int_{0}^{0.3} \left(x + \frac{x^4}{4}\right) dx$$

$$y_2 = 0 + \frac{x^2}{2} + \frac{x^5}{20} = 0 + \frac{0.3^2}{2} + \frac{0.3^5}{20} = 0.0451$$

3. Third iteration

$$\frac{dy}{dx} = x + y_2^2$$

$$\frac{dy}{dx} = x + \left(\frac{x^2}{2} + \frac{x^5}{20}\right)^2 = x + \frac{x^4}{4} + \frac{2x^7}{40} + \frac{x^{10}}{400}$$

Therefore

$$y_3 = 0 + \int_0^{0.3} \left(x + \frac{x^4}{4} + \frac{x^7}{20} + \frac{x^{10}}{400}\right) dx$$

$$y_3 = \frac{x^2}{2} + \frac{x^5}{20} + \frac{x^8}{160} + \frac{x^{11}}{4400} = \frac{0.3^2}{2} + \frac{0.3^{0.5}}{20} + \frac{0.3^8}{160} + \frac{0.3^{11}}{4400} = 0.0451$$

Accordingly, at $x = 0.3$, $y = 0.0451$, it is correct up to four decimal places.

Simulink and MATLAB Solution

The Simulink block diagram for Example 7.8 is shown in Figure 7.27. The default solver is used for RK4 as the Picard's method is not built in MATLAB solver.

The alternative way to solve Example 7.8 is using the Simulink function, as shown in Figure 7.28, to output the MATLAB code.

The following MATLAB codes correspond to the Picard's method planted in the Simulink function for the solution of the following ODE solution (Example 7.8).

```
function [y1,y2,y3] = Picard(a)
%Picard's Method
%
f1 = @(x) x^2/2;
```

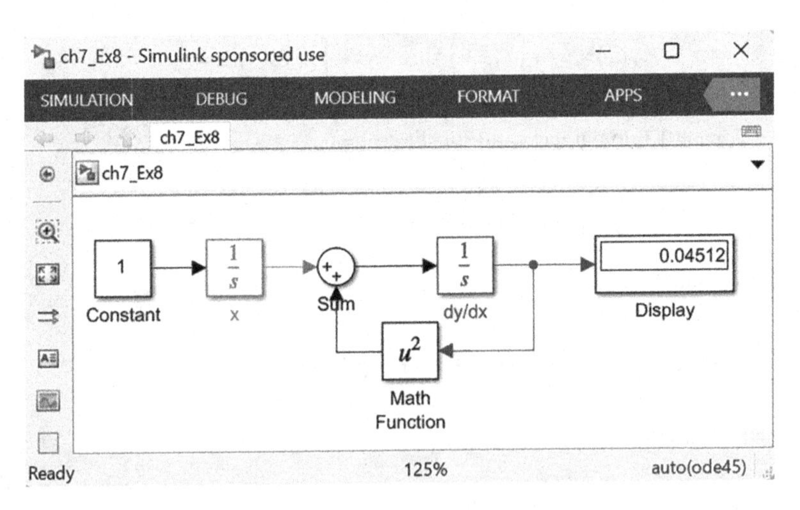

FIGURE 7.27 Simulink's solution, with downtime equal to 0.3, of the equation given in Example 7.8.

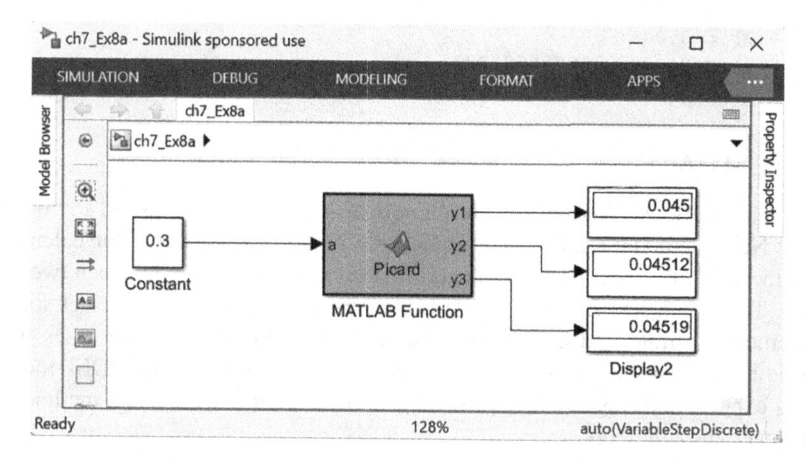

FIGURE 7.28 The block diagram of the simulation using the Picard method with a downtime of 0.2 represents the solution of the equation given in Example 7.8.

```
f2 = @(x) x^2/2 + x^5/20;
f3 = @(x) x^2/2 + x^5/20+ x^8/160 + x^1/4400;
%
y1=f1(a);
y2=f2(a);
y3=f3(a);
end
```

Python Solution

The approximate numerical solution using Picard's approximation method followed by its solution is shown below in Python format.

```
#Picard's Method
#
def y1(x):
Y1 = x**2/2
return Y1
def y2(x):
Y2 = x**2/2 + x**5/20
return Y2
def y3(x):
Y3 = x**2/2 + x**5/20+ x**8/160 + x**1/4400
return Y3
#
print (' ')
print (' === Solution ===')
print ('for dy(x) = x + y**2 the results are :')
print ('--------------------------------------')
print ('y1(0.3) = ', y1(0.3))
print ('y2(0.3) = ', y2(0.3))
print ('y3(0.3) = ', y3(0.3))
```

The results obtained from the Python code are as follows:

```
=== Solution ===
for dy(x) = x + y**2 the results are :
--------------------------------------
```

```
y1(0.3) = 0.045
y2(0.3) = 0.045121499999999995
y3(0.3) = 0.04519009188068l814
```

7.8 SUMMARY

Numerical methods are used to find numerical approximations to the solutions of ODEs. Numerical integration includes a wide range of algorithms for calculating the numerical value of a definite integral (the area under a curve between two fixed terms). The numerical method with a minor error compared to the exact solution is the most accurate and easiest to solve. The problem lies in those cases where there is no exact solution. The approximate numerical solution of ODEs includes Euler's explicit and implicit methods, the Midpoint method, the Heun method, and the Runge-Kutta method.

7.9 PROBLEMS

1. Using the Euler method, find an approximation for the solution of the differential equation

$$\frac{dy}{dx} = f(x, y)$$

With the initial condition

$$y(x_o) = y_o$$

2. The dilution of the concentrated orange in homemade orange juice is given as a function of time by

$$\frac{dy}{dt} = 37.5 - 3.5y$$

Initially, the orange concentration in the tank was 50 g/L. Using explicit Euler's method and a step size of $h = 1.5$ min, what is the orange concentration after 3 minutes? Compare the predicted result with the analytical solution. Find the best step size that gives the closest result to the exact solution. The exact solution of the ODE is given by

$$y(t) = 10.714 + 39.286e^{-3.5t}$$

Answer: (10.7154 g/L, with h = 0.1)

3. Using the RK4 method, find the approximate solution of the following initial value first-order differential equation within the interval [1–4].

$$\frac{dy}{dx} = 1 + \frac{y}{x}$$

With the initial condition, $y(1)=1$, the approximate numerical answer is (3.38).

4. Solve the following ODE using the Picard iterative method with an initial condition, $y(0)=100$, calculate y(50)

$$\frac{dy}{dt}=50-0.01y$$

Answer (1732)

5. Using the Midpoint method, solve the following ODE with an initial condition, $y(0)=1$, and step size, $h=0.1$, calculate $y(0.5)$

$$\frac{dy}{dx}=yx^3-0.01y,\quad y(0)=1.0$$

The exact solution is $y(0.5)=1.01068$.

6. Using the Heun method, solve the following ODE with an initial condition, $y(0)=1$, and step size, $h=0.1$, calculate $y(0.5)$

$$\frac{dy}{dx}=yx^3-0.01y,\quad y(0)=1.0$$

The exact solution is $y(0.5)=1.01068$ and the approximate answer is (1.0113).

7. Using the Euler's method within the range $t=0$ to 2, solve the following first-order initial value ODE, using a step size, $h=0.5$

$$\frac{dy}{dt}=t^2y-1.2y,\quad y(0)=1$$

The exact solution is $y(2)=1.30561$, and the approximate answer is (0.2882).

8. Using the Midpoint's method within the range $t=0$ to 2, solve the following first-order initial value ODE, using a step size, $h=0.5$

$$\frac{dy}{dt}=t^2y-1.2y,\quad y(0)=1$$

The exact solution is $y(2)=1.30561$, and the approximate answer is (1.16).

9. Using the RK4 method within the range $t=0$ to 2, solve the following first-order initial value ODE, using a step size, $h=0.5$

$$\frac{dy}{dt}=t^2y-1.2y,\quad y(0)=1$$

The exact solution is $y(2)=1.30561$, and the approximate answer is (1.29855).

10. Using the RK4 method within the range $x=0$ to 2, solve the following first-order initial value ODE, using a step size, $h=0.5$

$$\frac{dy}{dx}=x-1.2y,\quad y(0)=1$$

The exact solution is $y(2) = 1.12594$, and the approximate RK4 answer is (1.12594).

REFERENCES

1. Gupta, S.K., 2014, *Numerical Methods for Engineers.* London: New Academic Science.
2. Kharab, A. and Guenther, R., 2018, *An Introduction to Numerical Methods: AMatlab Approach.* Boca Raton, FL: CRC Press.
3. Kong, Q., Siauw, T., and Bayen, A., 2020, *Python Programming and Numerical Methods: A Guide for Engineers and Scientists.* Amsterdam: Academic Press Inc.
4. Hill, C., 2020, *Learning Scientific Programming with Python.* Cambridge: Cambridge University Press.

8 Simultaneous Systems of Differential Equations

In this chapter, we will learn how to simultaneously solve a system of binary, linear, independent differential equations of the first order. Such systems arise when the model includes two or more variables. This chapter grants numerical methods programming for generating simple and effective Python and MATLAB codes along with Simulink to output numerical solutions with the required accuracy. It uses the plotting functions of Matplotlib in Python and the Plot function in MATLAB to present the required results graphically.

LEARNING OBJECTIVES

1. Solve binary ODEs simultaneously using the Euler method.
2. Employ the Midpoint method to solve a system of ODEs simultaneously.
3. Use the fourth-order Runge-Kutta (RK4) method to solve binary ODEs.
4. Solve ODEs simultaneously using the Picard method.

8.1 INTRODUCTION

Numerical methods for solving ordinary differential equations (ODEs) are used to find approximate numerical solutions for ODE. Their use is also known as a numerical integral, although this term can also refer to the computation of integrals [1]. A first-order differential equation with only the first derivative of the dependent variable (e.g., y) without higher derivatives is considered an initial value problem (IVP),

$$\frac{dy}{dx} = y'(x) = f(x, y(x)\,,\; y(x_o) = y_o \tag{8.1}$$

Several differential equations cannot be solved using symbolic computation. However, a numerical approximation of the solution is often sufficient in engineering and for practical purposes. The algorithms considered here can be used to calculate such an approximation. ODEs occur in various scientific disciplines, including physics, chemistry, biology, and economics. In addition, specific scalar partial differential equation methods convert the partial differential equation into an ordinary one, which must then be solved. The following sections use various methods to solve two ODEs simultaneously [2].

DOI: 10.1201/9781003360544-8

8.2 SIMULTANEOUS NUMERICAL SOLUTION OF TWO ODEs

We will now consider systems of simultaneous linear differential equations that have one independent variable.

8.2.1 Euler Method

Considering that the following ODEs are to be solved simultaneously using the Euler method,

$$\frac{dy_1}{dt} = f_1\left(t, y_1, y_2\right), \quad y_1(0) = y_{1,0} \tag{8.2}$$

$$\frac{dy_2}{dt} = f_2\left(t, y_1, y_2\right), \quad y_2(0) = y_{2,o} \tag{8.3}$$

The Euler algorithms for solving the stated two ODEs are as follows:

$$y_{1,i+1} = y_{1,0} + f_1\left(t_i, y_{i,1}, y_{i,2}\right) \tag{8.4}$$

$$y_{2,i+1} = y_{2,o+}\ f_2\left(t_i, y_{i,1}, y_{i,2}\right) \tag{8.5}$$

The following example illustrates the application of the explicit Euler method in solving two ODEs simultaneously.

Example 8.1 Application of Explicit Euler Method

Solve the following two ODEs simultaneously using the Euler method and a step size of 0.1.

$$\frac{dy_1}{dt} = -y_1 e^{1-t} + 0.8y_2, \quad y_1(0) = 0$$

$$\frac{dy_2}{dt} = y_1 - y_2^3, \quad y_2(0) = 2$$

Manually calculate the values of the dependent variables (y_1, y_2) within the interval [0, 3]. Compare the manual calculations with the approximate solutions using Simulink (the graphical programming of MATLAB) and Python programming of the Euler method.

Solution

Apply the formulas of the explicit Euler method

$$\frac{dy_1}{dx} = f_1(t, y_1, y_2) = -y_1 e^{1-x} + 0.8y_2$$

$$\frac{dy_2}{dx} = f_2(t, y_1, y_2) = y_1 - y_2^3$$

The explicit Euler algorithm

$$y_{1,i+1} = y_{1,i} + hf_1(t_i, y_{i,1}, y_{i,2})$$

$$y_{2,i+1} = y_{2,i} + hf_2(t_i, y_{i,1}, y_{i,2})$$

The initial conditions are at $t = 0$, where $y_1(0) = 0,\ y_2(0) = 2$

At $t = 0.1$,

$$y_{1,1} = 0 + hf_1(0,\ 0{,}2) = 0 + 0.1\left(-0e^{1-0} + 0.8(2)\right) = 0.16$$

$$y_{2,1} = 2 + hf_2(0{,}0{,}2) = 2 + 0.1\left(0 - 2^3\right) = 1.2$$

At $t = 0.2$,

$$y_{1,2} = 0.16 + hf_1(0.1,\ 0.16{,}1.2) = 0.16 + 0.1\left(-0.16e^{1-0.1} + 0.8(1.2)\right) = 0.2166$$

Simplify

$$y_{2,2} = 1.2 + hf_2(0.1{,}0.16{,}1.2) = 1.2 + 0.1\left(0.16 - (1.2)^3\right) = 1.043$$

The rest of the calculations are listed in Table 8.1.

TABLE 8.1
Euler's Method Solution of Example 8.1

Time (t)	i	y_1	y_2
0.000	0	0.000	2.000
0.100	1	0.160	1.200
0.200	2	0.217	1.043
2.800	28	1.067	0.953
2.900	29	1.126	0.973
3.000	30	1.187	0.994

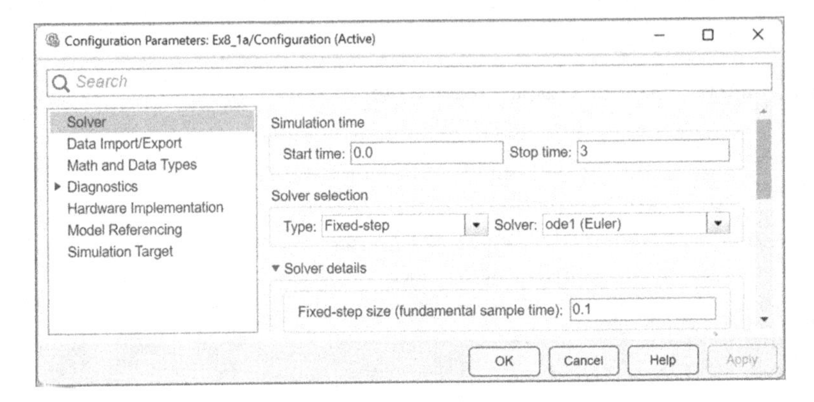

FIGURE 8.1 Simulink solver configuration parameters for selecting the Euler method to solve the equations defined in Example 8.1.

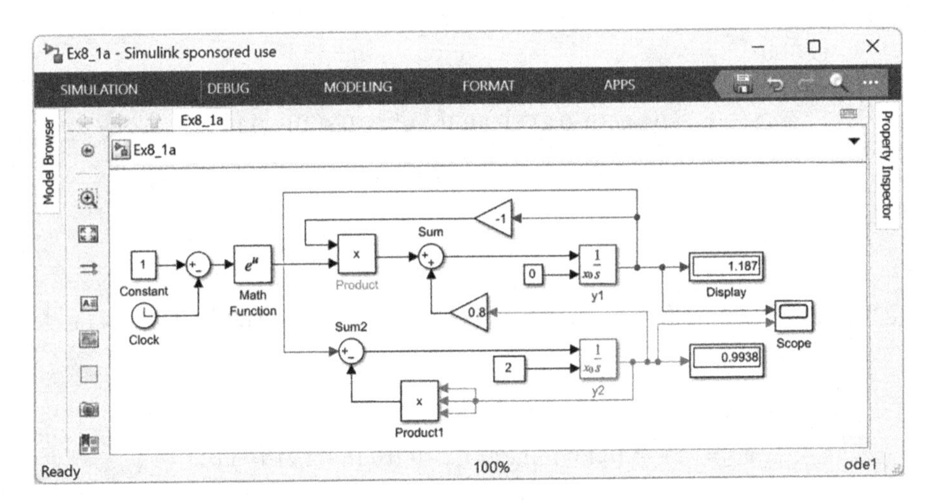

FIGURE 8.2 Numerical solution of the two equations defined in Example 8.1 applying the explicit Euler method in Simulink.

Simulink Solution

To use the integrator block function from the Simulink library, the configuration parameters such as start time, stop time, type, and solver should be changed in the modeling set. Therefore, to do so, while on the Simulink simulation page, click on the modeling tab, and select the modeling setting. Change the start and stop times to 0.1 and 3, respectively. For the solver selection, indicate the type: fixed step and the solver: Euler, as shown in Figure 8.1. The entire Simulink graphical solution of the two ODEs defined in Example 8.1 using the explicit Euler method is presented in Figure 8.2. The values in the display blocks are the values of y_1 and y_2 with a stop time of three.

Individuals may want to change the background colors of the scope before copying the scale image. To do so, click on the 'scope parameters' icon (the

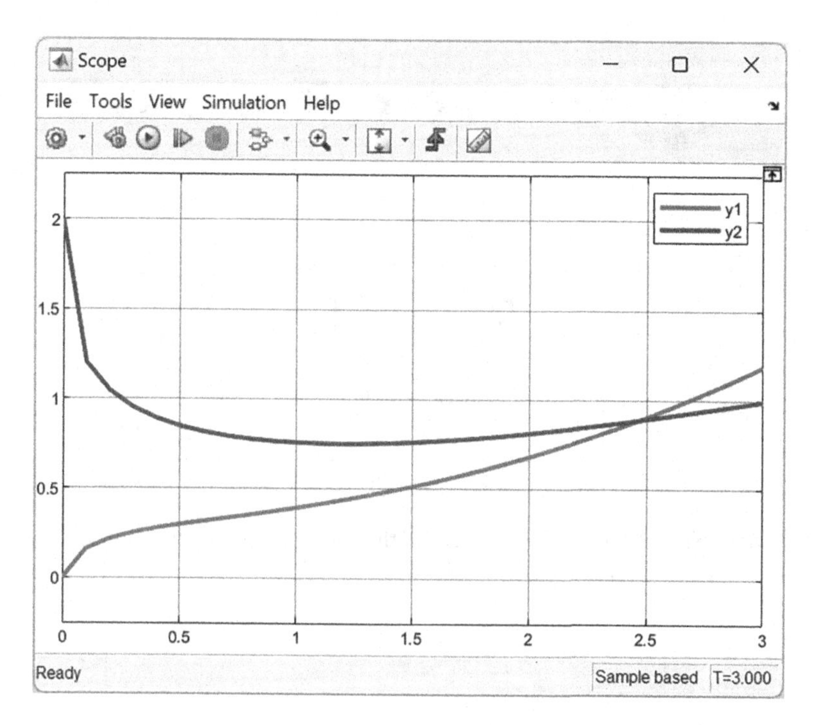

FIGURE 8.3 Simulink plot of the numerical solution utilizing the explicit Euler method of the model equations defined in Example 8.1.

second icon) and go to the 'style' tab. Change the shape color to black, the axes colors to white background and black writing, and the font color to black. After the modifications, the plot of the two ODEs generated by Simulink is shown in Figure 8.3. There is no option to automatically set the axis or name labels in the plot generated by the scope simulated graph. However, after the simulation, a person can manually add axis names from the Graphical User Interface (GUI) shape or the command window. Select the scope graph from Simulink and run the following commands into the command window.

```
set(0,'ShowHiddenHandles', 'on')
set(gcf, 'menubar', 'figure')
```

These commands will enable the toolbar in the scope graph shape window. One can add axis names from Insert>> X Label and Insert>> Y Label. To change the legend, while in the scope window, select the View > Legend option to display different variables to show the legend. If one double-clicks on a legend, a small text entry field will appear, allowing one to name it. After the scope amendment, the graph should look like that in Figure 8.3.

An alternative way to solve the equations defined in Example 8.1 is to use the Simulink MATLAB function (Figure 8.4). The MATLAB function requires writing a MATLAB code. Figure 8.4 shows the Simulink MATLAB function with five input ports: the lower limit interval (*a*), the upper limit interval (*b*), the step size (*h*), and the initial values of $y_1(y_o)$ and $y_2(z_o)$. The two displays release the values of y_1 and y_2 at the simulation stop time [3].

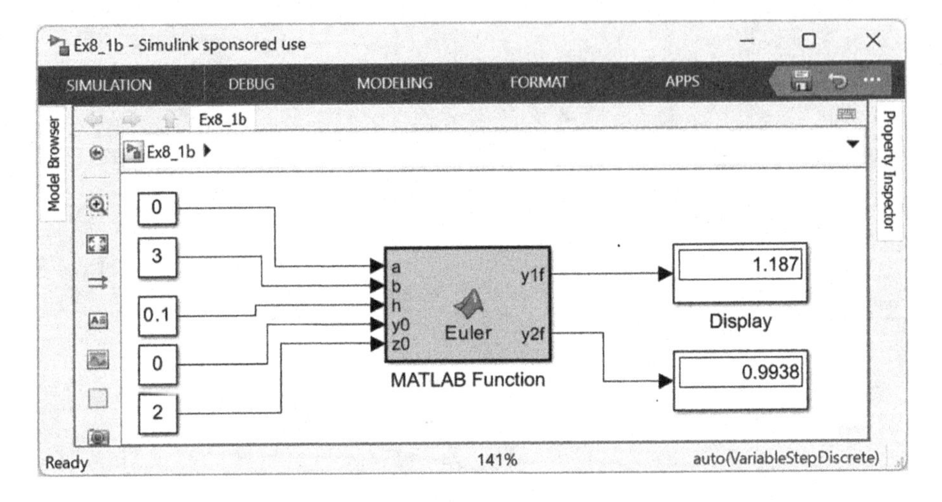

FIGURE 8.4 Simulink numerical solution of the two equations described in Example 8.1 using the Euler method. The stop time is three.

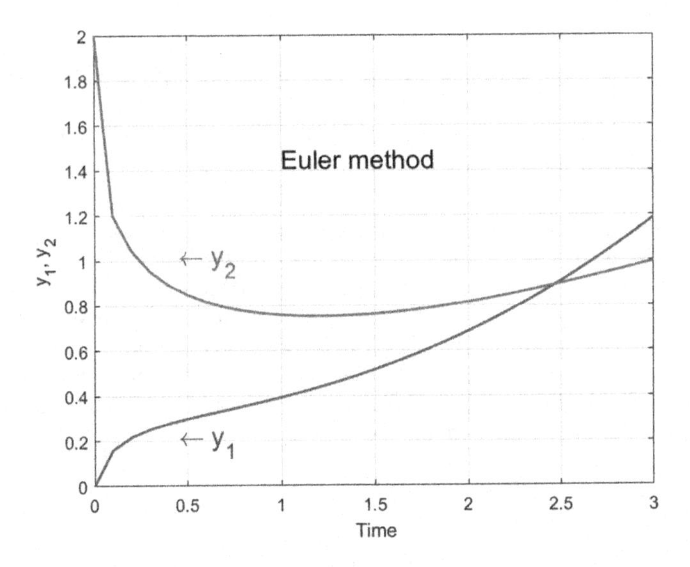

FIGURE 8.5 Simulink generated a plot of numerical solution using the explicit Euler method of the equations described in Example 8.1.

The following program is the MATLAB code implanted in the Simulink MATLAB function shown in Figure 8.4. When running the Simulink graphical program, it generates the plot presented in Figure 8.5. The figure describes the change in the dependent variables (y_1, y_2) with time within the desired interval [0, 3].

```
% Example 8.1 using the Euler method
function [y1f, y2f] = Euler(a, b, h, y0,z0)
%Euler integration method
```

```
% Routine starts here
f1=@(x, y, z)(-y*exp(1-x)+ 0.8*z); % Enter RHS of the first ODE
f2=@(x, y, z) (y-z^3); % Enter RHS of the first ODE

x=a:h:b;
y=zeros(1,length(x));
z=zeros(1,length(x));
y(1)=y0;
z(1)=z0;
x(1)=a;
for i=1:(length(x)-1)

    m1 = f2(x(i), y(i), z(i));
    k1 = f1(x(i), y(i), z(i));

    y(i+1) = y0 + h*k1;
    y0=y(i+1);
    z(i+1) = z0 + h*m1;
    z0=z(i+1);
end
y1f=y0;
y2f=z0;
plot(x, y, x, z, 'linewidth', 1.5);
text(0.45,1, '\leftarrow z', 'Color', 'r', FontSize=15)
text(0.45,0.2, '\leftarrow y', 'Color', 'b', FontSize=15)
text(1.0,1.45, 'Euler method', FontSize=15)
xlabel('x');
ylabel('y, z');
grid on
```

Python Solution

The following is the Python code that utilizes the Euler method simultaneously for solving the two ODEs defined in Example 8.1. Numerical methods with Python programming allow the creation of simple and efficient Python code that produces numerical solutions with the required degree of accuracy. Use of matplotlib plotting functions to present the results graphically [3]. The implementation of the program resulted in Figure 8.6.

```
# Euler method for solving 2 ODEs simultaneously
import matplotlib.pyplot as plt
import numpy as np
from numpy import arange
# y1' = -y1*np.exp(1-x)+0.8*y2
# y2' = y1-y2**3
def F1(y1, y2, x):
    return -y1*np.exp(1-x)+0.8*y2
def F(y1, y2,x):
    return y1-y2**3
a = 0
b = 3
N =30
h = (b-a)/N
xp = arange(a, b, h)
yp = []
up = []
```

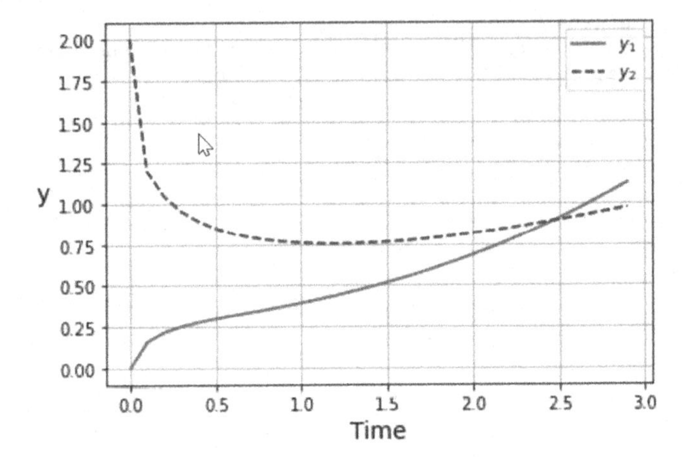

FIGURE 8.6 Python plot represents the numerical solution using the Euler method of the equations defined in Example 8.1.

```
# initial conditons
y1 = 0.0
y2 = 2
for x in xp:
    yp.append(y1)
    up.append(y2)
    m1 = h*F1(y1, y2, x)
    k1 = h*F(y1, y2, x)
    y1+= m1
    y2+= k1
# plot section
plt.plot(xp, yp, 'r-', label=r'$y_{1}$')
plt.plot(xp, up, 'b--', label=r'$y_{2}$')
plt.xlabel("Time", fontsize=14)
plt.ylabel('y ', fontsize=14,rotation=0)
plt.legend()
plt.grid(True)
```

8.2.2 Midpoint Method

The Midpoint method is a type of second-order Rung-Kutta method. It is used to solve ODEs with a given initial condition. This technique is called the Midpoint method because it uses the tangent at the midpoint and checks where the tangent cuts for the vertical lines. Considering h as the step size and the following two ODEs:

$$\frac{dy_1}{dx} = f_1(x, y_1, y_2) \tag{8.6}$$

$$\frac{dy_2}{dx} = f_2(x, y_1, y_2) \tag{8.7}$$

The Midpoint algorithm for solving the two equations at once is as follows:

$$k_{1,1} = f_1\left(x_i, y_{i,1}, y_{i,2}\right) \tag{8.8}$$

$$k_{1,2} = f_2\left(x_i, y_{i,1}, y_{i,2}\right) \tag{8.9}$$

$$k_{2,1} = f_1\left(x_i + \frac{h}{2},\ y_{i,1} + \frac{h}{2}k_{1,1},\ y_{i,2} + \frac{h}{2}k_{2,1}\right) \tag{8.10}$$

$$k_{2,2} = f_2\left(x_i + \frac{h}{2},\ y_{i,1} + \frac{h}{2}k_{2,1},\ y_{i,2} + \frac{h}{2}k_{2,2}\right) \tag{8.11}$$

Finally,

$$y_{i,1} = y_{i-1,1} + h\ k_{21} \tag{8.12}$$

$$y_{i,2} = y_{i-1,2} + h\ k_{22} \tag{8.13}$$

Example 8.2 Applying the Midpoint Method to Solve Two Simultaneous Equations

Apply the Midpoint method to solve the following two ODEs using a step size of 0.1.

$$\frac{dy_1}{dt} = -y_1 e^{1-t} + 0.8y_2, \quad y_1(0) = 0$$

$$\frac{dy_2}{dt} = y_1 - y_2^3, \quad y_2(0) = 2$$

Calculate y_1 and y_2 manually within the interval [0, 3] and confirm your manual calculations with Python and Simulink graphical programming.

Manual Solution

Following the Midpoint formulas (8–13),

$k_{1,1}$

$$k_{1,1} = f_1\left(t_i, y_{i,1}, y_{i,2}\right) = f_1(0,0,2) = -0\mathrm{e}^{1-0} + 0.8 * 2 = 1.6$$

$k_{1,2}$

$$k_{1,2} = f_2\left(t_i, y_{i,1}, y_{i,2}\right) = f_2\left(t_0, y_{o,1}, y_{o,2}\right) = f_2\left(0,0,2\right) = 0 - 2^3 = -8$$

$k_{2,1}$

$$k_{2,1} = f_1\left(t_o + \frac{h}{2},\ y_{o,1} + \frac{h}{2}k_{1,1},\ y_{o,2} + \frac{h}{2}k_{1,2}\right)$$

$$k_{2,1} = f_1\left(0 + \frac{0.1}{2},\ 0 + \frac{0.1}{2}(1.6),\ 2 + \frac{0.1}{2}(-8)\right)$$

$$k_{2,1} = f_1\left(0.05,\ 0.08, 1.6\right) = -0.08e^{(1-0.05)} + 0.8(1.6) = 1.07314$$

$k_{2,2}$

$$k_{2,2} = f_2\left(t_o + \frac{h}{2},\ y_{o,1} + \frac{h}{2}k_{1,1}, y_{o,2} + \frac{h}{2}k_{1,2}\right)$$

$$k_{2,2} = f_2\left(0.05,\ 0.08, 1.6\right) = y_1 - y_2^3 = 0.08 - 1.6^3 = -4.016$$

Accordingly

$$y_1 = y_{1,0} + h\left(k_{2,1}\right)$$

Substitute the values

$$y_1 = 0 + 0.1 \times \left(1.07314\right) = 0.1107314$$

Following the same procedure to calculate, y_2

$$y_2 = y_{2,0} + h \times k_{2,2}$$

Substitute the known values

$$y_2 = 2 + 0.1 \times \left(-4.016\right) = 1.598$$

The remaining values are calculated using excel and tabulated in Table 8.2.

Figure 8.7 shows the excel plot of the numerical solution utilizing the Midpoint method. It shows the change in the dependent variables (y_1 and y_2) versus time (t).

TABLE 8.2
Solution of Binary Equation Defined in Example 8.2 Using the Midpoint Method

i	x	$k_{1,1}$	$k_{1,2}$	$k_{2,1}$	$k_{2,2}$	y_1	y_2
1	0.000					0.000	2.000
2	0.100	1.600	–8.000	1.073	–4.016	0.107	1.598
3	0.200	1.015	–3.976	0.750	–2.583	0.182	1.340
29	2.900	0.597	0.193	0.609	0.195	1.196	1.000
30	3.000	0.621	0.197	0.633	0.198	1.260	1.020

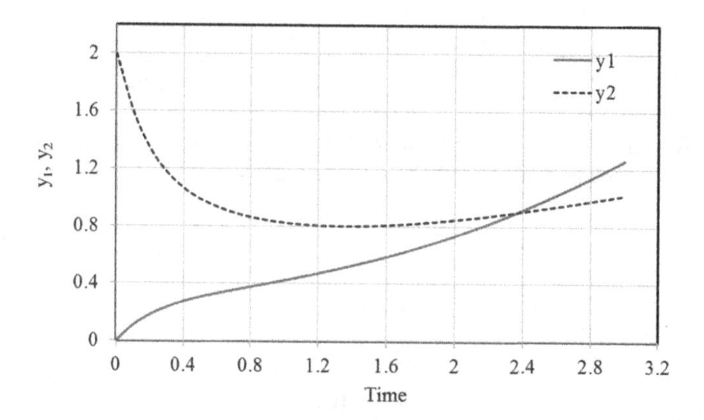

FIGURE 8.7 Excel plot of the manual calculations, utilizing the Midpoint method, of the equations defined in Example 8.2.

Simulink Solution

The Simulink numerical solution of the two differential equations defined in Example 8.2 is shown in Figure 8.8. The Simulink graphical program uses the MATLAB function dragged from the Simulink library (under the user-defined functions). The Simulink MATLAB function relates five input constants represent the lower limit interval (*a*), upper limit interval (*b*), step size (*h*), and the initial values of y_1 (y_o) and initial value of y_2 (z_o). The two output displays present the values of y_1 and y_2 at the simulation stop time [3].

The following is the MATLAB code of the Midpoint method implanted in the Simulink MATLAB function present in Figure 8.8. The generated plot of the dependent variables versus time is revealed in Figure 8.9.

```
% Example 8.2 using Midpoint
function [y1f, y2f] = MidPoint(a, b, h, y0,z0)
% Midpoint integration method

% Enter RHS of the first ODE
f1=@(x, y, z)(-y*exp(1-x)+ 0.8*z);
```

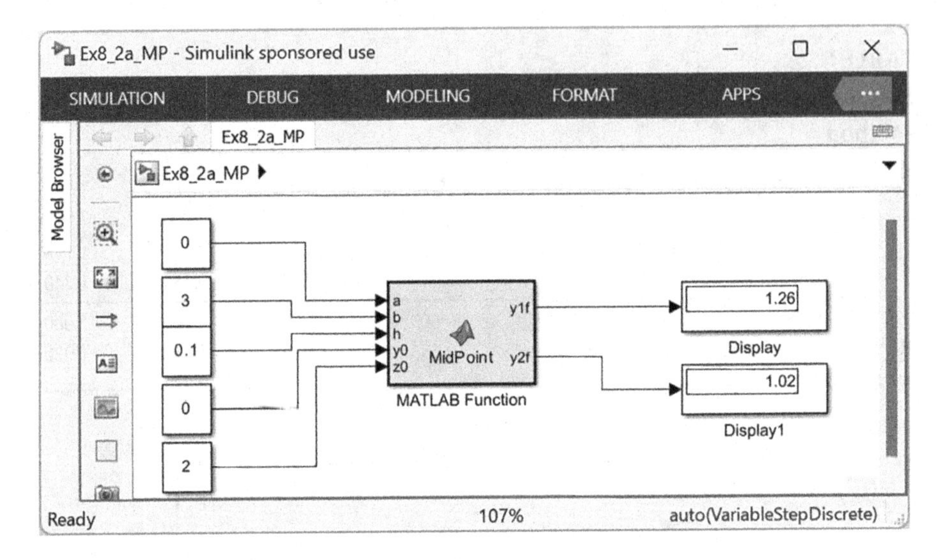

FIGURE 8.8 Simulink solution utilizing MATLAB function and the Midpoint method of the two ODEs defined in Example 8.2, stop time equal three.

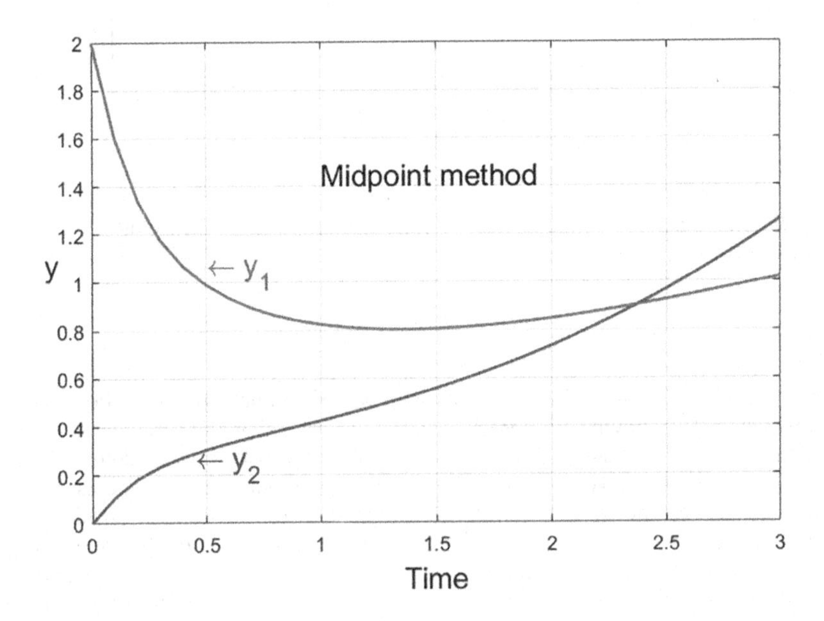

FIGURE 8.9 Simulink plot generated using the Midpoint method of the two equations defined in Example 8.2.

```
% Enter RHS of the first ODE
f2=@(x, y, z) (y-z^3);

x=a:h:b;
y=zeros(1,length(x));
z=zeros(1,length(x));
```

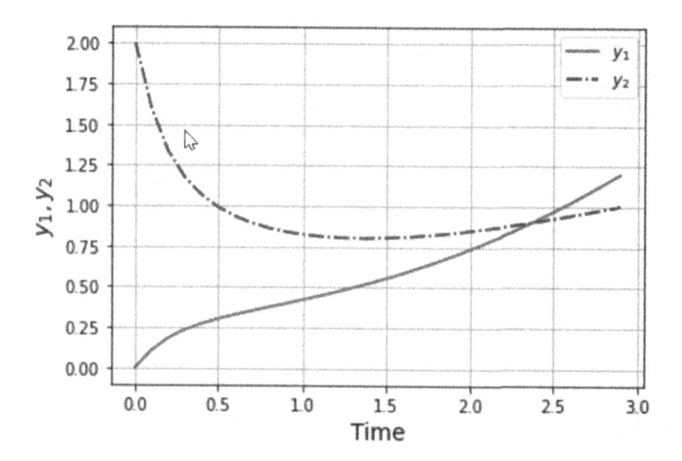

FIGURE 8.10 Python-generated plot using the Midpoint method represents the solution of the equations defined in Example 8.2.

```
y(1)=y0;
z(1)=z0;
x(1)=a;
    for i=1:(length(x)-1)
    k1 = f1(x(i), y(i), z(i));
    m1 = f2(x(i), y(i), z(i));
    k2 = f1(x(i)+0.5*h, y(i)+0.5*h*k1,z(i)+0.5*h*m1 );
    m2 = f2(x(i)+0.5*h, y(i)+0.5*h*k1,z(i)+0.5*h*m1 );

    y(i+1) = y0 + h*k2;
    y0=y(i+1);

    z(i+1) = z0 + h*m2;
    z0=z(i+1);
end
y1f=y0;
y2f=z0;

% Plot sections
plot(x, y, x, z, 'linewidth', 1.5);
text(0.5,1.05, '\leftarrow z', 'color', 'r', FontSize=15)
text(0.45,0.25, '\leftarrow y', 'color', 'b', FontSize=15)
text(1.0,1.45, 'Midpoint method', FontSize=15)
xlabel('Time', FontSize=15);
ylabel('y_1, y_2', FontSize=15);
grid on
```

The following is the Python program that employed the Midpoint method to solve two ODEs defined in Example 8.2. Running the program releases the plot of the dependent variables versus the independent variable, as shown in Figure 8.10.

```
# Example 8.2
# Midpoint method for solving two ODEs
import matplotlib.pyplot as plt
import numpy as np
from numpy import arange
# Enterthe two functions
```

```
def F1(y, u, x):
return -y*np.exp(1-x)+0.8*u
def F2(y, u, x):
    return y-u**3
a = 0
b = 3
N =30
h = (b-a)/N
xp = arange(a, b, h)
yp = []
up = []
# initial conditions
y = 0.0
u = 2

for x in xp:
yp.append(y)
up.append(u)

m1 = h*F1(y, u, x)
k1 = h*F2(y, u, x)
m2 = h*F1(y+0.5*m1, u+0.5*k1, x+0.5*h)
k2 = h*F2(y+0.5*m1, u+0.5*k1, x+0.5*h)

y = y + m2
u = u + k2

# plot section
plt.plot(xp, yp, 'r-', label="y")
plt.plot(xp, up, 'b-.', label="x")
plt.xlabel("Time", fontsize=14)
plt.ylabel("y, x", fontsize=14)
plt.legend()
plt.grid(True)
```

The generated solution of the Python code after execution is represented in Figure 8.10.

8.2.3 Fourth-Order Runge-Kutta

The Runge-Kutta method is a mathematical algorithm used to solve systems of ODEs of the following forms:

$$\frac{dy_1}{dt} = f_1(t, y_1, y_2) \tag{8.14}$$

$$\frac{dy_2}{dt} = f_2(t, y_1, y_2) \tag{8.15}$$

The general form of these equations is as follows:

$$k_{1,1} = f_1(t_i, y_{i,1}, y_{i,2}) \tag{8.16}$$

$$k_{1,2} = f_2(t_i, y_{i,1}, y_{i,2}) \tag{8.17}$$

$$k_{2,1} = f_1\left(t_i + \frac{h}{2},\ y_{i,1} + \frac{h}{2}k_{1,1}, y_{i,2} + \frac{h}{2}k_{1,2}\right) \tag{8.18}$$

$$k_{2,2} = f_2\left(t_i + \frac{h}{2},\ y_{i,1} + \frac{h}{2}k_{1,1}, y_{i,2} + \frac{h}{2}k_{1,2}\right) \tag{8.19}$$

$$k_{3,1} = f_1\left(t_i + \frac{h}{2},\ y_{i,1} + \frac{h}{2}k_{2,1}, y_{i,2} + \frac{h}{2}k_{2,2}\right) \tag{8.20}$$

$$k_{3,2} = f_2\left(t_i + \frac{h}{2},\ y_{i,1} + \frac{h}{2}k_{2,1}, y_{i,2} + \frac{h}{2}k_{2,2}\right) \tag{8.21}$$

$$k_{4,1} = f_1\left(t_i + h,\ y_{i,1} + hk_{3,1},\ y_{3,2} + h\ k_{3,2}\right) \tag{8.22}$$

$$k_{4,2} = f_2\left(t_i + h,\ y_{i,1} + hk_{3,1},\ y_{3,2} + h\ k_{3,2}\right) \tag{8.23}$$

where $h > 0$ is a step size parameter, $I = 1, 2, 3, \ldots$, the RK4 method is given by:

$$y_{i,1} = y_{i-1,1} + \frac{h}{6}\{k_{1,1} + 2k_{2,1} + 2k_{3,1} + k_{4,1}\} \tag{8.24}$$

$$y_{i,2} = y_{i-1,2} + \frac{h}{6}\{k_{1,2} + 2k_{2,2} + 2k_{3,2} + k_{4,2}\} \tag{8.25}$$

The Runge-Kutta method provides greater precision than multiplying each function in ODE by the step size factor and adding the results to the current values in y (such as the Euler method).

Example 8.3 Applying RK4 to Solve Two ODEs Simultaneously

Solve the following two ODEs simultaneously using RK4 and a step size of 0.1.

$$\frac{dy_1}{dt} = -y_1 e^{1-t} + 0.8y_2, \quad y_1(0) = 0$$

$$\frac{dy_2}{dt} = y_1 - y_2^3, \quad y_2(0) = 2$$

Calculate the two dependent variables (y_1, y_2) manually as a function of time (t) within the interval [0, 3]. Verify the manual calculations with Simulink (the graphical programming of MATLAB) and Python programming.

Solution

Following the RK4 formulas (16–25),
Calculate $k_{1,1}$ and $k_{1,2}$

$$k_{1,1} = f_1\left(t_i, y_{i,1}, y_{i,2}\right) = f_1\left(0,0,2\right) = -0\mathrm{e}^{1-0} + 0.8 * 2 = 1.6$$

$$k_{1,2} = f_2\left(t_i, y_{i,1}, y_{i,2}\right) = f_2\left(t_0, y_{o,1}, y_{o,2}\right) = f_2\left(0,0,2\right) = 0 - 2^3 = -8$$

Calculate $k_{2,1}$ and $k_{2,2}$

$$k_{2,1} = f_1\left(t_o + \frac{h}{2}, y_{o,1} + \frac{h}{2}k_{1,1}, y_{o,2} + \frac{h}{2}k_{1,2}\right)$$

$$k_{2,1} = f_1\left(0 + \frac{0.1}{2}, 0 + \frac{0.1}{2}(1.6), 2 + \frac{0.1}{2}(-8)\right)$$

$$k_{2,1} = f_1\left(0.05, 0.08, 1.6\right) = -0.08e^{(1-0.05)} + 0.8(1.6) = 1.07314$$

$$k_{2,2} = f_2\left(t_o + \frac{h}{2}, y_{o,1} + \frac{h}{2}k_{1,1}, y_{o,2} + \frac{h}{2}k_{1,2}\right)$$

$$k_{2,2} = f_2\left(0.05, 0.08, 1.6\right) = y_1 - y_2^3 = 0.08 - 1.6^3 = -4.016$$

Calculate $k_{3,1}$ and $k_{3,2}$

$$k_{3,1} = f_1\left(t_o + \frac{h}{2}, y_{o,1} + \frac{h}{2}k_{2,1}, y_{o,2} + \frac{h}{2}k_{2,2}\right)$$

$$k_{3,1} = f_1\left(0 + \frac{0.1}{2}, 0 + \frac{0.1}{2}(1.07314), 2 + \frac{0.1}{2}(-4.016)\right)$$

$$k_{3,1} = f_1\left(0.05, 0.053657, 1.799\right) = -y_1e^{1-t} + 0.8y_2$$

$$k_{3,1} = -0.0536576 * e^{(1-0.05)} + 0.8(1.799) = 1.3004$$

$$k_{3,2} = f_2\left(t_o + \frac{h}{2}, y_{o,1} + \frac{h}{2}k_{2,1}, y_{o,2} + \frac{h}{2}k_{2,2}\right)$$

$$k_{3,2} = f_2\left(0 + \frac{0.1}{2}, 0 + \frac{0.1}{2}(1.07314), 2 + \frac{0.1}{2}(-4.016)\right)$$

$$k_{3,2} = f_2\left(0.05, 0.053657, 1.799\right) = y_1 - y_2^3$$

$$k_{3,2} = 0.053657 - 1.799^3 = -5.77$$

Calculate $k_{4,1}$ and $k_{4,2}$

$$k_{4,1} = f_1\left(t_i + h,\ y_{i,1} + hk_{3,1},\ y_{3,2} + h\ k_{3,2}\right)$$

$$k_{4,1} = f_1\left(0 + 0.1,\ 0 + 0.1(1.3),\ 2 + 0.1(-5.77)\right)$$

$$k_{4,1} = f_1(0.1,\ 0.13,\ 1.423) = -0.13e^{1-0.1} + 0.8(1.423) = 0.81865$$

$$k_{4,2} = f_2\left(t_i + h,\ y_{i,1} + hk_{3,1},\ y_{3,2} + h\ k_{3,2}\right)$$

$$k_{4,2} = f_2(0.1,\ 0.13,\ 1.423) = 0.13 - 1.423^3 = -2.751$$

calculate $y_{1,1}$

$$y_{1,1} = y_{1,0} + \frac{h}{6}\left\{k_{1,1} + 2k_{2,1} + 2k_{3,1} + k_{4,1}\right\}$$

$$y_{1,1} = 0 + \frac{0.1}{6}\left\{1.6 + 2(1.073) + 2(1.3) + 0.8186\right\} = 0.1194$$

Calculate $y_{1,2}$

$$y_{2,1} = y_{2,0} + \frac{h}{6}\left\{k_{1,2} + 2k_{2,2} + 2k_{3,2} + k_{4,2}\right\}$$

$$y_{2,1} = 2 + \frac{0.1}{6}\left\{-8 + 2(-4.016) + 2(-5.77) - 2.751\right\} = 1.4946$$

Table 8.3 summarizes the manual calculations of the first step (i.e., $x_1 = 0.1$). The remaining values are calculated using Excel, listed in Table 8.4. Figure 8.11 is a plot of the data obtained from the excel calculations.

Simulink Solution

Before performing the Simulink block diagram solution of the equations described in Example 8.3, we must change the configuration parameters to RK4 (through the modeling/model setting). The changes include the start time, the stop time, the type, the solver selection, and the fixed step size (Figure 8.12). The Simulink graphical programming (utilizing the RK4 method) of the equations defined in Example 8.3 is illustrated in Figure 8.13. The figure shows the values of the dependent

TABLE 8.3
Manual Calculation Using RK4 and Step Size, h = 0.1

		k_1		k_2		k_3		k_4			
i	x	$k_{1,1}$	$k_{1,2}$	$k_{2,1}$	$k_{2,2}$	$k_{3,1}$	$k_{3,2}$	$k_{4,1}$	$k_{4,2}$	y_1	y_2
0	0									0	2
1	0.1	1.6	−8	1.073	−4.016	1.3	−5.77	0.81865	−2.751	0.119	1.4946

TABLE 8.4
The Following Is the Excel Solution

i	t	k_{11}	$k_{1,1}$	$k_{1,2}$	$k_{2,2}$	$k_{3,1}$	$k_{3,2}$	$k_{4,1}$	$k_{4,2}$	y_1	y_2
0	0									0	2
1	0.1	1.600	−8.000	1.073	−4.016	1.301	−5.771	0.818	−2.751	0.119	1.495
2	0.2	0.902	−3.219	0.682	−2.207	0.748	−2.499	0.563	−1.734	0.192	1.255
3	0.3	0.578	−1.786	0.466	−1.364	0.495	−1.457	0.402	−1.125	0.240	1.113
4	0.4	0.407	−1.137	0.346	−0.916	0.361	−0.957	0.311	−0.776	0.275	1.018
5	0.5	0.313	−0.780	0.279	−0.648	0.287	−0.669	0.260	−0.557	0.304	0.952
29	2.9	0.595	0.195	0.607	0.196	0.607	0.197	0.619	0.198	1.186	0.996
30	3	0.619	0.198	0.632	0.199	0.632	0.200	0.644	0.201	1.249	1.016

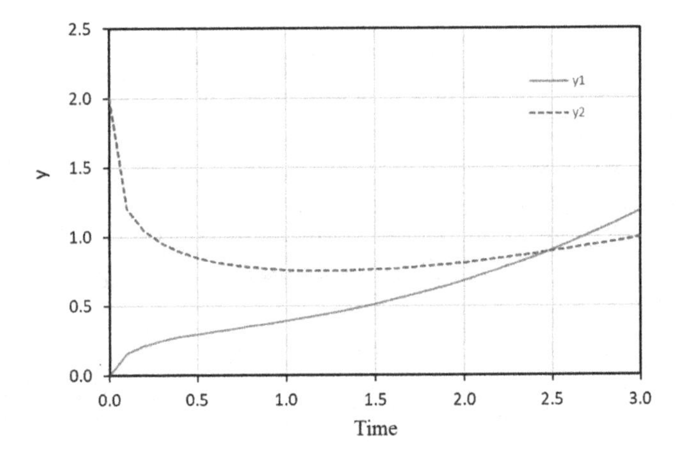

FIGURE 8.11 Plot of the manual calculations utilizing the RK4 method for solving the two ODEs defined in Example 8.3.

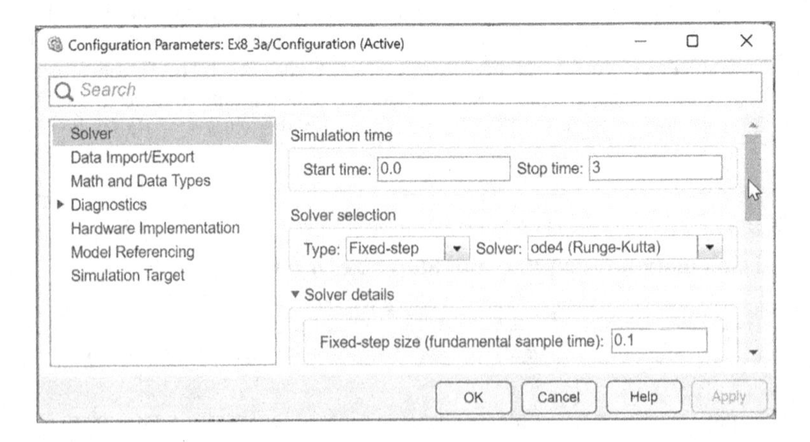

FIGURE 8.12 Simulink solver setting to RK4 method for solving the equations defined in Example 8.3.

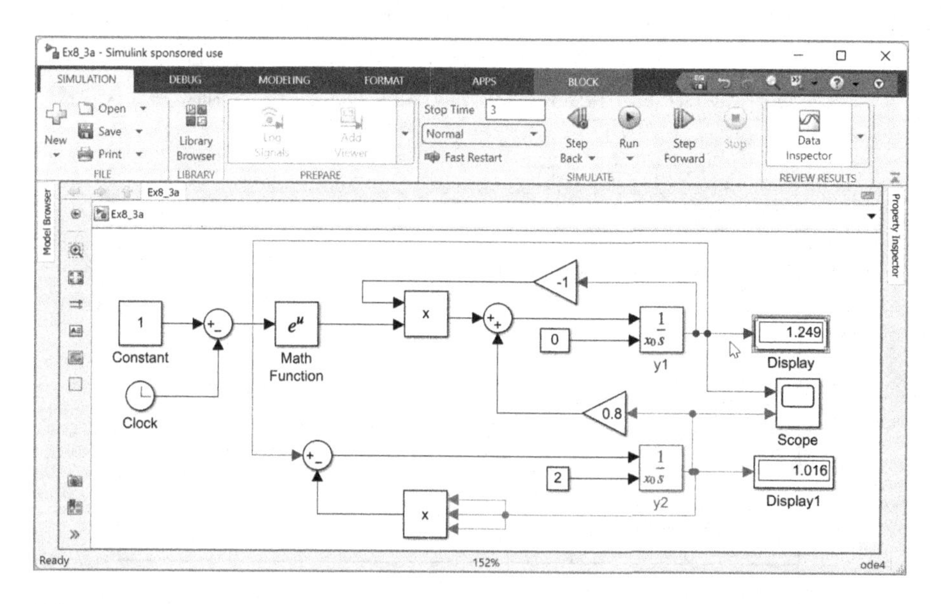

FIGURE 8.13 Simulink solution using the RK4 method of the two ODEs specified in Example 8.3.

variables (y_1, y_2) at the simulation stop time (i.e., 3). Figure 8.14 shows the Simulink graphical programming solution utilizing the RK4 method of the twin ODEs specified in Example 8.3.

Figure 8.15 shows the graphical programming in Simulink using the MATLAB function from the Simulink library as an alternative way. Following the figure is the MATLAB code implanted in the MATLAB function. The following MATLAB program is linked to the Simulink MATLAB function (Figure 8.15). Figure 8.16 shows the plot of the two ODEs described in Example 8.3. The Simulink MATLAB function is fed with five input constants representing the lower limit interval (a), upper limit interval (b), step size (h), and the initial values of y_1 (y_o) and the initial value of y_2 (z_o). The two output displays the values of the dependent variables (y_1 and y_2) at the simulation stop time (3).

```
% Example 8.3 using RK4
function [y1f, y2f] = Rk(a, b, h, y0,z0)

% Fourth-order Runge-Kutta integration routine
% Routine starts here
f1=@(x, y, z)(-y*exp(1-x)+ 0.8*z); % Enter RHS of the first ODE
f2=@(x, y, z) (y-z^3); % Enter RHS of the first ODE

x=a:h:b;
y=zeros(1,length(x));
z=zeros(1,length(x));
y(1)=y0;
z(1)=z0;
x(1)=a;
for i=1:(length(x)-1)

    m1 = f2(x(i), y(i), z(i));
    k1 = f1(x(i), y(i), z(i));

    m2 = f2(x(i)+0.5*h, y(i)+0.5*h*k1,z(i)+0.5*h*m1 );
    k2 = f1(x(i)+0.5*h, y(i)+0.5*h*k1,z(i)+0.5*h*m1 );
```

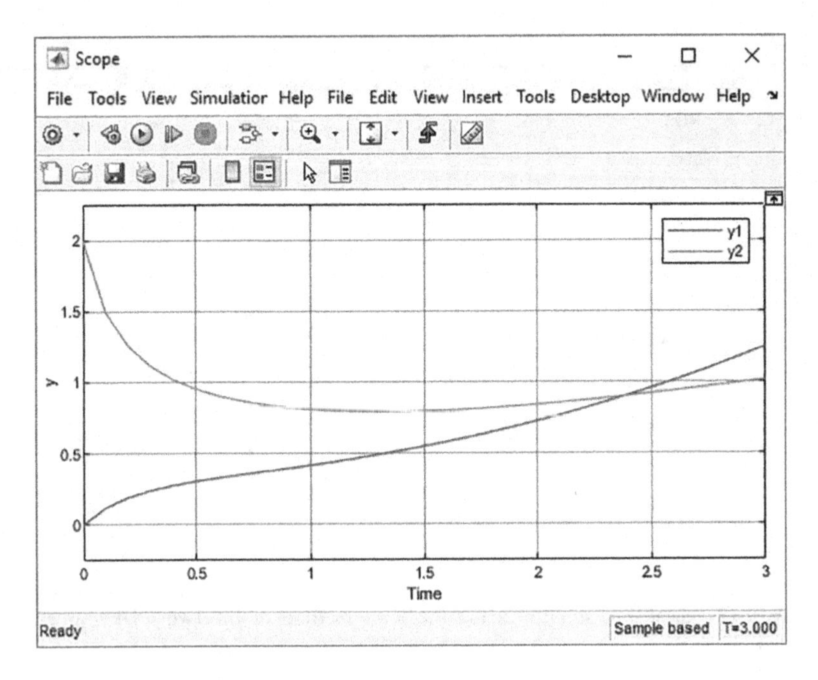

FIGURE 8.14 Simulink solution utilizing RK4 to solve the two ODEs defined in Example 8.3.

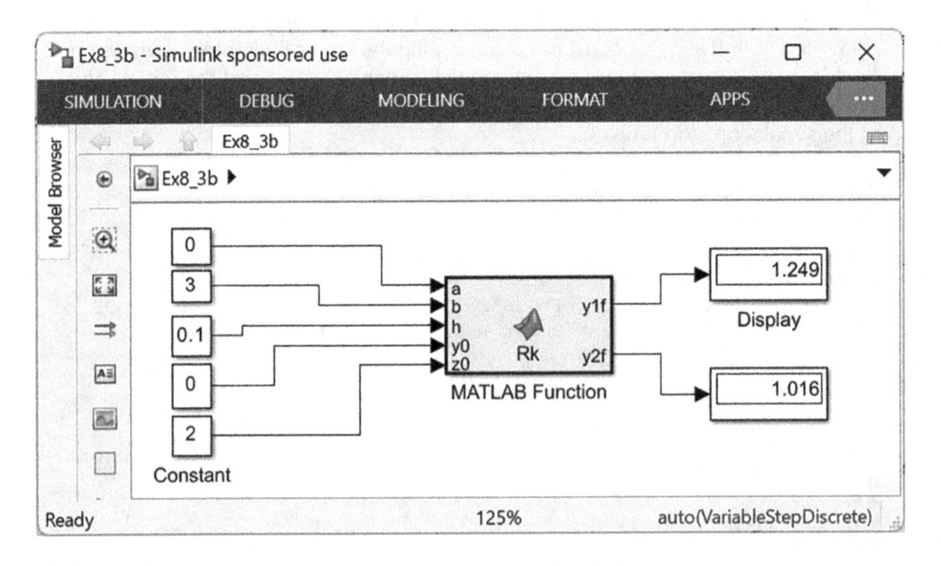

FIGURE 8.15 Simulink simultaneous solution using RK4 of the two ODEs described in Example 8.3. Stop time is equal to three.

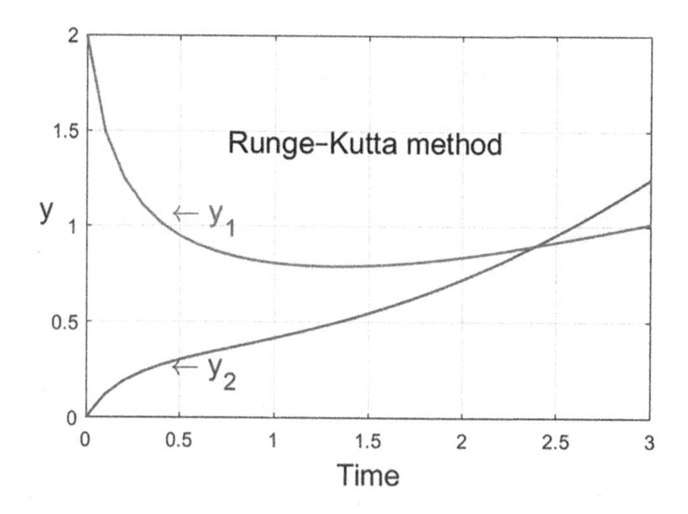

FIGURE 8.16 Simulink plot generated utilizing RK4 of the twin equations defined in Example 8.3.

```
    m3 = f2((x(i)+0.5*h), (y(i)+0.5*h*k2), z(i)+0.5*h*m2);
    k3 = f1((x(i)+0.5*h), (y(i)+0.5*h*k2), z(i)+0.5*h*m2);

    m4 = f2((x(i)+h), (y(i)+k3*h), z(i)+ h*m3);
    k4 = f1((x(i)+h), (y(i)+k3*h), z(i)+ h*m3);

y(i+1) = y0 + (1/6)*(k1+2*k2+2*k3+k4)*h;
y0=y(i+1);
z(i+1) = z0 + (1/6)*(m1+2*m2+2*m3+m4)*h;
z0=z(i+1);
end
y1f=y0;
y2f=z0;

% Plot section
plot(x, y, x, z, 'linewidth', 1.3);
text(0.45,1.05, '\leftarrow y_1', 'color', 'r', FontSize=15)
text(0.45,0.25, '\leftarrow y_2', 'color', 'b', FontSize=15)
text(0.75,1.45, 'Runge-Kutta method', FontSize=15)
xlabel('x', FontSize=15);
ylabel('y_1, y_2', FontSize=15);
grid on
```

Python Solution

The 'SciPy' method is used for solving the required set of equations defined in Example 8.3, and it now has '*solve_ivp* '. This solves the system on the interval (0,3) with an initial value (0,2). The result has an independent value (t in the notation) as 'res.t'. These values were chosen automatically. One can provide 't_{eval}' to have the solution evaluated at desired points: for example,

$$t_{eval} = \text{np.linspace}(0,0.1)$$

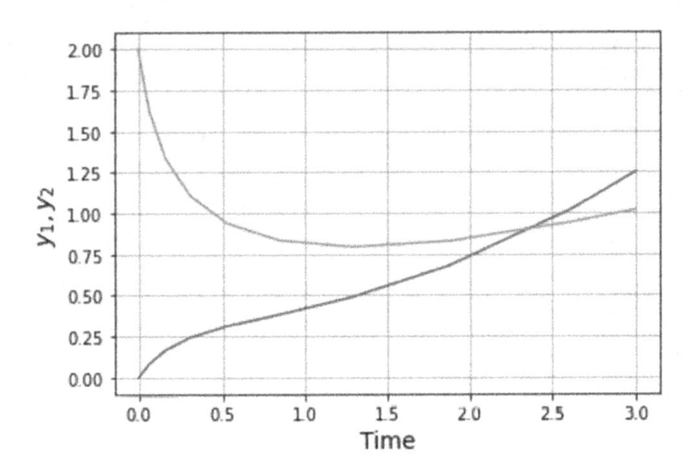

FIGURE 8.17 Plot of dependent variables (y_1 and y_2) against the time generated by the Python program to solve Example 8.3.

While the dependent variable, the function we are solving, is in res.y

Solve_ivp function built-in Python library (SciPy. Integrate) is employed to solve the two ODEs simultaneously. The Python code simultaneously solved the two ODEs described in Example 8.3, using the solve_ivp function. Running the Pyhone code generates Figure 8.17.

```
#Example 8.3
import numpy as np
import matplotlib.pyplot as plt
from scipy.integrate import solve_ivp
# enter the model equations
def rhs(t, y):
    return [-y[0]*np.exp(1-t)+0.8*y[1], y[0]-y[1]**3]
# enter the initial conditions as shown below
res = solve_ivp(rhs, (0, 3), [0,2])
print (res.t)
t_eval=np.linspace(0, 3)
print (res.y)
plt.plot(res.t, res.y.T)
plt.ylabel(r'$y_{1}, y_{2}$', fontsize=14)
plt.grid(True)
plt.xlabel('Time', fontsize=14)
```

The following Python code is programmed using RK4 syntax to solve these two types of ODEs defined in Example 8.3. The results are in good agreement with those obtained from the solver of the IVP ($solve_{ivp}$). Execution of the Python program results in the plot disclosed in Figure 8.18. The Python code presented here is for the RK4 method in two dimensions.

```
# Example 8.3
# Runge Kutta method
import matplotlib.pyplot as plt
import numpy as np
from numpy import arange
```

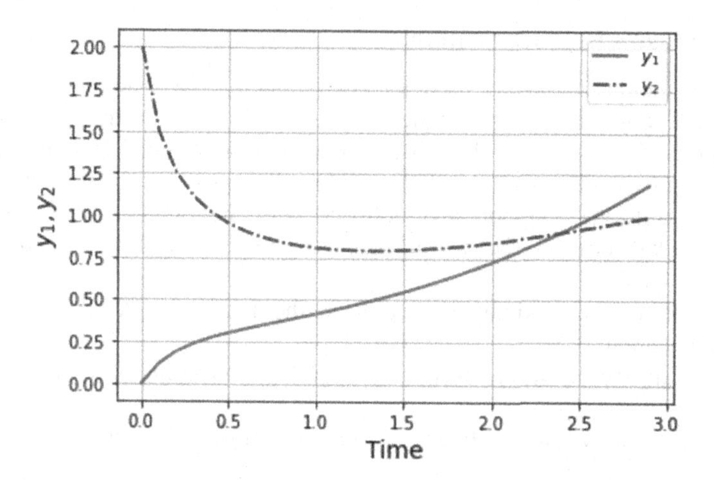

FIGURE 8.18 Python solution using RK4 of the two ODEs given in Example 8.3.

```
# y' = -y*np.exp(1-x)+0.8*u
# u' = y-u**3
def F1(y, u, x):
    return -y*np.exp(1-x)+0.8*u
def F(y, u, x):
    return y-u**3
a = 0
b = 3
N =30
h = (b-a)/N
xp = arange(a, b, h)
yp = []
up = []
# initial conditons
y = 0.0
u = 2

for x in xp:
    yp.append(y)
    up.append(u)
    m1 = h*F1(y, u, x)
    k1 = h*F(y, u, x) #(x, v, t)
    m2 = h*F1(y+0.5*m1, u+0.5*k1, x+0.5*h)
    k2 = h*F(y+0.5*m1, u+0.5*k1, x+0.5*h)
    m3 = h*F1(y+0.5*m2, u+0.5*k2, x+0.5*h)
    k3 = h*F(y+0.5*m2, u+0.5*k2, x+0.5*h)
    m4 = h*F1(y+m3, u+k3, x+h)
    k4 = h*F(y+m3, u+k3, x+h)
    y += (m1 + 2*m2 + 2*m3 + m4)/6
    u += (k1 + 2*k2 + 2*k3 + k4)/6
# plot section
plt.plot(xp, yp, 'r-', label='$y_{1}$')
plt.plot(xp, up, 'b-.', label='$y_{2}$' )
plt.xlabel('Time', fontsize=14)
plt.ylabel(r'$y_{1}, y_{2}$', fontsize=14)
plt.legend()
plt.grid(True
```

8.2.4 Picard's Iterative Method

Picard's method is iteratively used for approximating the solution to differential equations. The method gives a sequence of approximations to the solution of differential equations such that the nth approximation is obtained from one or more previous approximations $(y_1(x), y_2(x),....y_n(x))$. The more iterations used, the more numerical results become more and more accurate. The following steps are used in solving simultaneous differential equations using Picard's iterative method:

1. Substitute an approximate value of y into the right-hand side of the differential equation.
2. Integrate the equation concerning x, giving y in terms of x as a second approximation.
3. Iterate until two consecutive numerical solutions are the same.

Let

$$\frac{dy}{dx} = f(x,y,z), \text{ with } y(x_o) = y_o \tag{8.26}$$

and

$$\frac{dz}{dx} = g(x,y,z), \text{ with } z(x_o) = z_o \tag{8.27}$$

where z and y are dependent variables on x (independent). Using the Picard method, the value of y_n and z_n is given as

$$y_n = y_o + \int_{x_o}^{x} f(x, y_{n-1}, z_{n-1})\,dx \tag{8.28}$$

and

$$z_n = z_o + \int_{x_o}^{x} g(x, y_{n-1}, z_{n-1})\,dx \tag{8.29}$$

$n = 1$, The first approximation

$$y_1 = y_o + \int_{x_o}^{x} f(x, y_o, z_o)\,dx \tag{8.30}$$

and

$$z_1 = z_o + \int_{x_o}^{x} g(x, y_o, z_o)\,dx \tag{8.31}$$

$n = 2$, The second approximation

$$y_2 = y_o + \int_{x_o}^{x} f(x, y_1, z_1)\,dx \tag{8.32}$$

and

$$z_2 = z_o + \int_{x_o}^{x} g(x, y_1, z_1)\,dx \tag{8.33}$$

$n = 3$, The second approximation

$$y_3 = y_o + \int_{x_o}^{x} f(x, y_2, z_2)\,dx \tag{8.34}$$

and

$$z_3 = z_o + \int_{x_o}^{x} g(x, y_2, z_2)\,dx \tag{8.35}$$

Example 8.4 Application of Picard's Method

Using the Picard method, find an approximate value of y and z at $x = 0.1$ for the following two ODEs

$$\frac{dy}{dx} = x + z, \quad y(0) = 2$$

$$\frac{dz}{dx} = x - y^2, \quad z(0) = 1$$

Solve manually and using Python and Simulink

Solution

Applying Picard's method for solving the following two ODEs together is as follows:

At $x = 0$

$$\frac{dy}{dx} = x + z = f(x, y, z)$$

$$\frac{dz}{dx} = x - y^2 = g(x, y, z)$$

Employing the Picard method,

$$y_n = y_o + \int_{x_o}^{x} f(x, y_{n-1}, z_{n-1})\,dx$$

$$z_n = z_o + \int_{x_o}^{x} g(x, y_{n-1}, z_{n-1})\,dx$$

$n = 1$, the first approximation of y and z

$$y_1 = y_o + \int_{x_o}^{x} f(x, x_o, z_o)\,dx = 2 + \int_{x_o}^{x} f(x + z_o)\,dx$$

$$y_1(x) = 2 + \int_{x_o}^{x} (x + 1)\,dx = 2 + x + \frac{x^2}{2}$$

Substitute $x = 0.1$

$$y_1(0.1) = 2 + \int_{0}^{0.1} (x + 1)\,dx = 2 + x + \frac{x^2}{2} = 2 + 0.1 + \frac{0.1^2}{2} = 2.105$$

Calculate z at $x = 0.1$

$$z_1 = z_o + \int_{x_o}^{x} g(x, y_o, z_o)\,dx = 1 + \int_{x_o}^{x} (x - y_o^2)\,dx$$

$$z_1(x) = 1 + \int_{0}^{x} (x - 2^2)\,dx = 1 + \frac{x^2}{2} - 4x$$

Substitute $x = 0.1$

$$z_1(0.1) = 1 + \int_0^{0.1} \left(x - 2^2\right) dx = 1 + \frac{x^2}{2} - 4x = 1 + \frac{0.1^2}{2} - 4(0.1) = 0.605$$

($n = 2$) The second approximation of y and z

$$y_2 = y_o + \int_{x_o}^{x} f(x, y, z)\, dx$$

$$y_2 = 2 + \int_0^x (x + z_1)\, dx = 2 + \int_0^x \left(x + 1 - 4x + \frac{x^2}{2}\right) dx$$

$$y_2(x) = 2 + x - 3\frac{x^2}{2} + \frac{x^3}{6}$$

Substitute $x = 0.1$

$$y_2(0.1) = 2 + 0.1 - 3\frac{0.1^2}{2} + \frac{0.1^3}{6} = 2.085$$

Now we will calculate z_2

$$z_2 = z_o + \int_{x_o}^{x} g(x, y_1, z_1)\, dx$$

$$z_2 = z_o + \int_{x_o}^{x} g(x, y_1, z_1)\, dx$$

$$= 1 + \int_0^x \left(x - y_1^2\right) dx$$

$$= 1 + \int_0^x \left(x - \left(2 + x + \frac{x^2}{2}\right)^2\right) dx$$

$$= 1 + \int_0^x \left(x - 4 - x^2 - \frac{x^4}{4} - x^3 - 2x^2\right) dx$$

$$= 1 + \frac{x^2}{2} - 4x - \frac{x^3}{3} - \frac{x^5}{20} - \frac{x^4}{4} - \frac{2x^3}{3}$$

Substitute $x = 0.1$

$$z_2(0.1) = 1 + \frac{0.1^2}{2} - 4(0.1) - \frac{0.1^3}{3} - \frac{0.1^5}{20} - \frac{0.1^4}{4} - \frac{2(0.1)^3}{3} = 0.603975$$

3rd approximation ($n = 3$)

$$y_3 = y_o + \int_{x_o}^{x} f(x, y_2, z_2)\,dx$$

$$y_3 = 2 + \int_{x_o}^{x} (x + z_2)\,dx$$

$$y_3 = 2 + \int_{0}^{x} \left[x + 1 - 4x - \frac{3x^2}{2} - x^3 - \frac{x^4}{4} - \frac{x^5}{20} \right] dx$$

Rearrange and integrate

$$y_3 = 2 + x - \frac{3x^2}{2} - \frac{x^3}{2} - \frac{x^4}{4} - \frac{x^5}{20} - \frac{x^6}{120}$$

Substitute $x = 0.1$

$$y_3(0.1) = 2 + 0.1 - \frac{3(0.1)^2}{2} - \frac{(0.1)^3}{2} - \frac{(0.1)^4}{4} - \frac{(0.1)^5}{20} - \frac{(0.1)^6}{120}$$

$$y_3(0.1) = 2.085$$

For z

$$z_3 = z_o + \int_{x_o}^{x} g(x, y_2, z_2)\,dx$$

$$z_3 = 1 + \int_{0}^{x} \left(x - y_2^2\right) dx$$

$$z_3 = 1 + \int_{0}^{x} \left[x - \left(2 + x - \frac{3x^2}{2} + \frac{x^3}{6} \right)^2 \right] dx$$

Integrate

$$z_3 = 1 - 4x - \frac{3x^2}{2} + \frac{5x^3}{2} + \frac{7x^4}{12} - \frac{31x^5}{60} + \frac{x^6}{12} - \frac{x^7}{252}$$

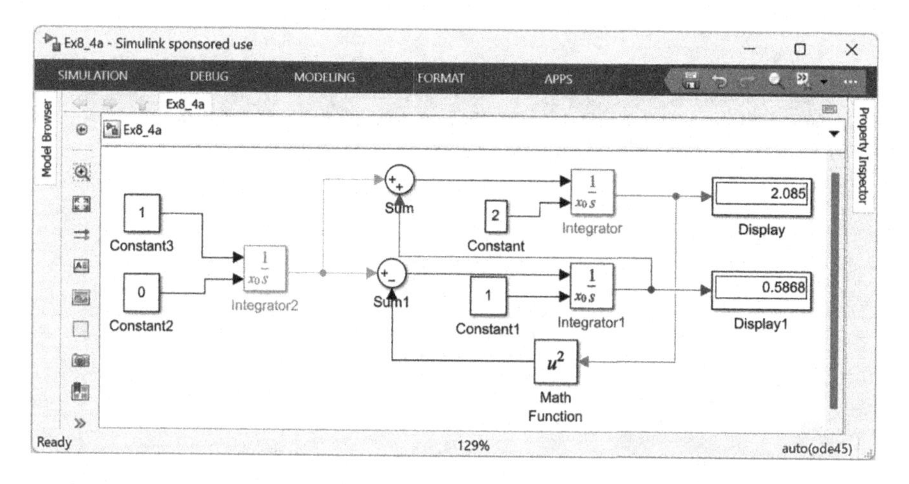

FIGURE 8.19 Simulink solution using the Simulink integrator with default solver of the equation defined in Example 8.4, the stop time is 0.1.

Finally,

$$z_3(0.1) = 0.587$$

Simulink Solution

The Simulink solution of Example 8.4 is shown in Figure 8.19.

An alternative solution is the Simulink MATLAB function embedded with the MATLAB code. The MATLAB code describes the solution of the two ODEs defined in Example 8.4. The input port to the MATLAB function, the constant (*a*), represents the value at which the dependent variables are required to find (Figure 8.20).

```
% Example 8.4 using the Picards method
function [y, z]= fcn(a)
%x=a;res
y1=@(x) 2+x+x^2/2;
z1=@(x) (1 + x^2/2 - 4*x) ;
y2=@(x) 2 + x - 3*x^2/2 + x^3/6;
z2=@(x) 1+x^2/2 -4*x-x^3/3-x^5/20 - x^4/4 - 2*x^3/3;
y3=@(x) 2+x-3*x^2/2-x^3/2-x^4/4 -x^5/20-x^6/120;
z3=@(x) 1-4*x - 3*x^2/2 +5*x^3/2+7*x^4/12-31*x^5/60+x^6/12-x^7/252;


y =y3(a);
z =z3(a);
```

Python Solution

The following is the Python code utilized by the Picard method to solve the two ODEs defined in Example 8.3.

```
#Picard's Method
#
```

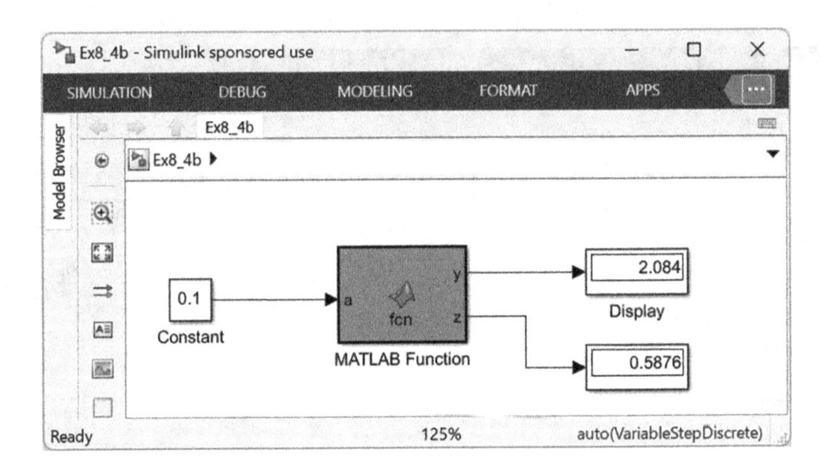

FIGURE 8.20 Simulink solution using Picard's method of the equations defined in Example 8.4 and a stop time of three.

```
def y1(x):
Y1 = 2+x+x**2/2
   return Y1
def z1(x):
   z1 = 1 + x**2/2 - 4*x
   return z1
def y2(x):
   Y2 = 2 + x - 3*x**2/2 + x**3/6
   return Y2
def z2(x):
   z2 = 1+x**2/2 -4*x-x**3/3-x**5/20 - x**4/4 - 2*x**3/3
   return z2
def y3(x):
   Y3 = 2 + x - 3*x**2/2-x**3/2-x**4/4 -x**5/20-x**6/120
   return Y3
def z3(x):
   z3 = 1-4*x - 3*x**2/2 +5*x**3/2+7*x**4/12-31*x**5/60+x**6/12-x**7/252
   return z3
#
print (' ')
print (' === Solution ===')
print ('for dy(x) = x + y^2 , dz(x)=x-y^2, the results are :')
print ('-------------------------------------')
print ('y1(0.1) = ', "%.3f"%y1(0.1))
print ('y2(0.1) = ', "%.3f"%y2(0.1))
print ('y3(0.1) = ', "%.3f"%y3(0.1))
print ('z1(0.1) = ', "%.3f"%z1(0.1))
print ('z2(0.1) = ', "%.3f"%z2(0.1))
print ('z3(0.1) = ', "%.3f"%z3(0.1))
```

The Python execution results are displayed as follows:

```
=== Solution ===
for dy(x) = x + y^2 , dz(x)=x-y^2, the results are :
--------------------------------------
y1(0.1) = 2.105
y2(0.1) = 2.085
```

```
y3(0.1) = 2.084
z1(0.1) = 0.605
z2(0.1) = 0.604
z3(0.1) = 0.588
```

8.3 SUMMARY

Numerical methods of ODEs are used to find a numerical approximation of solutions simultaneously to two ODEs. The chapter goes through the numerical integration methods, including Euler, Midpoint, the RK4, and the Picard as appropriate methods for solving two ODEs simultaneously. The numerical integration methods were graphically programmed using Simulink and coded using Python.

8.4 PROBLEMS

1. Solve the following two first-order initial value ODEs using the Euler method. Using a step size of 0.1, find the value of x and y at $t=0.3$. verify the solution using Simulink and Python.

$$\frac{dx}{dt} = 3x + 8y, \quad x(0) = 6$$

$$\frac{dy}{dt} = -x - 3y, \quad y(0) = -2$$

Answer: 6.782, –2.060

2. Applying the Midpoint method, solve the following two first-order initial value ODEs. Using a step size of 0.1, find the value of y_1 and y_2 at $x = 0.3$. Verify the manual solution using Simulink and Python.

$$\frac{dy_1}{dx} = 3y_1 - 4y_2, \quad y_1(0) = 1$$

$$\frac{dy_2}{dx} = y_1 - y_2, \quad y_2(0) = 0$$

Answer: 2.155, 0.4029

3. Utilizing the RK4 method, solve the following two first-order initial value ODEs using a step size of 0.1 to find the value of x and y at $t=0.5$. Confirm the manual solution using Simulink and Python programming.

$$\frac{dy}{dt} = 3y + 4x, \quad y(0) = 0$$

$$\frac{dx}{dt} = -4y + 3x, \quad x(0) = 1$$

Answer: –1.87, 4.075

4. Using the RK4 method, solve the following two first-order initial value ODEs using a step size of 0.1 to find the value of x and y at $x = 0.3$. Validate the manual results using Simulink and Python programming.

$$\frac{dy_1}{dt} = y_2,\ y_1(0) = 0$$

$$\frac{dy_2}{dx} = 10\left(1 - y_1^2\right)y_2 - y_1,\ y_2(0) = 1$$

Answer: 1.308, 6.518

5. Utilizing the RK4 method, solve the following two first-order initial value ODEs representing the concentration of components A and B in a batch reactor using a step size of 0.1 to find the value of C_A and C_B at $t = 0.5$. Confirm the manual solution using Simulink and Python programming. The values of the rate constants, $k_1 = 0.1$, $k_2 = 0.2$.

$$\frac{dC_A}{dt} = -k_1 C_A,\ C_A(0) = 1$$

$$\frac{dC_B}{dt} = k_1 C_A - k_2 C_B,\quad C_B(0) = 0$$

Answer: 0.95, 0.046

6. Solve the following two ODEs by Picard's iteration method up to the second approximation. Verify the manual solution using Python programming and Simulink graphical programming.

$$\frac{dy_1}{dx} = y_2,\quad y_1(0) = 1$$

$$\frac{dy_2}{dx} = x^3\left(y_1 + y_2\right),\quad y_2(0) = 0.5$$

Answer: 1.4896, 0.8500

REFERENCES

1. Atkinson, K., Han, W., and Stewart, D., 2009, *Numerical Solution of Ordinary Differential Equations*. Newyork: John Wily.
2. Griffiths, D.F., and Higham, D., 2010, *Numerical Methods for Ordinary Differential Equations: Initial Value Problems*. London, UK: Springer.
3. Kong, Q., Siauw, T., and Bayen, A., 2021, *Python Programming and Numerical Methods: A Guide for Engineers and Scientists*. California, CA: Academic Press.

9 Boundary Value Problems of Ordinary Differential Equations

A boundary value problem (BVP) is a system of ordinary differential equations (ODEs) with a solution and derivative values defined at more than one point. This chapter uses finite differences and shooting methods to convert BVPs into algebraic equations. Furthermore, the Thomas algorithm is presented and explained in this chapter. Manual calculations are compared to Python and Simulink programming predictions, where the code is given for each case.

LEARNING OBJECTIVES

1. Identify a BVP.
2. Implement the shooting method to solve a BVP.
3. Apply the finite difference method.
4. Use the Thomas algorithm to solve a BVP.
5. Utilize Python and Simulink/MATLAB to tackle a BVP.

9.1 INTRODUCTION

A differential equation is defined as an equation that relates one or more unknown functions and their derivatives in mathematics. ODEs can only be solved given additional information, called boundary conditions (BCs). If the BCs are specified at the same independent variable value (e.g., x), the problem is an initial value problem (IVP). Some BCs are specified at different values for the independent variable, such as initial and final positions of x; these are called BVPs. The differential equation contains one or more terms and the derivatives of one dependent variable concerning another independent variable [1]. We often see them for problems where conditions are specified at various locations in space rather than time.

$$\frac{dy}{dx} = f(x) \tag{9.1}$$

Here, 'x' is an independent variable and 'y' is a dependent variable, such as

$$\frac{dy}{dx} = 5x \tag{9.2}$$

DOI: 10.1201/9781003360544-9

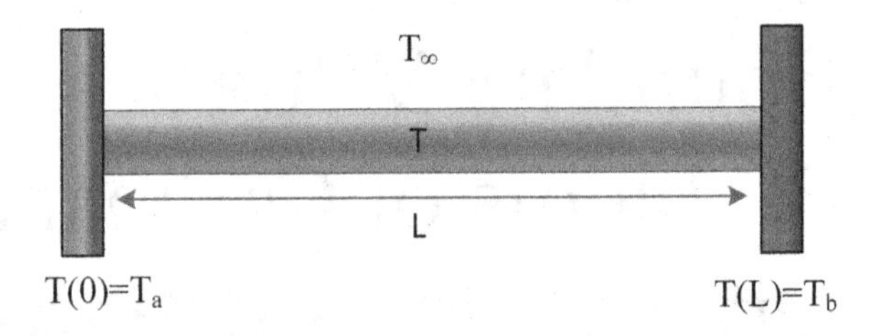

FIGURE 9.1 Schematic of a rod fixed between two plats with different temperatures.

A differential equation contains derivatives that are either partial derivatives or ordinary derivatives. BVPs commonly arise in many engineering subjects, such as heat transfer, mass transfer, and reactor design. An example of thc BVP is the steady-state temperature distribution along a rod hung between two plates of different temperatures (Figure 9.1). A BVP comprises a differential equation subject to a given set of BCs. BCs are prescribed for the unknown variable and its derivatives at more than one point [2].

An example of a second-order BVP is shown below:

$$\frac{d^2y}{dx^2} = y(x) \tag{9.3}$$

9.2 BOUNDARY CONDITIONS

BCs are constraints to solve a BVP. It is the conditions that are satisfied on the boundaries of the spatial domain of the problem. Considering the following BCs,

$$y(0) = 0, \; y(1) = 1$$

Note that the number of BCs equals the order of the ODE. In this example, the given interval is [0, 1]. Then the number of subintervals, n,

$$n = \frac{b-a}{h}$$

For $h = 0.25$

$$n = \frac{1-0}{0.25} = 4$$

The aim is to find $y(0.25)$, $y(0.5)$, $y(0.75)$, if $h = 0.25$. From the BCs, $y(0)$ and $y(1)$ are known (Figure 9.2).

Every ordinary or partial derivative occurring in the equation and the BCs is replaced by finite difference approximation. Accordingly, the equation is reduced to

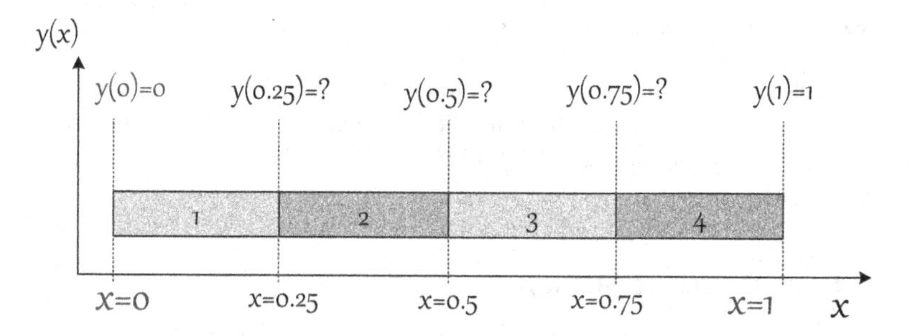

FIGURE 9.2 Schematic dividing the main interval [0, 1] into four subintervals.

a set of linear systems of equations, which are then solved using standard methods. The following sections describe initial and BCs.

9.2.1 Initial Value Problems

In mathematics, an IVP is an ODE with an initial condition that defines the value of the unknown function at a given point in the domain. Modeling a system in physics or other sciences often amounts to solving an IVP. The auxiliary conditions are at one point of the independent variable

$$\frac{d^2y}{dx^2} + 2\frac{dy}{dx} + y = e^{-2x} \tag{9.4}$$

The initial conditions are at one point of the independent variable ($x = 0$),

$$y(0) = 1, \ \frac{dy}{dx}(0) = 2.5$$

9.2.2 Boundary Value Problem

The BVP is a differential equation with additional constraints called BCs. In the BVPs, the auxiliary conditions are not at one point of the independent variable. BVP is more difficult to solve compared to the IVPs. By contrast, the initial conditions in the IVP are at one point of the independent variable. A BVP is a system of ODEs with a solution and derivative values specified at more than one point. Usually, the solution and derivatives are specified at the boundaries (two points) that define the two-point BVP. Equation (9.5) is a second-order BVP with BCs at two different locations of the independent variable, x.

$$\frac{d^2y}{dx^2} + 2\frac{dy}{dx} + y = e^{-2x} \tag{9.5}$$

With the BCs: $y(0) = 1, y(2) = 1.5$

9.3 SOLUTION OF BVPs

The following sections present two methods of solving BVP: the finite difference method and shooing methods. These methods transform BVPs into algebraic equation problems. The generated set of equations is linear when the differential equation is linear for either method.

9.3.1 Finite Difference Method

The finite difference method replaces derivatives in a differential equation with approximation, leading to an extensive linear equation system that must be solved using numerical techniques. The following formulas are used in solving BVPs:

The forward finite difference approximation formula for $y'(x)$

$$\frac{dy}{dx} = y' = \frac{y_{i+1} - y_i}{h} \tag{9.6}$$

Backward finite difference formula for $y'(x)$

$$\frac{dy}{dx} = y' = \frac{y_i - y_{i-1}}{h} = \frac{y(x) - y(x-h)}{h} = \frac{\text{Present} - \text{Previous}}{h} \tag{9.7}$$

Central finite difference, also called the central difference approximation for $y'(x)$

$$\frac{dy}{dx} = y' = \frac{y_{i+1} - y_{i-1}}{2h} = \frac{y(x+h) - y(x-h)}{2h} = \frac{\text{Next} - \text{Previous}}{2 \times h} \tag{9.8}$$

The formula for the second derivative, $y''(x)$

$$\frac{d^2y}{dx^2} = y'' = \frac{y_{i+1} - 2y_i + y_{i-1}}{h^2} = \frac{y(x+h) - 2y(x) + y(x-h)}{h^2} \tag{9.9}$$

$$\frac{d^2y}{dx^2} = \frac{\text{Next} - 2 \times \text{Present} + \text{Previous}}{h^2} \tag{9.10}$$

Figure 9.3 describes the forward finite difference methods and central and backward formulas.

Follow the steps below to solve the BVP using the finite differences method:

1. Divide the interval between the initial and boundary value into n intervals.
2. Use the finite difference approximation method to replace the derivatives.
3. Solve the generated set of algebraic equations resulting from the finite difference approximation to obtain the solution to the BVP.

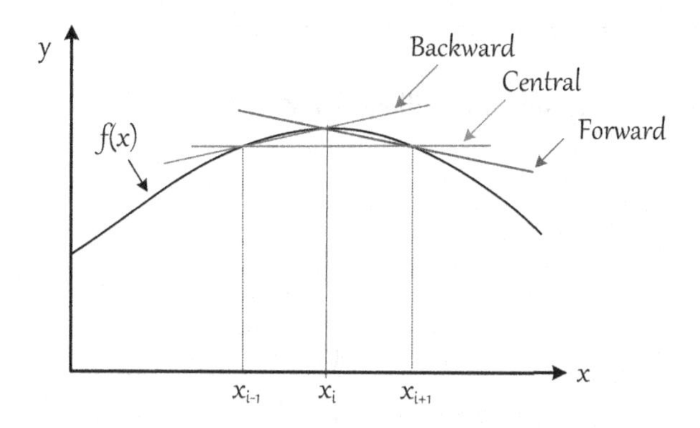

FIGURE 9.3 Representation of the finite difference method approximation.

The solution method can be summarized by substituting the finite difference approximation for the derivatives occurring in the equation and the BCs and then solving the algebraic equations by substitution or using any known solver.

9.3.2 Thomas Algorithm

The Thomas algorithm can be used to obtain the solutions for the tridiagonal matrix system. The tridiagonal matrix has three non-zero elements in all its rows, except in the first and last rows. The first and last rows have one element on the right and left of the diagonal elements. The tridiagonal systems are standard when using the finite difference second-order method to solve BVPs. Having the algebraic equation,

$$a_i \; y_{i-1} - b_i \; y_i + c_i \; y_{i+1} = r_i$$

where a_i is the coefficient of y_{i-1}, b_i is the coefficient of y_i, c_i is the coefficient of y_{i+1}, and r_i is the right-hand side of the equation. The equations describing the Thomas algorithm are listed in Table 9.1.

Example 9.1 Application of Finite Difference Method

Solve the following BVP using the finite difference approximation approach applying four subintervals ($n = 4$).

$$\frac{d^2y}{dx^2} + 2\frac{dy}{dx} + y = x^2$$

The BCs are as follows:

$$y(0) = 0.2, \; y(1) = 0.8$$

Solve the BVP manually, using Python and Simulink graphical programming language.

TABLE 9.1
Thomas Algorithm for Solvent Tridiagonal Matrix

$$\beta_1 = b_1$$

$$\gamma_1 = \frac{\gamma_1}{\beta_1}$$

For $i = 2,\ldots,n$

$$\beta_i = b_i - a_i\left(\frac{c_i - 1}{\beta_i - 1}\right)$$

$$\gamma_i = \left(\frac{\gamma_i - a_i\gamma_{i-1}}{\beta_i}\right)$$

$$y_n = \gamma_n$$

End

For $j = 1,\ldots,n-1$

$$y_{n-j} = \gamma_{n-j} - c_{n-j}\frac{y_{n-j+1}}{\beta_{n-j}}$$

End

Solution

Divide the interval [0, 1] into four subintervals ($n = 4$), and hence $h = 0.25$.
Using finite central difference formulas,

$$\frac{d^2y}{dx^2} = \frac{y_{i+1} - 2y_i + y_{i-1}}{h^2}$$

Central method for the first derivatives

$$\frac{dy}{dx} = \frac{y_{i+1} - y_{i-1}}{2h}$$

The second-order ODE becomes

$$\frac{y_{i+1} - 2y_i + y_{i-1}}{h^2} + 2\frac{y_{i+1} - y_{i-1}}{2h} + y_i = x_i^2$$

Multiply both sides by h^2

$$y_{i+1} - 2y_i + y_{i-1} + h\left(y_{i+1} - y_{i-1}\right) + h^2 y_i = h^2 x^2$$

Rearrange

$$\left(1-h\right)y_{i-1} + \left(h^2 - 2\right)y_i + \left(1+h\right)y_{i+1} = h^2 x^2$$

Divide by $(1-h)$

$$0.75\ y_{i-1} - 1.9375 y_i + 1.25\ y_{i+1} = 0.0625 x_i^2$$

Solve for $i = 1$ to 3

$$i=1: -1.9375 y_1 + 1.25\ y_2 + 0 = 0.0625 x_1^2 - 0.75 y_o$$

$$i=2:\ \ 0.75\ y_1 - 1.9375 y_2 + 1.25\ y_3 = 0.0625 x_2^2$$

$$i=3:\ \ 0 + 0.75\ y_2 - 1.9375 y_3 = 0.0625 x_3^2 - 1.25\ y_4$$

Rearrange (previous y_1, present y_2, next y_3) and consider the substitution of the known initial y_0 and boundary value y_4

$$-1.9375 y_1 + 1.25\ y_2 + 0 = 0.0625 x_1^2 - 0.75 y_0$$

$$0.75\ y_1 - 1.9375 y_2 + 1.25\ y_3 = 0.0625 x_2^2$$

$$0 + 0.75\ y_2 - 1.9375 y_3 = 0.0625 x_3^2 - 1.25\ y_4$$

From BC 1, $y(0) = y_0 = 0.2$ to BC 2, $y(3) = y_4 = 0.8$

$$\begin{bmatrix} -1.9375 & 1.25 & 0 \\ 0.75 & -1.9375 & 1.25 \\ 0 & 0.75 & -1.9375 \end{bmatrix} \begin{bmatrix} y_1 \\ y_2 \\ y_3 \end{bmatrix} = \begin{bmatrix} -0.146 \\ 0.0156 \\ -0.9648 \end{bmatrix}$$

In matrix form,

$$A.Y = B$$

The solution was obtained using the Python code shown below:

$$y_1 = 0.5166,\ y_2 = 0.684,\ y_3 = 0.7627$$

Python provides a straightforward method to calculate the inverse of a matrix. The function linalg.inv(), available in the Python NumPy module, is used to compute the inverse of a matrix. The inverse of a matrix (*A*) is multiplied by the constants (*B*). The set of linear algebraic equations is solved as shown in the following Python codes.

```
# Solving a set of linear algebraic equations
import numpy as np
# Matrix coefficient
A1=[-1.9375, 1.25,0]
A2=[0.75, -1.9375,1.25]
A3=[0, 0.75,-1.9375]

A = np.array([A1,A2 , A3])
B = np.array([-0.146, 0.0156, -0.9648])
```

```
X = np.linalg.inv(A).dot(B)
print("")
print(X)
```

The following calculated values of y_1, y_2, and y_3 are generated from the Python program after execution.

```
[0.51668057 0.68405489 0.76275673]
```

Simulink Solution

Using Math Operations/Algebraic Constraints

First, the linear algebraic equations are arranged and solved by Simulink using the 'Math Operations/Algebraic Constraints' function in the Simulink library.

$$-1.9375y_1 + 1.25y_2 + 0.146 = 0$$

$$0.75y_1 - 1.9375y_2 + 1.25y_3 - 0.0156 = 0$$

$$0.75y_2 - 1.9375y_3 + 0.9648 = 0$$

The BCs

$$y(0) = 0.2, \quad y(1) = 0.8$$

The Simulink predicted results (Figure 9.4) agree with the manual calculations used in solving the three linear algebraic equations.

Using the Simulink MATLAB function, an alternative Simulink solution uses the finite difference method by discretizing the ODE. The second-order BVP

$$\frac{d^2y}{dx^2} + 2\frac{dy}{dx} + y = x^2$$

Discretizing the second-order ODE into a tridiagonal matrix by using the finite difference formula for the second derivative,

$$\frac{d^2y}{dx^2} = \left(\frac{y(i+1) - 2y(i) + y(i-1)}{h^2}\right)$$

Using the central method for the first derivative,

$$\frac{dy}{dx} = \frac{y(i+1) - y(i-1)}{2h}$$

The discretized ODE into a tridiagonal matrix

$$\frac{y(i+1) - 2y(i) + y(i+1)}{h^2} + 2\times\frac{y(i+1) - y(i+1)}{2h} + y(i) = x(i)^2$$

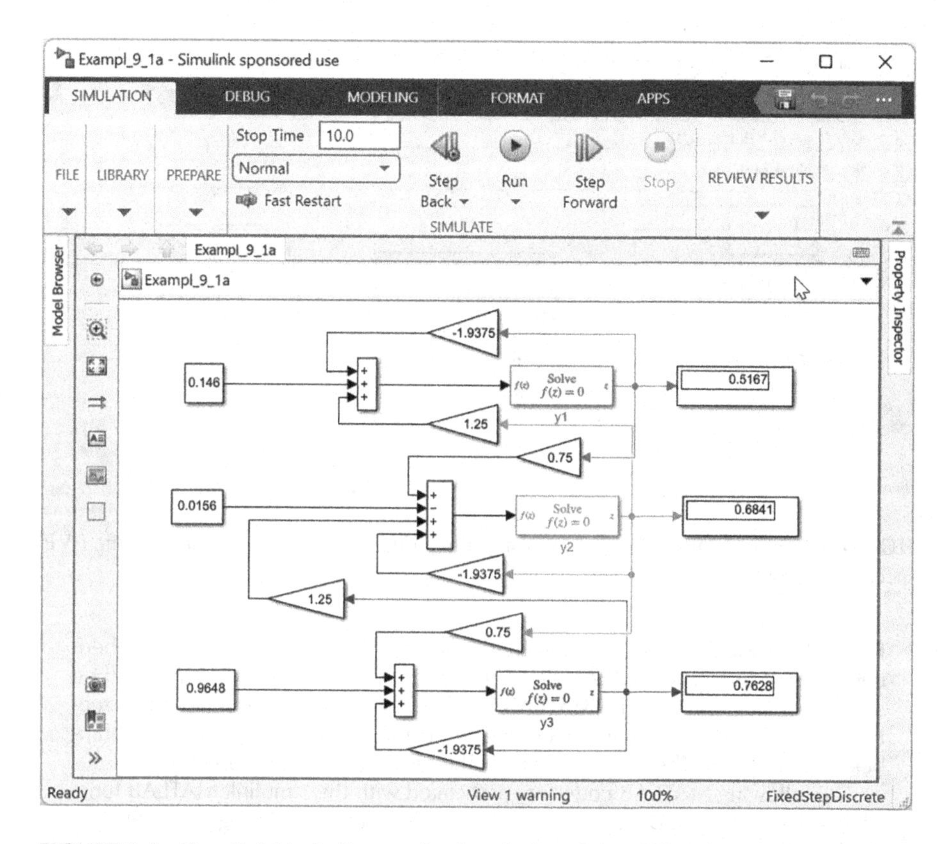

FIGURE 9.4 Simulink block diagram for the solution of the BVP defined in Example 9.1.

Arranging,

$$y(i+1)\left[\frac{1}{h^2}+\frac{1}{h}\right]+y(i)\left[-\frac{2}{h^2}+1\right]+y(i+1)\left[1-\frac{1}{h}\right]=x(i)^2$$

The supper diagonal is the coefficient of $y(i+1)$

$$\text{supd}=\frac{1}{h^2}+\frac{1}{h}$$

The main diagonal is the coefficient of $y(i)$

$$\text{maind}=\frac{2}{h^2}+1$$

The sub diagonal is the coefficient of $y(i-1)$

$$\text{supd}=1-\frac{1}{h}$$

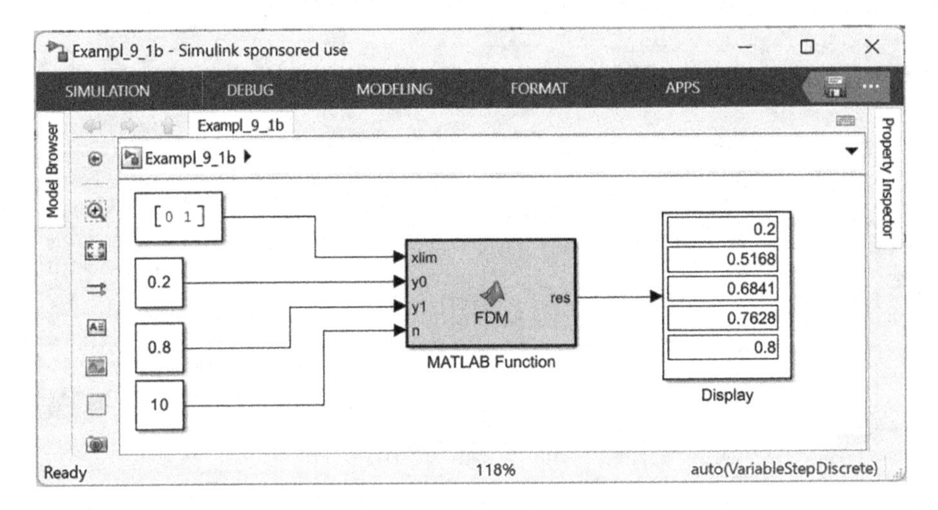

FIGURE 9.5 Simulink block diagram using finite difference method for solving the BVP specified in Example 9.1.

Figure 9.5 is the Simulink block diagram followed by the MATLAB codes embedded into the Simulink MATLAB function, illustrating the finite difference method solution of Example 9.1. The MATLAB function contains four input ports that represent the interval lower and upper limit (xlim), BC 1 (y_0), BC 2 (y_1), and the number of intervals (n). The exit port displays the results at each grid point.

The following MATLAB codes are associated with the Simulink MATLAB function and solve the ODE using an indicated discretization. Using MATLAB, the results are drawn in Figure 9.6.

```
function res = FDM(xlim, y0,y1,n)
% y"+2y'+y=x^2
% xlim: is the limit interval [a, b]
% y0, y1 are the boundary conditions at a, and b
% n is the number of elements in the final result

function [y, x] =ODEsolve(xlim, y0,y1,n)
     x = linspace(xlim(1), xlim(2), n)';
     % stride
     h = x(2) - x(1);

     % The main diagonal has coefficients
     maind = [1;repmat(-2/h^2 + 1,n - 2,1);1];
     % The sub diagonals has coefficients
     subd = [repmat(1/h^2 - 1/h, n-2,1);0];
     % The super diagonals has coefficients
     supd = [0;repmat(1/h^2 + 1/h, n-2,1)];

     % The ODE matrix
     ODEmat = diag(maind) + diag(subd, -1) + diag(supd, 1);

     % create the right hand side for the solve:
     rhs = [y0;x(2:end-1).^2;y1];
```

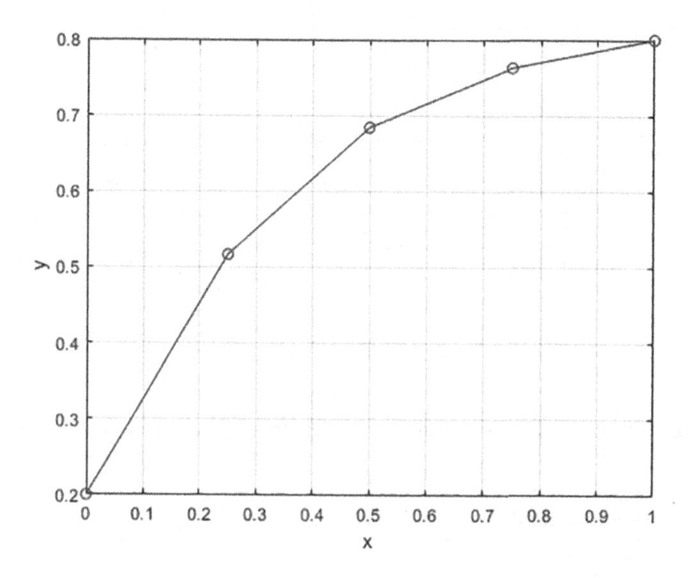

FIGURE 9.6 Solution of the BVP defined in example 9.1 using finite difference method and the Simulink MATLAB function.

```
    % solve the matrixa a call to backslash
    y =ODEmat\rhs;
    end
[y, x] = ODEsolve(xlim, y0,y1,5);
res =y;
plot(x, y, 'bo-'), xlabel("x"), ylabel("y");
grid on
end
```

Python Solution

In the finite difference method, the derivatives of the differential equation are approximated using the finite difference equations. Python solution generates Figure 9.7 while using the finite difference method discretizing the BVP of Example 9.1. The following Python program uses the finite difference method to solve the BVP equation using the finite difference approximation method. The values of y are shown below, and the entire data are plotted in Figure 9.7.

```
import numpy as np
import matplotlib.pyplot as plt
# (1-h)y_(i-1)+(h^2-2)y_i+(1+h)y_(i+1)=h^2 x^2
a=0
b=1
n = 4
h = (b-a) / n
# Matrix coefficient
A = np.zeros((n+1, n+1))
A[0, 0] = 1
A[n, n] = 1
```

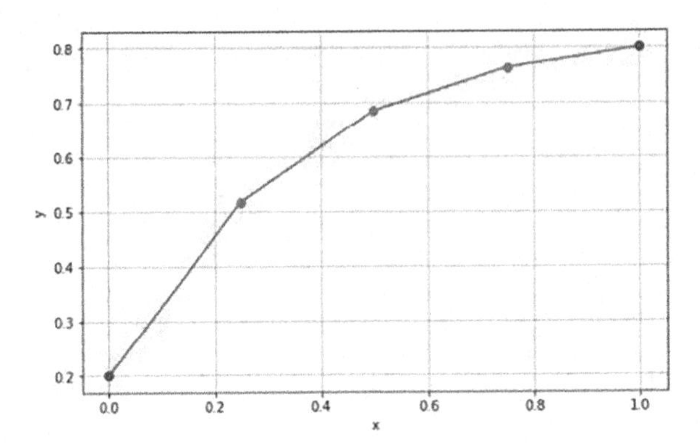

FIGURE 9.7 Python solution using the BVP finite difference method described in Example 9.1.

```
# matrix coefficient
x = np.linspace(a, b, n+1)
for i in range(1, n):
     A[i, i-1] = 1-h
     A[i, i] = h*h-2
     A[i, i+1] = 1+h
#print(A)
x = np.linspace(a, b, n+1)
b = np.zeros(n+1)
b[0]=0.2 # boundary conditon 1
for i in range(1, n):
     b[i] = h**2*x[i]**2
b[n] = 0.8 # boundary condtion 2
#print(b)
# solve the linear equations
y = np.linalg.solve(A, b)
plt.figure(figsize=(8,5))
plt.plot(x, y, 'ro-')
plt.plot(a, b[0], 'bo'), plt.plot(1, 0.8, 'bo')
plt.xlabel('x'), plt.ylabel('y')
plt.grid(True)
print (y)
```

Execution

```
[0.2 0.51675525 0.68409563 0.76279508 0.8 ]
```

Example 9.2 Application of the Finite Difference Method

Solve the following second-order BVP using the finite difference approximation method with $h=0.5$ within the interval [1, 3].

$$\frac{d^2y}{dx^2} = \left[1-\frac{x}{5}\right]y + x; y(1) = 2, y(3) = -1$$

Solve manually using Python, Simulink, and MATLAB.

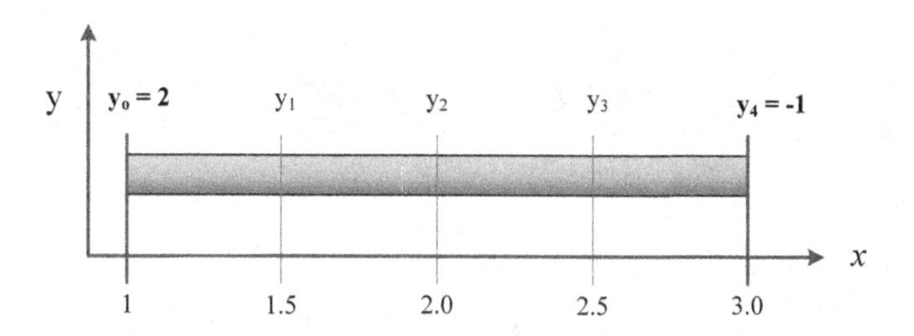

FIGURE 9.8 Describes the subintervals used to solve the BVP defined in Example 9.2.

Solution

Using a step size of 0.5, as stated in the problem statement ($h=0.5$), resulting in four subintervals, as shown in Figure 9.8.

There are five nodes with $h=0.5$: y_0, y_1, y_2, y_3, and y_4. However, there are only three unknowns since y_0, y_4 are known:

$$y_0 = y(1) = 2 \text{ and } y_4 = y(3) = -1$$

Discretizing the equation in the finite difference form, the ODE becomes

$$\frac{y_{i+1} - 2y_i + y_{i-1}}{h^2} - \left[1 - \frac{x_i}{5}\right] y_i = x_i \;\; i = 1, 2, 3, 4$$

Multiply both sides by h^2 and rearranging,

$$y_{i+1} - 2y_i - h^2\left[1 - \frac{x_i}{5}\right] y_i + y_{i+1} = h^2 x_i$$

Rearrange the equation (previous $(i-1)$, present (i), next $(i-1)$):

$$y_{i+1} - \left(2 + h^2\left[1 - \frac{x_i}{5}\right]\right) y_i + y_{i+1} = h^2 x_i$$

Substitute, $i=1, 2, 3$

$i = 1, x_1 = 1.5$

$$y_0 - \left(2 + 0.5^2\left[1 - \frac{1.5}{5}\right]\right) y_1 + y_2 = 0.5 \times 1.5$$

$i = 2, x_2 = 2.0$

$$y_1 - \left(2 + 0.5^2\left[1 - \frac{2}{5}\right]\right) y_2 + y_3 = 0.5^2 \times 2.0$$

$i = 3, x_3 = 2.5$

$$y_2 - \left(2 + 0.5^2\left[1 - \frac{2.5}{5}\right]\right) y_3 + y_4 = 0.5^2 \times 2.5$$

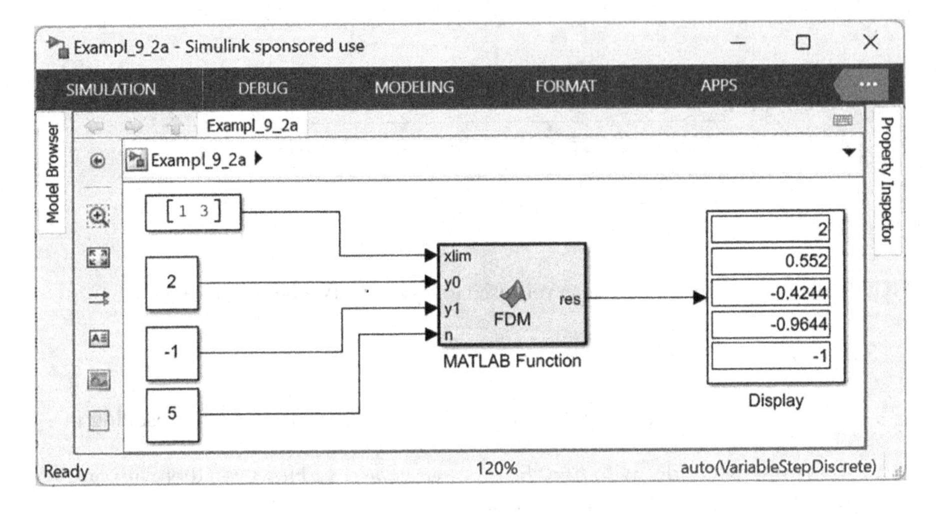

FIGURE 9.9 Simulink block diagram for the solution of BVP stated in Example 9.2.

After simplifying and substituting the BC, $y_0 = 2, y_4 = -1$, the following three linear algebraic equations arise:

$$-2.175y_1 + y_2 + 0 = -1.625$$

$$y_1 - 2.15y_2 + y_3 = 0.5$$

$$0 + y_2 - 2.125y_3 = 1.625$$

Solve the three unknowns in the linear algebraic equations by any means. The answer is as follows:

$$y_1 = 0.552, y_2 = -0.424, y_3 = -0.9644$$

Simulink Solution

The linear algebraic equations are solved using Simulink and the Thomas algorithm shown in Figure 9.9, followed by the MATLAB code embeds the Simulink function block. Figure 9.9 discloses the results plotted in Figure 9.10.

$$\frac{d^2y}{dx^2} - \left(1 - \frac{x}{5}\right)y = x;\ y(1) = 2, y(3) = -1$$

After the substitution of the finite difference formulas, the discretized equation is

$$\frac{y_{i+1} - 2y_i + y_{i-1}}{h^2} - \left(1 - \frac{x_i}{5}\right)y_i = x_i$$

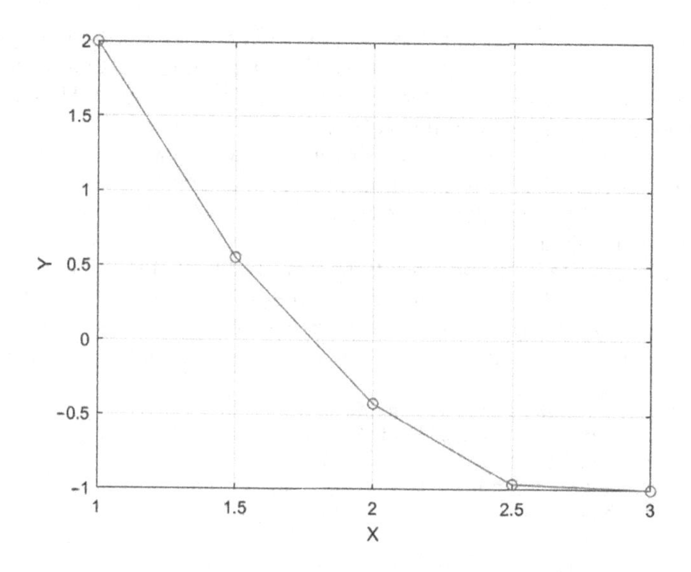

FIGURE 9.10 Simulink solution of the BVP defined in Example 9.2.

Rearranging,

$$y(i+1)\{1\}+y(i)\left[-2-2\left(1-\frac{x_i}{5}\right)h^2\right]+y(i-1)=h^2x(i)$$

The supper diagonal is the coefficient of $y(i+1)$

$$\text{supd}=1$$

The main diagonal is the coefficient of $y(i)$

$$\text{maind}=-2-2\left(1-\frac{x_i}{5}\right)h^2$$

The subdiagonal is the coefficient of $y(i-1)$

$$\text{subd}=1$$

On the right-hand side, RHS

$$rhs=xh^2$$

Simulink Solution

Finite difference methods (FDMs) are numerical methods for solving Partial Differential Equations (PDE) by approximating them with difference equations, where finite differences approximate the derivatives. Thus, FDMs are discretionary methods. FDMs convert linear or nonlinear ODE/ Partial Differential Equation (PDE) into a system of linear (nonlinear) equations, which matrix algebra techniques can solve. The reduction of the differential equation to a system of algebraic equations makes the problem of finding a solution to a particular ODE ideally suited to modern computers, hence the widespread use of FDM in modern numerical analysis. The Simulink block diagram (Figure 9.9) is a solution to Example 9.2. The MATLAB function is connected with four input ports, *xlim*, y_0, y_1, and n, represent the lower and upper limit intervals, the lower BC 1, the higher BC 2, and the number of intervals, respectively. The exit display port contains the values of the five nodes.

The following MATLAB code inserts the Simulink block MATLAB function, generates the result displayed in Figure 9.9, and the plotted the values of y at various values of x with step size, $h=0.2$. The plotted results of y versus x are shown in Figure 9.10. The curve could be smoother if applying a smaller step size,

```
function res = FDM(xlim, y0,y1,n)
% y(i-1) - 2yi - h^2(1-xi/5)yi + y(i+1)=h^2 xi
% xlim: is the limit interval [a, b]
% y0, y1 are the boundary conditions at a, and b
% n is the number of elements in the final result

    function [y, x] =ODEsolve(xlim, y0,y1,n)
    x = linspace(xlim(1), xlim(2), n)';
    % stride
    h = x(2) - x(1);

    % The main diagonal has coefficients
    maind = [1;-(2+h^2*(1-x(2:end-1)/5));1];
    % The sub diagonals has coefficients
    subd = [repmat(1,n-2,1);0];
    % The super diagonals has coefficients
    supd = [0;repmat(1,n-2,1)];

    % The ODE matrix
    ODEmat = diag(maind) + diag(subd, -1) + diag(supd, 1);

    % create the right hand side for the solve:
    rhs = [y0;x(2:end-1)*(h^2);y1];

    % solve the matrixa a call to backslash
    y =ODEmat\rhs;
    end
[y, x] = ODEsolve(xlim, y0,y1,5);
res =y;
plot(x, y, 'ro-'), xlabel("X"), ylabel("Y");
grid on
end
```

The Thomas algorithm is utilized to obtain the solution for the tridiagonal matrix system.

The equations are solved using Simulink and the Thomas algorithm (Figure 9.11), followed by the MATLAB code describing the Thomas algorithm, followed by the executed results illustrated in Figure 9.12.

By double-clicking the MATLAB function block, the following MATLAB code is appended to the Simulink block:

```
% Thomas Algorithm, Example 9.2
function Res = Thomas(a0,b0,h, ya, yb)
```

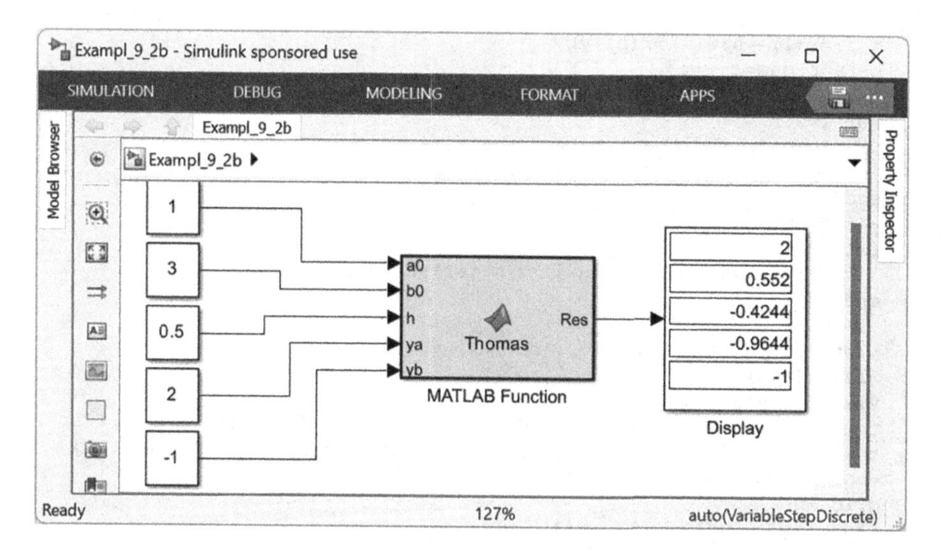

FIGURE 9.11 Simulink block diagram using Thomas algorithm for the solution of the BVP presented in Example 9.2.

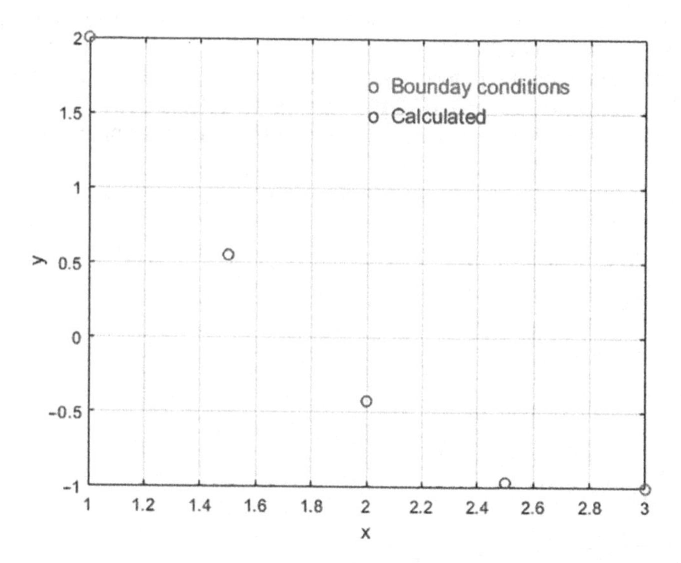

FIGURE 9.12 Simulink plot for the solution of Example 9.2, red circle is the BCs; blue circles are calculated y values at the three nodes.

```
% yi-1+[2 + h^2(1-xi/5)yi + yi+1 = h^2xi
a1=a0+h;
b1=b0-h;
x=a1:h:b1;
n=length(x);
b=-(2+h*h*(1-x/5));
c=ones(1,n);a=c;
r=h*h*x;
```

```
r(1)=r(1)-ya;r(n)=r(n)-yb;
beta=c;gam=c;y=c;
beta(1)=b(1);gam(1)=r(1)/beta(1);
for i=2:n
      beta(i)=b(i)-a(i)*c(i-1)/beta(i-1);
      gam(i)=(r(i)-a(i)*gam(i-1))/beta(i);
end
y(n)=gam(n);
for j=1:n-1
      y(n-j)=gam(n-j)-c(n-j)*y(n-j+1)/beta(n-j);
end

plot (a0,ya, 'ro', b0,yb, 'ro', x, y, 'bo');
xlabel('x');
ylabel ('y');
text(2,1.5,'o Calculated', 'Color', 'blue', 'FontSize', 12)
text(2,1.7,'o Bounday conditions', 'Color', 'red', 'FontSize', 12)

grid on
Res = [ya, y(1), y(2), y(3), yb];
```

Python Solution

Figure 9.13 presents the python solution of the set of equations generated using the finite difference method of Example 9.2. The four y values followed the python program are at the studied x values. The y values shown below the Python program results after executing the python program are the same as those obtained manually and by Simulink. Figure 9.13 displays the plotted data bounded by the BCs at $x(1)=2$ and $x(2)=-1$. The curve could be smoother if decreasing the step size.

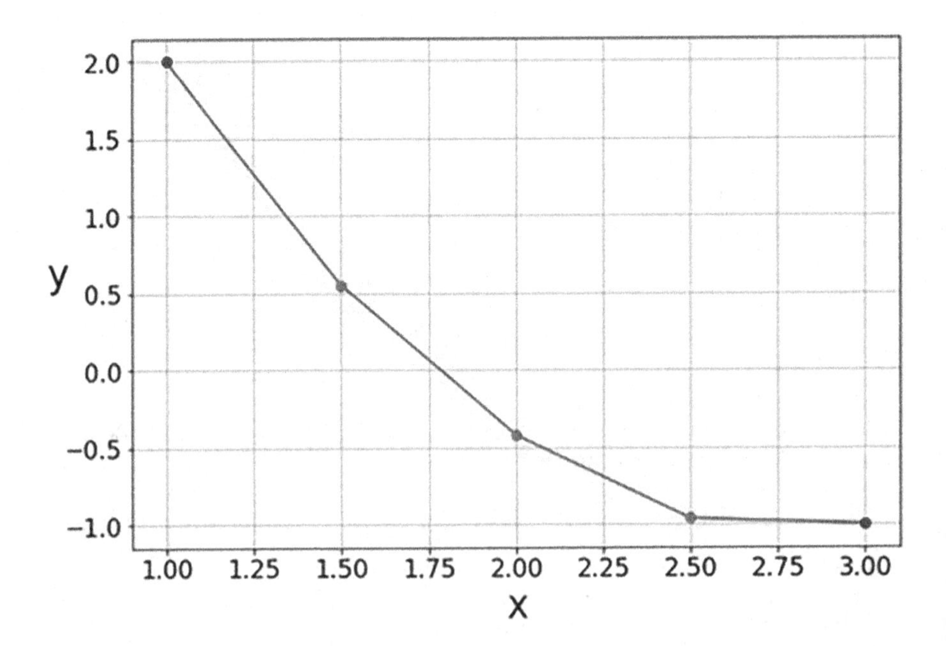

FIGURE 9.13 Python solution for the BVP defined in Example 9.2.

```
import numpy as np
import matplotlib.pyplot as plt
a=1
b=3
n = 4
h = (b-a) / n
# Matrix coefficient
A = np.zeros((n+1, n+1))
A[0, 0] = 1
A[n, n] = 1
# matrix coefficient
x = np.linspace(1, 3, n+1)
for i in range(1, n):
    A[i, i-1] = 1
    A[i, i] = -(2+h**2*(1-x[i]/5))
    A[i, i+1] = 1
print(A)
x = np.linspace(1, 3, n+1)
b = np.zeros(n+1)
b[0]=2 # boundary conditon 1
for i in range(1, n):
    b[i] = h**2*x[i]
b[n] = -1 # boundary condtion 2
print(b)
# solve the linear equations
y = np.linalg.solve(A, b)
plt.figure(figsize=(9,6))
plt.plot(x, y, 'ro-')
plt.plot(1,2,'bo'), plt.plot(3, -1, 'bo')
plt.xlabel('x', fontsize=24)
plt.ylabel('y', fontsize=24,rotation=0)
plt.yticks(fontsize=16)
plt.xticks(fontsize=16)
plt.savefig('image.jpg')
plt.grid()
print (y)
```

Execution results

```
[ 2. 0.55201375 -0.4243701 -0.96440946 -1]
```

Example 9.3 Applying Finite Difference Method to Second-Order BVP

Solve the following second-order differential equation using the finite difference approximation method,

$$\frac{d^2y}{dx^2} = 1 - y,$$

Boundary conditions

$$y(0) = 0,\ y(1) = 0$$

Solve manually using Python, Simulink, and MATLAB.

Solution

Using the finite difference approximation formulas, replace the derivatives with the finite difference as follows:

$$\frac{y_{i+1} - 2y_i + y_{i-1}}{h^2} + y_i - 1 = 0$$

Multiply by h^2

$$y_{i+1} - 2y_i + y_{i-1} + h^2 y_i - h^2 = 0$$

Rearrange by collecting identical terms

$$y_{i+1} + (h^2 - 2)y_i + y_{i-1} - h^2 = 0$$

Substitute step size, $h = 0.25$

$$y_{i+1} - 1.875y_i + y_{i-1} - 0.0625 = 0$$

Substitute $i = 0$ (known BC 1)

$$y_o = 0$$

Substitute $i = 1$

$$y_2 - 1.938y_1 + y_0 - 0.0625 = 0$$

Substitute $i = 2$

$$y_3 - 1.938y_2 + y_1 - 0.0625 = 0$$

Substitute $i = 3$

$$y_4 - 1.938y_3 + y_2 - 0.0625 = 0$$

Substitute $i = 4$ known BC 2

$$y_4 = 0$$

After substituting the BCs, solve the set of three linear algebraic equations to fined y_1, y_2, y_3

$$y_2 - 1.938y_1 + 0 - 0.0625 = 0$$

$$y_3 - 1.938y_2 + y_1 - 0.0625 = 0$$

$$0 - 1.938y_3 + y_2 - 0.0625 = 0$$

There are three equations and three unknowns (the degree of freedom is zero). Polymath and any other solver can quickly solve the equations:

$$y_1 = -0.105,\ y_2 = -0.140,\ y_3 = -0.105$$

Simulink Solution

The three linear algebraic equations can be solved using the algebraic constraint function in Simulink under Math Operations. The Simulink block diagram is shown in Figure 9.14.

An alternative Simulink solution uses the MATLAB function (Figure 9.15). Using finite difference approximation formulas, replace the derivatives with the finite difference.

$$\frac{y_{i+1} - 2y_i + y_{i-1}}{h^2} + y_i - 1 = 0$$

Multiply by h^2

$$y_{i+1} - 2y_i + y_{i-1} + h^2 y_i - h^2 = 0$$

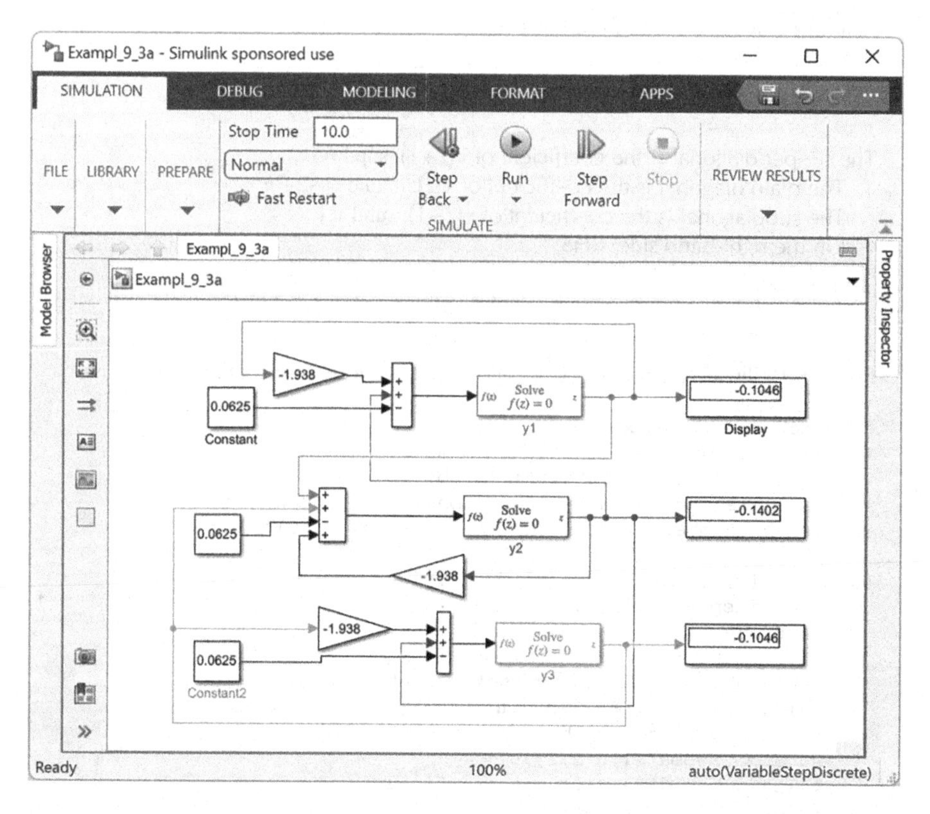

FIGURE 9.14 Solution of three linear algebraic equations using the algebraic constraints function in Simulink (Example 9.3).

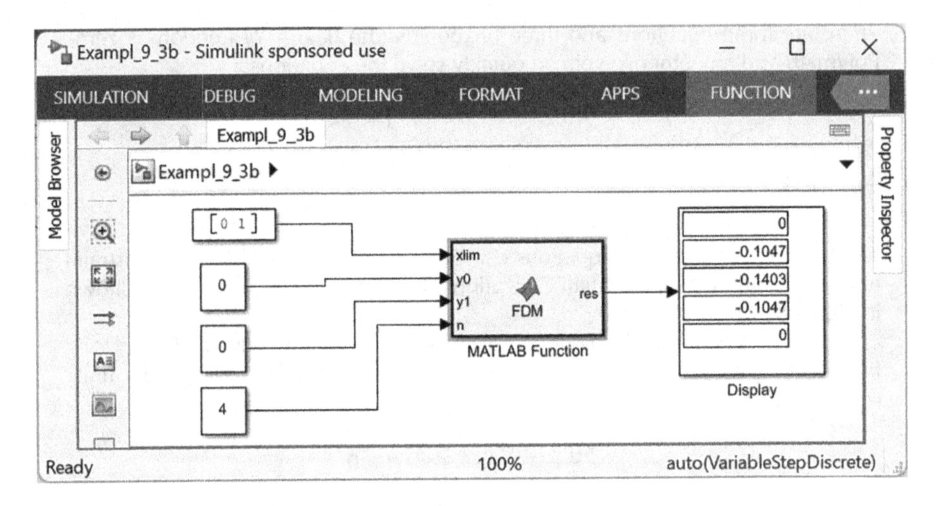

FIGURE 9.15 Simulink block diagram represents the solution using the finite difference method of the equation defined in Example 9.3.

Rearrange by collecting the same terms

$$y(i+1)\{1\}+y(i)\{-2+h^2\}+y(i-1)=h^2$$

The supper diagonal is the coefficient of $y(i+1)$: supd = 1

The main diagonal is the coefficient of $y(i)$: maind $= -2+h^2$

The subdiagonal is the coefficient of $y(i-1)$: subd = 1

On the right-hand side, RHS

$$rhs = h^2$$

Figure 9.16 illustrates the values of y along with various values of x.

```
function res = FDM(xlim, y0,y1,n)
% y(i+1) - 2yi + y(i-1) + h^2 yi= h^2
% xlim: is the limit interval [a, b]
% y0, y1 are the boundary conditions at a, and b
% n is the number of elements in the final result

function [y, x] =ODEsolve(xlim, y0,y1,n)
    x = linspace(xlim(1), xlim(2), n)';
    % stride
    h = x(2) - x(1);
    % The main diagonal has coefficients
    maind = [1;repmat(-2+h^2,n - 2,1);1];
    % The sub diagonals has coefficients
    subd = [repmat(1,n-2,1);0];
    % The super diagonals has coefficients
    supd = [0;repmat(1,n-2,1)];
    % The ODE matrix
    ODEmat = diag(maind) + diag(subd, -1) + diag(supd, 1);
    % create the right hand side for the solve:
```

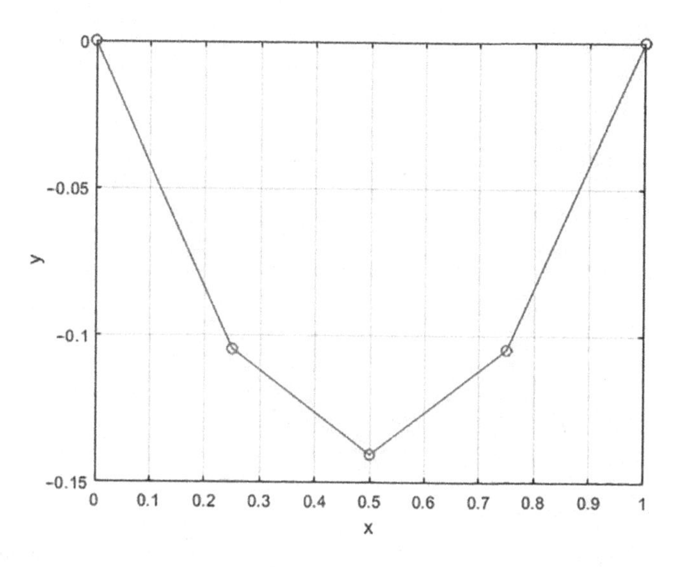

FIGURE 9.16 Simulink-generated plot using the finite difference method represents the equation defined in Example 9.3.

```
    rhs = [y0;repmat(h^2, n-2,1);y1];
    % solve the matrixa a call to backslash
    y =ODEmat\rhs;
    end

[y, x] = ODEsolve(xlim, y0,y1,5);
res =y;
hold on;
plot(x, y, '-ro')
plot(xlim(1), y0, 'bo')
plot(xlim(2), y1, 'bo')
grid on, xlabel("x"), ylabel("y");
end
= [y0;repmat(h^2, n-2,1);y1];
```

Python Solution

The following results were generated by Python of the set of algebraic equations produced by the finite difference method:

[0.0 – 0.10467706 – 0.1403118 – 0.10467706 0.0]

The curve depicted in Figure 9.17 could be smoother if more grids were used (increase n). The python code is shown as follows:

```
import numpy as np
import matplotlib.pyplot as plt
plt.style.use('seaborn-poster')
a=0
b=1
n = 4
h = (b-a) / n
# Matrix coefficient
A = np.zeros((n+1, n+1))
```

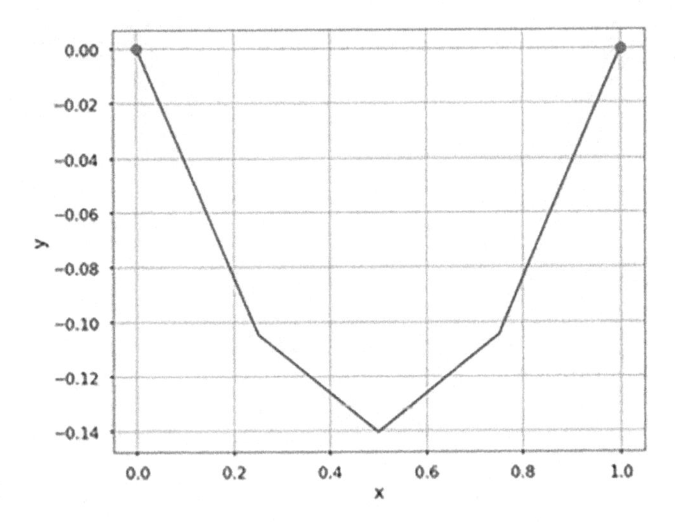

FIGURE 9.17 Python solution using the finite difference method of the equation described in Example 9.3.

```
A[0, 0] = 1
A[n, n] = 1
# matrix coefficient
x = np.linspace(0, 1, n+1)
for i in range(1, n):
    A[i, i-1] = 1
    A[i, i] = h**2-2
    A[i, i+1] = 1
print(A)
x = np.linspace(0, 1, n+1)
b = np.zeros(n+1)
b[0]=0 # boundary conditon 1
for i in range(1, n):
    b[i] = h**2
b[n] = 0 # boundary condtion 2
print(b)
# solve the linear equations
y = np.linalg.solve(A, b)
plt.figure(figsize=(9,6))
plt.plot(x, y, 'ro-')
plt.plot(0,0,'bo')
plt.plot(1, 0, 'bo')
plt.xlabel('x')
plt.ylabel('y', rotation=0)
plt.grid()
```

9.3.3 Shooting Method for BVPs

In numerical analysis, the shooting method solves a BVP by reducing it to an IVP. Then we can solve it using the methods we learned in the previous chapters. The shooting method is iterative and can efficiently solve BVPs. The shooting method

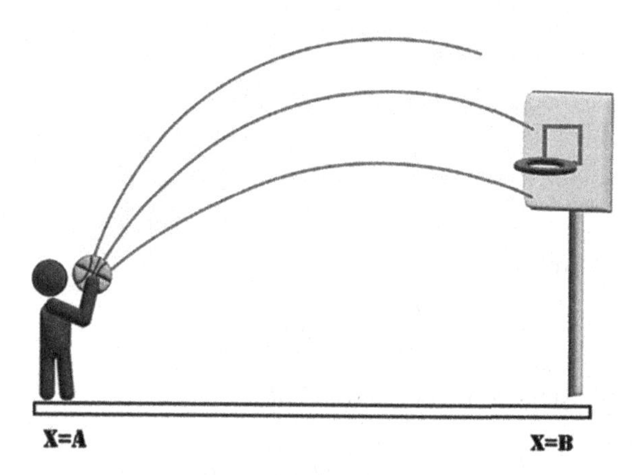

FIGURE 9.18 Presentation of the shooting method principles.

converts a BVP into an equivalent IVP. Then, the IVP is solved via a trial-and-error approach. This technique is called a 'shooting' method, by analogy to the procedure of shooting an object at a stationary target (Figure 9.18). The following procedure is recommended to follow in solving a BVP using the shooting method:

1. Start with a guess value for the auxiliary conditions at one point in time.
2. Solve the IVP using the numerical integration methods, Euler, and Runge-Kutta methods.
3. Check if the BCs are satisfied, otherwise modify the guess, and resolve the problem.
4. Use interpolation in updating the guess.

Example 9.4 Applying the Shooting Method for Heating Rod

Demonstrate the shooting method for the steady-state heat transfer in a rod fixed between two plates of differential temperatures. The rod is a length of 10 cm. The following second-order ODE describes the model equation.

$$\frac{d^2T}{dx^2} = -0.05(200 - T)$$

With the following BCs

$$T(0) = 200K,\ T(10) = 400K$$

Solve manually using Python and Simulink.

Solution

1. Apply the shooting method and divide second-order BVP into two IVPs. Let

$$z = \frac{dT}{dx}$$

It can be written as

$$\frac{dT}{dx} = z, \quad T(0) = 300$$

2. Take the first derivative of z; it will be dz/dx

$$\frac{dz}{dx} = \frac{d^2T}{dx^2} = -0.05(200 - T), \quad z(0) = \text{guess}$$

Applying the Euler method for solving the two initial value ODEs simultaneously.

$$T_{i+1} = T_i + hf_1(x_i, T_i, z_i)$$

$$z_{i+1} = z_i + hf_2(x_i, T_i, z_i)$$

The entire solution is described in Table 9.2.

Using excel and by utilizing the data/goal seek. The rest of the data can be found in Table 9.3.

Simulink Solution

The same data can be obtained using Simulink. Since we are using the Euler method, the solver in Simulink should be changed to a fixed value step size and the Euler method, as shown in Figure 9.19. Click on the modeling tab, then the setting, then fill in the start, stop time, the type to fixed step, and for the solver, select Euler, and enter the step as stated in the problem statement.

Figure 9.20 shows the Simulink block diagram. The initial guess found by excel (–10.8232) is used for the same purpose as an initial guess for z integrator. The temperature profile is shown in Figure 9.21.

TABLE 9.2
Solution Using the Shooting Method of Equation Defined in Example 9.4

i	*x*	*T*	*z*
0	0	300	–7 (Initial guess)
1	1	300 + 1 * (–7) = 293	–7 + 1 * (–0.05 * (200–300) = –2
2	2	293 + 1 * (–2) = 291	–2 + 1 * (–0.05 * (200–293) = 2.65
3	3		
9	9		
10	10	400 (Target)	

TABLE 9.3
Solution of Two IVPs Simultaneously Using the Euler Method

i	x	T	z
0	0	300.00	−10.82
1	1	289.18	−5.82
2	2	283.35	−1.36
3	3	281.99	2.80
4	4	284.79	6.90
5	5	291.70	11.14
6	6	302.84	15.73
7	7	318.56	20.87
8	8	339.43	26.80
9	9	366.23	33.77
10	10	400.00	42.08

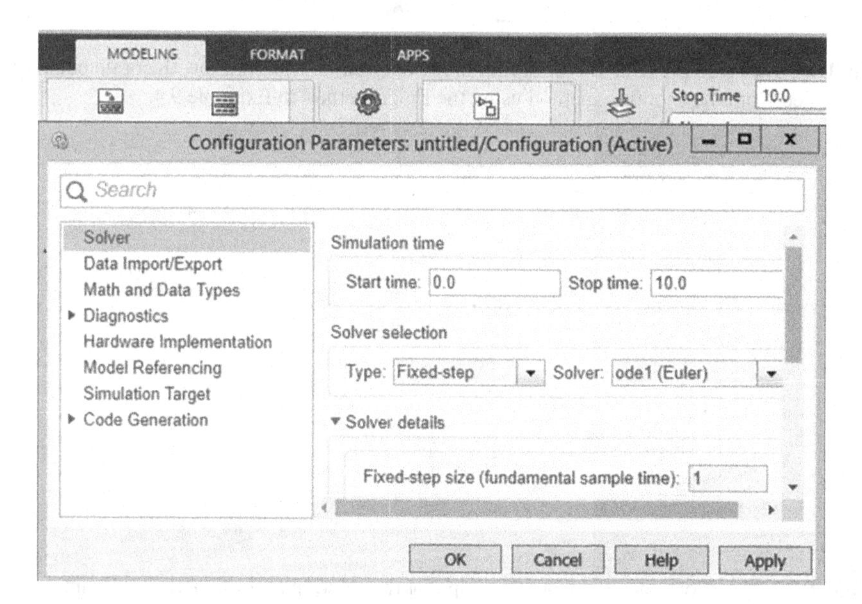

FIGURE 9.19 Setting the Simulink solver to the Euler method and step size, $h=1$.

Python Solution

The solution of Example 9.4 using Python is plotted as shown in Figure 9.22. A trial and error of the initial guess value of z are required. The shooting methods are developed to convert ODE BVPs into equivalent IVPs, and then we can solve them using the methods we learned from the previous chapter. In the IVPs, we can start from the initial value and move forward to get the solution.

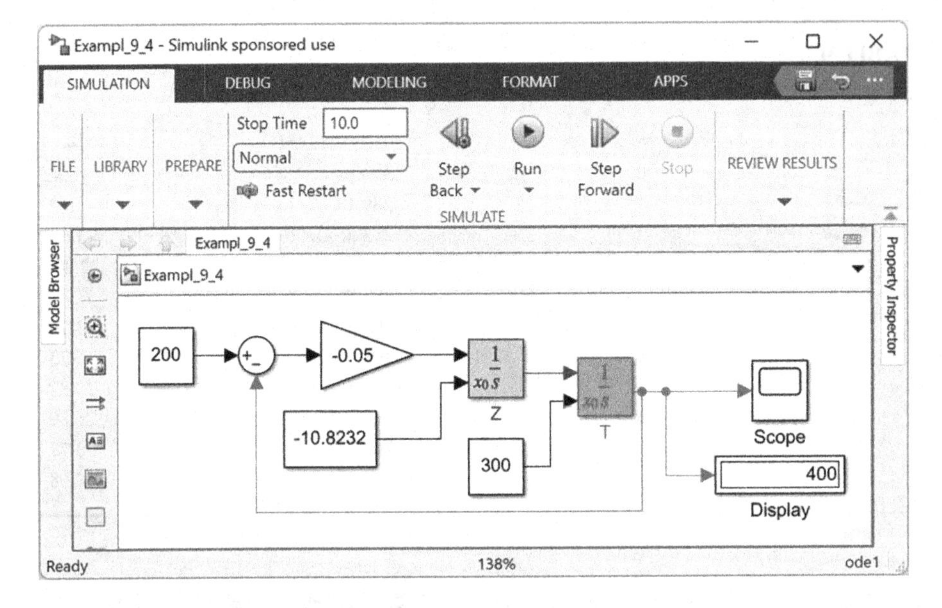

FIGURE 9.20 Represents the Simulink block diagram for solving the algebraic equations generated from the shooting method using the Euler method in Example 9.4.

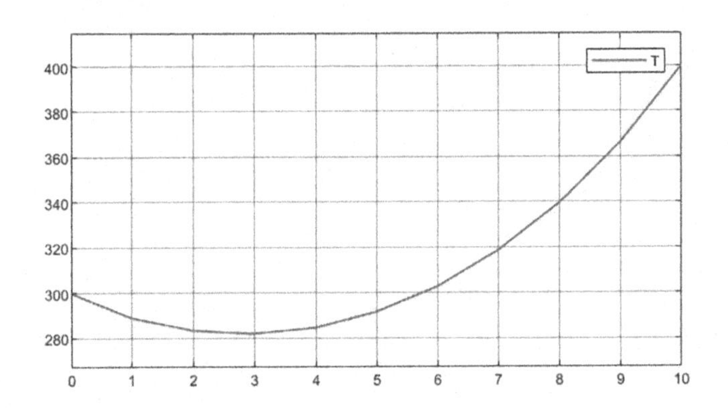

FIGURE 9.21 Plot generated from the scope for the solution of the equation in Example 9.4.

```
import numpy as np
import matplotlib.pyplot as plt
from scipy.integrate import solve_ivp
x = np.linspace(0, 10, 100)
def equations(x, y):
    yprime = np.zeros(2)
    # Shooting method, desired equations
    yprime[0] = y[1]
    yprime[1] = -0.05*(200- y[0])
    return yprime
tol = 1e-6
```

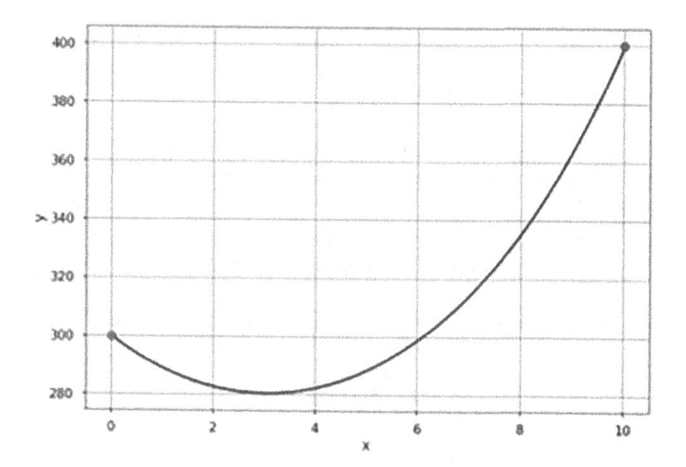

FIGURE 9.22 Python solution of equation described in Example 9.4.

```
max_iters = 1000
low = -10
high = 10
count = 0
while count <= max_iters:
      count = count + 1
      xspan = (x[0], x[-1])
      # Set the initial condition
      y0 = [300, -13.24]
      # Solve the system using our guess
      sol = solve_ivp(equations, xspan, y0, t_eval = x)
      # Extract the function values from the solution object.
      y_num = sol.y[0, :]
      # Plot the solution
plt.plot(x, y_num, 'b-', label='Numeric')
plt.plot([0, 10], [300,400], 'ro')
print(y_num[-1])
plt.grid(True)
plt.xlabel('x', fontsize=24)
plt.ylabel('y', fontsize=24,rotation=0)
plt.yticks(fontsize=16)
plt.xticks(fontsize=16)
plt.xlim((0,10))
plt.ylim((250,400))
```

Example 9.5 Applying the Shooting Method to Solve BVP

Use the shooting method and solve the following second-order BVP within the interval [1, 2].

$$\frac{d^2y}{dx^2} = 6x, \ \ y(1) = 2, \ \ y(2) = 9$$

Solve manually using Python, Simulink, and MATLAB

Solution

We will follow the following steps:

1. Convert the second-order ODE to a system of first-order ODEs.
2. Guess the initial conditions for the new ODE that are not available.
3. Solve the IVP.
4. Check if the known BCs are satisfied.
5. Modify the guess and resolve the problem again when needed.

First, we will convert the second-order ODE to two first-order ODEs IVPs, let

$$\frac{dy}{dx} = z, \quad y(1) = 2$$

$$\frac{dz}{dx} = 6x, \quad z(1) = 2 \text{ (Guess)}$$

Now we have two IVPs, and as a first approximation, we used the Euler method.

$$\frac{dy}{dx} = f_1(x, y, z) = z$$

$$\frac{dz}{dx} = f_2(x, y, z) = 6x$$

The initial conditions: $x_0 = 1,\ y_0 = 2,\ z_0 = 2\text{(guess)}$

$$x_1 = x_0 + h = 1 + 0.5 = 1.5$$

We are implementing the Euler method for solving two ODEs simultaneously.

$$y_{i+1} = y_i + hf_1(x_i, y_i, z_i)$$

$$z_{i+1} = z_i + hf_2(x_i, y_i, z_i)$$

Substitute

$$x_o = 1,\ y_o = 2,\ z_o = 2$$

$$y_1 = y_o + hf_1(1,2,2) = 2 + 0.5(2) = 3$$

$$z_1 = z_o + hf_2(1,2,2) = 2 + 0.5(6(1)) = 5.$$

$$x_1 = 1.5,\ y_1 = 3,\ z_1 = 5$$

$$y_2 = y_1 + hf_1(x_1, y_1, z_1) = y_1 + f_1(1.5,3,5) = 3 + 0.5(5) = 5.5$$

$$z_2 = z_1 + hf_2(x_1, y_1, z_1) = 5 + hf(1.5,\ 3,\ 6) = 5 + 0.5(6(1.5)) = 9.5$$

It is better to be arranged in a table, as shown in Table 9.4.

TABLE 9.4
Assume $z(0) = 0$ as an Initial Guess

i	x_i	y_i	z_i
0	1	2	2
1	1.5	3	5
2	2.0	5.5	9.5

TABLE 9.5
Assume $z_0 = 5.5$ as an Initial Guess

i	x_i	y_i	z_i
0	1.0	2.0	5.5
1	1.5	4.75	8.5
2	2.0	9.0	13.0

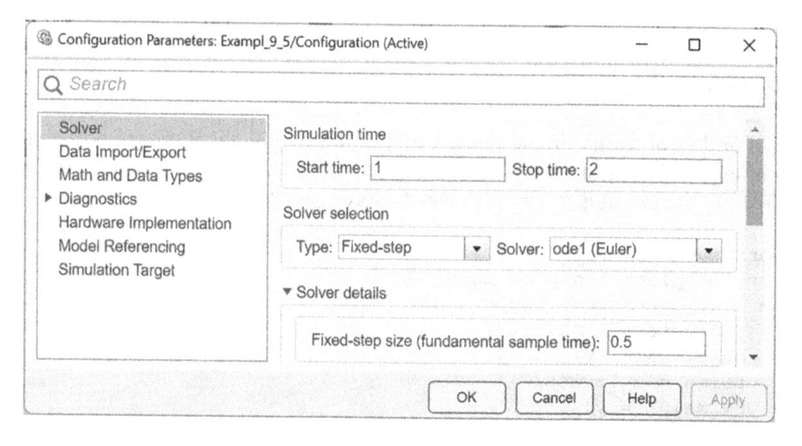

FIGURE 9.23 Solver setting of the configuration parameters page, for Example, 9.5 (Euler, fixed step size).

Since $y(2) = 5.5$, the first trial initial guess ($z_o = 2$) is leading to not correct BC ($y(2) = 9$). A second trial guess, $z_o = 5.5$, based on this initial value guess for z, the results are listed in Table 9.5. The shooting method converged at $y(2) = 9$.

Simulink Solution

First, from the modeling page in the toolbar, click on model/setting. Change the default setting of the configuration parameters. Set for the type: select fixed step; for the solver, select: mode1 (Euler) (Figure 9.23). The Simulink block diagram describing the solution of Example 9.5 is presented in Figure 9.24.

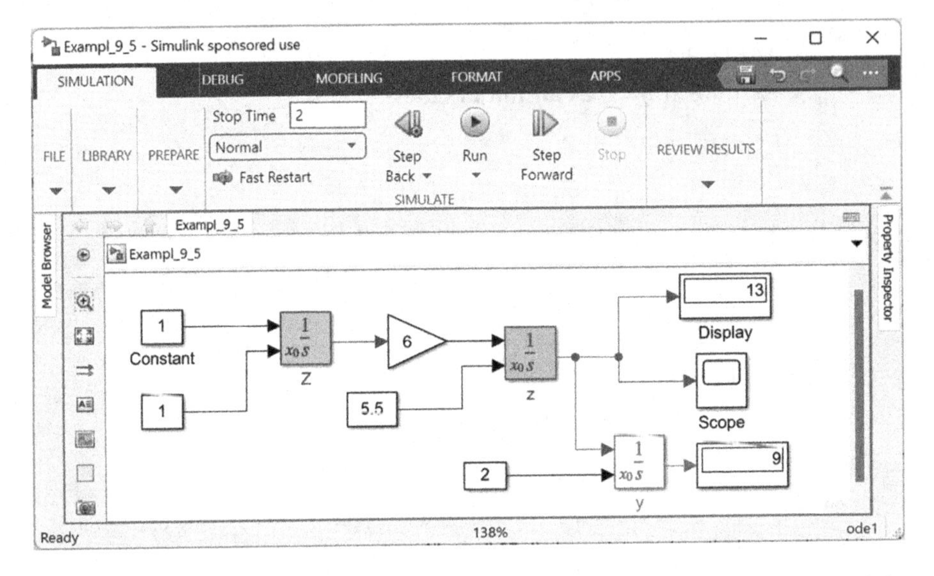

FIGURE 9.24 Simulink block diagram using Euler and the desired fixed step size of the equations defined in Example 9.5.

Python Solution

A trial and error for the initial guess value of the hypothetical model equation in Example 9.5. The results are plotted in Figure 9.25. The shooting method reduces the BVP to the IVP. It is a common way to solve two-point BVPs.

```
import numpy as np
import matplotlib.pyplot as plt
from scipy.integrate import solve_ivp
x = np.linspace(1, 2, 100)
def equations(x, y):
    yprime = np.zeros(2)
    yprime[0] = y[1]
    yprime[1] = 6*x
return yprime
tol = 1e-6
max_iters = 100
low = -10
high = 10
count = 0
while count <= max_iters:
count = count + 1
xspan = (x[0], x[-1])
    # Set the initial condition vector to be passed into the solver
    y0 = [2, 2.9]
    # Solve the system using our guess
    sol = solve_ivp(equations, xspan, y0, t_eval = x)
    # For ease of use, extract the function values from the
    solution object.
    y_num = sol.y[0, :]
```

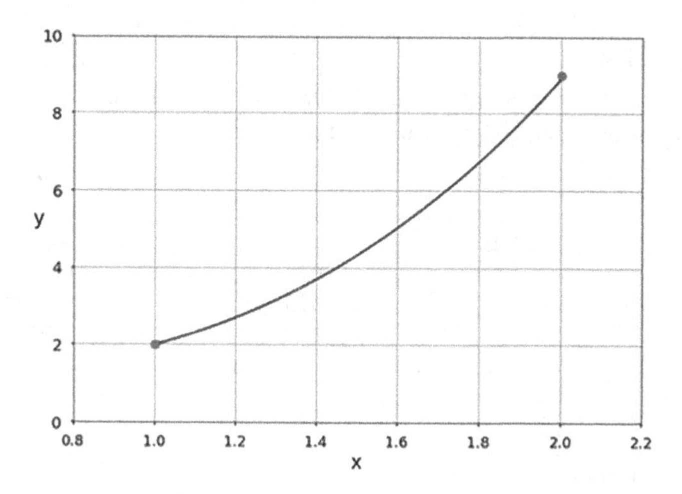

FIGURE 9.25 Python solution for equation defined in Example 9.5.

```
    # Plot the solution
plt.plot(x, y_num, 'b-')
plt.plot([1, 2], [2,9], 'ro')
print(y_num[-1])
plt.grid(True)
plt.xlabel('x'), plt.ylabel('y')
plt.xlabel('x', fontsize=24)
plt.ylabel('y', fontsize=24,rotation=0)
plt.yticks(fontsize=20)
plt.xticks(fontsize=18)
plt.xlim((0.8,2.2))
plt.ylim((0,10))
```

Example 9.6 Applying the Shooting Method

Use the shooting method and the explicit Euler method to approximate the solution to the BVP:

$$\frac{d^2y}{dx^2} = 2y, \ \ y(0) = 1.2, \ y(1) = 0.9, \ \ \ h = 0.25$$

Confirm the manual results with Python programming.

Solution

The basic idea of the shooting method is that we take the second-order ODE and write it as a system of the first-order ODE. Accordingly, split the second-order ODE BVP into two IVPs.

$$\frac{dy}{dx} = z = f_1(x, y, z), \ \ y(0) = 1.2$$

$$\frac{dz}{dx} = 2y = f_2(x, y, z), \ \ z(0) = \text{guess}$$

TABLE 9.6
Solution Using Shooting and Euler's Methods of the Equations Defined in Example 9.6

i	x_i	y	z
0	0	1.2	−1.0834
1	0.25	0.92915	−0.4834
2	0.5	0.8083	−0.01883
3	0.75	0.803594	0.385325
4	1	0.899925	0.787122

Using the Euler method and $h = 0.25$,

At $x = 0.25$

$$y_1 = y_0 + 0.25 f_1(0, 1.2, -1.0834)$$

$$y_1 = 1.2 - 0.25 \times (-1.0834) = 0.929$$

$$z_1 = z_o + 0.25 f_2(0, 1.2, -1.0834)$$

$$z_1 = -1.0834 + 0.25 \times (2 \times 1.2) = -0.4834$$

The rest of the values are listed in Table 9.6.

Python Solution

Figure 9.26 displays the results of a trial and error for the hypothetical ODE.

```
import numpy as np
import matplotlib.pyplot as plt
from scipy.integrate import solve_ivp
x = np.linspace(0, 1, 4)
def equations(x, y):
    yprime = np.zeros(2)
    # the desired ODE(BVP) d2y/dx2=2y, dy/dx=z, dz/dx = 2y
    yprime[0] = y[1]
    yprime[1] = 2*y[0]
    return yprime
tol = 1e-6
max_iters = 100
low = -10
high = 10
count = 0
while count <= max_iters:
    count = count + 1
    xspan = (x[0], x[-1])
    # Set the initial condition vector to be passed into the solver
    y0 = [1.2,-1.25]
    # Solve the system using our guess
    sol = solve_ivp(equations, xspan, y0, t_eval = x)
```

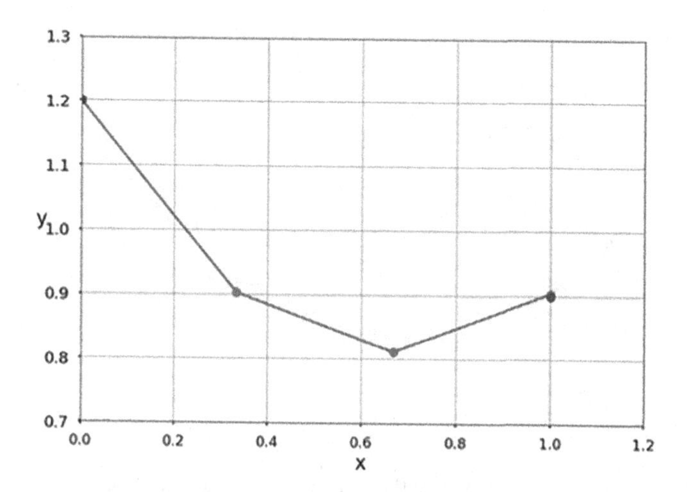

FIGURE 9.26 A trial-and-error solution of Example 9.6 using python.

```
# For ease of use, extract the function values from the solution
object.
y_num = sol.y[0, :]
# Plot the solution
plt.plot(x, y_num, 'ro-', label='Numeric')
plt.plot (0,1.2, 'bo'), plt.plot(1,0.9, 'bo')
print(y_num[-1])
plt.grid(True)
plt.xlabel('x'), plt.ylabel('y')
plt.xlabel('x', fontsize=24)
plt.ylabel('y', fontsize=24,rotation=0)
plt.yticks(fontsize=20)
plt.xticks(fontsize=18)
plt.xlim((0,1.2))
plt.ylim((0.7,1.3))
```

In the following solution, the initial guess of the z value is determined by Python without the initial guess trial-and-error solution (Figure 9.27).

```
from scipy.integrate import solve_bvp
import numpy as np
import matplotlib.pyplot as plt
# Solve the BVP of differential equations
# d2y/dx2 = 2y, y(0)=1.2 y(1)=0.9
def dydx(x, y):
    dy0 = y[1]
    dy1 = 2*y[0]
    return np.vstack((dy0, dy1))
# Calculation The boundary conditions
def boundCond(ya, yb):
    # The boundary conditions y(xa=0) = 1.2, y(xb=1) = 0.9
    fa = 1.2
    fb = 0.9
    return np.array([ya[0]-fa, yb[0]-fb])
# Indepedent variables, points (xa, xb)
```

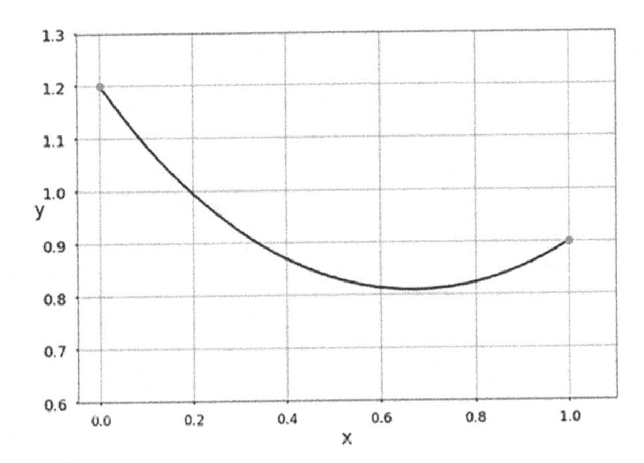

FIGURE 9.27 Python solution where the function found the initial guess.

```
xa, xb = 0, 1
xini = np.linspace(xa, xb, 11) # determine x Initial value of
yini = np.zeros((2, xini.size)) # determine y Initial value of
res = solve_bvp(dydx, boundCond, xini, yini) # solve BVP
xSol = np.linspace(xa, xb, 100) # Output mesh node
ySol = res.sol(xSol)[0] # At mesh nodes y value
# plot section
plt.plot(xSol, ySol, 'r-')
plt.plot(0,1.2, 'bo')
plt.plot(1, 0.9, 'bo')
plt.grid()
plt.xlabel('x', fontsize=24)
plt.ylabel('y', fontsize=24,rotation=0)
plt.yticks(fontsize=20)
plt.xticks(fontsize=18)
plt.xlim((-0.05,1.1))
plt.ylim((0.7,1.3))
```

9.4 SUMMARY

This chapter discusses the numerical solution of ODEs and BVPs. The BVPs specify the value of the dependent variable at the boundaries. Use strategies for IVPs to solve BVPs. The shooting method requires guessing an initial condition that substitutes for a BC and iterates to find the proper solution.

9.5 PROBLEMS

1. Solve the following BVP using the shooting method; after splitting the equation, use the explicit Euler method (step size, $h = 1$) to perform the numerical integration of the two IVPs simultaneously.

$$\frac{d^2y}{dx^2} = 2y - 8x + 72, \quad y(0) = 0, \; y(9) = 0$$

2. Use the shooting method to solve the following BVP

$$\frac{d^2y}{dx^2} = 2y, \quad y(0) = 1.2, \ y(1) = 0.9$$

Use the step size $h = 0.25$. Solve the equivalent system of IVP via explicit Euler's method.

3. Consider the following BVP

$$\frac{d^2y}{dx^2} = y, \ \ y(0) = 0, \ y(2) = 1$$

Solve it using the shooting method and perform the numerical integration with the simultaneous IVPs using Heun's method with step size, $h = 0.1$.

4. Solve the following BVP using the finite difference method

$$\frac{d^2y}{dx^2} = -y, \ \ y(0) = 0, \ y(1) = 1, \ h = 0.25$$

Answer: $y1 = 0.2943$, $y_2 = 0.5701$, $y_3 = 0.8108$

5. Solve the following ODE using a finite difference method,

$$\frac{d^2y}{dx^2} = 4(y - x)$$

With the following BCs and step size, $h = 0.25$

$$y(0) = 0, \ y(1) = 2$$

Answer: $y_1 = 0.3951$, $y_2 = 0.8265$, $y_3 = 1.3396$

6. Solve the following BVP using find difference approximation by dividing the interval into four equal portions.

$$\frac{d^2y}{dx^2} = y + x, \ \ y(0) = 0, \ y(1) = 0$$

Answer: $y_1 = -0.035$, $y_2 = -0.056$, $y_3 = -0.050$

7. Solve the following BVP using the shooting method; use any method to solve the ODE at $y = 10$.

$$\frac{d^2y}{dx^2} + 0.01(20 - y) = 0, \quad y(0) = 40, \ y(10) = 200$$

8. Use the shooting method to solve

$$\frac{d^2y}{dx^2} - \frac{2}{7}\frac{dy}{dx} - \frac{y}{7} + \frac{x}{7} = 0$$

With the BCs, $y(0)=5$ and $y(20)=8$. Use the implicit Euler method and a step size of 0.5 to find $y(10)$

9. Use the shooting method to solve the following BVP with the Euler method, a step size of five, and an initial guess of one for z. Calculate the temperature at $x=10$.

$$\frac{d^2T}{dx^2} + \frac{dT}{dx} - Tx - 7 = 0$$

The BCs are

$$T(0) = 20, \frac{dT}{dx}(10) = 55$$

10. Solve the following BVP using the shooting method. Split BVP into two IVPs and solve the generated system of the two first-order ODE that is equivalent to the second-order ODE using the explicit Euler method with $h=0.2$, use an initial guess, $z(0)=1.5$.

$$\frac{d^2T}{dx^2} + 2\frac{dT}{dx} - x^2\frac{dT}{dx} - x = 0$$

The BCs are

$$y(0) = 1,\ y(1) = 1.75$$

REFERENCES

1. Ascher, U.M., Mattheij, R.M., and Russell, R.D., 1995, *Classics in Applied Mathematics, Numerical Solution of Boundary Value Problems for Ordinary Differential Equations.* Philadelphia: Society for Industrial and Applied Mathematics.
2. Polyanin, A.D., and Zaitsev, V.F., 2003, *Handbook of Exact Solutions for Ordinary Differential Equations.* Boca Raton: Chapman & Hall/CRC Press.

Appendix A
Python Programming Code

A.1 WHY IS PYTHON FAMOUS AND ESSENTIAL?

It is a universal programming language that is easy and simple to learn and use in multiple fields. For example, web development, data analytics, data engineering, data science, machine learning, and artificial intelligence. Therefore, the wide use of Python gained popularity in all fields, and users preferred it over other programming languages. Unlike other programming languages, Python does not require a compiler (translator) that changes the program into a form the machine understands. Worldwide compatibility is essential, which makes integrating folders and utilizing a single language in different fields very beneficial.

A.2 INTRODUCTION

There are several rules that programmers follow in writing programming codes. The list of rules are as follows:

1. Aim for a smaller number of code lines; if 10 lines can do the same task as 20 lines, then why the extra complexity and lines of code to run?
2. Aim for simpler code; program codes should be easily read and understood by different programmers since different programmers could work on the same code.
3. Aim for explicit rather than implicit code; the code should be clearly and fully expressed, leaving nothing implied.
4. Aim for flat code rather than nested code; for example, in the 'if' function, it is better to keep the code flat than nested, and the line's starting point keeps getting further and further.
5. Aim for sparse lines of code; having dense code is not favorable, and a straightforward sparse program is easier to work with.
6. Aim for readability; take into consideration other programmers working on the same code.

Writing Python code with these guidelines in mind will help the programmer's code to be more efficient, simpler, easier to understand, and require less computational time.

A.3 SETTING UP

In this book, the program Spyder is used to write the code. Spyder is a free and open-source tool that is written in Python to run Python codes. It features an interactive user interface and beautiful visualization capabilities. Spyder has several components and can be updated with any plugin, thus increasing the possibility of adding more functions to the program. Spyder is available for Windows, MacOS, and Linux for free. The latest version of Python is 3.9.5, released on May 3, 2021.

In this book, Spyder is employed to write Python programs. Spyder is a powerful Interactive Development Environment (or IDE) for the Python programming language. Figure A.1 shows the Spyder user interface.

To check for the current Python version on windows as an example, a simple code is used:

```
import sys
print("User Current Version:-", sys.version)
```

That results in the following as an example:

```
User Current Version:-3.7.13 (default, Apr 24 2022, 01:04:09)
[GCC 7.5.0]
```

After setting up the program and learning the rules of coding, the basics of Python can be introduced. In every programming language, the first function taught is responsible for displaying "Hello world!" on the screen. This can be easily done in Python with one line of code as follows:

```
print("Hello world!")
```

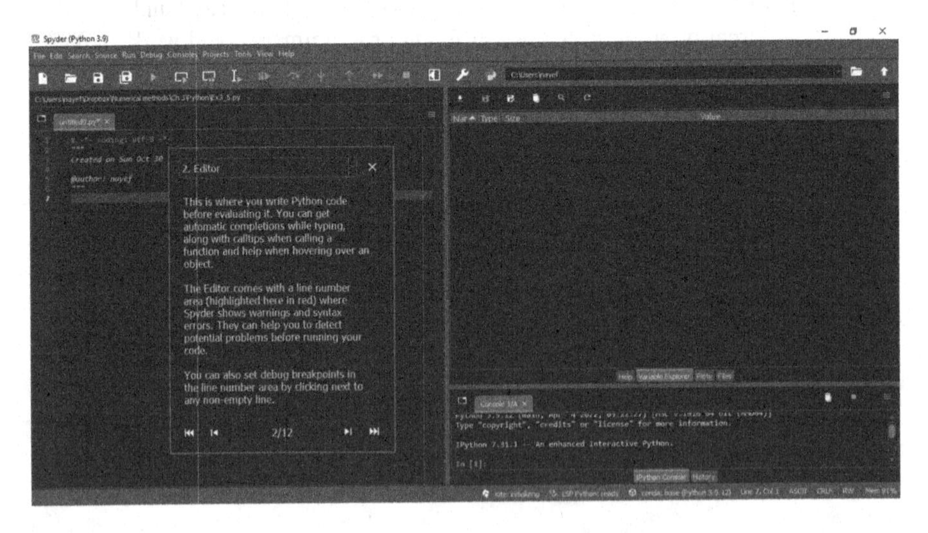

FIGURE A.1 Spyder interactive development environment for Python.

When running the code, you should see the following output:

```
Hello world!
```

As can be seen from the results, the parentheses are not displayed but are required when writing the code to indicate what is the required text. When the editor reads the function print, it will display whatever is inside the parenthesis. Getting started with Python a basic function is taught at first, the 'print' function. The print function can be used in any place where a message is to be displaced.

In writing comments:

'#' hashtag is used to indicate the start of a comment.

If a comment stretches over many lines:" is written at the start, then ''' indicates the comment end, and those lines will not be executed when command is run.

A.4 PART 1: PYTHON BASICS

Several expressions in mathematics are used in Python as follows:

'+ – * /' represent addition, subtraction, multiplication, division, respectively.

Example:

```
#in writing multiple lines of math, print function is required
print(4+5) #Addition
print(4+4*2) #Python follows precedence order when execusion
print(36/4) #Example of division
print(36/4+9/3+2*2) #Example on precedence (solves left to right)
#Results:
9
12
9.0
16.0
```

'%' represents the remainder

Example:

```
101%10 #finding the remainder from 101 using 10
#Results:
1 #the solution is 1
```

Python deals with integers and real numbers.

Example:

```
print(1.123) #real number with multiple digits
print(3/2) #operator results a real number
print(-1.123) #positive and negative values are possible
#Results:
1.123
1.5
-1.123
```

Other mathematical functions are also present in Python, but to activate them, the math class must be imported into the Python code. The functions can then be used as follows:

```
from math import * #to import class
print(cos(90)) #to solve cosine value in radians
print(pi) #pi represent pi value of a circle
print(e) #e represent euler's number
#Results:
-0.4480736161291701
3.141592653589793
2.718281828459045
```

There are other mathematical functions in Python, as shown in the following table.

Command Name	Description
Cos (value)	cosine, angle unit in radians
Sin (value)	sine, angle in radians
Log (value)	Logarithm, with base value of *e*
log[] (value)	Logarithm, with base value of []
Max (value1, value2)	Finds maximum value
Min (value1, value2)	Finds minimum value
Int (value)	Converts value into an integer
Float (value)	Converts value into a floating number
Abs (value)	Outputs absolute value
Round (value)	Outputs number to the nearest whole number
Sqrt (value)	Square root

Python can store variables and assign a specific value to them as follows:

```
x = 2 #defining variable x
y = 2 #defining variable y
z = 'hey' #variable can be defined as string
#variables can be utilized in anyway possible
print(x+y)
print(x*y)
print(x/y)
print(z)
#Results:
4
4
1.0
Hey
```

An important function of coding is the ability to input information; this can be done with 'input' function. When 'input' function is used with text, the text is displayed, and the program waits for an input from the user to continue. An example utilizing 'input' function is as follows:

```
age = input("How old are you:") #input function to assign to
varibale
print('Your age is ', age) #adding coma then variable displays it
print('You were born on ', 2022 - int(age)) #variable is defined
as integer to allow for substraction
#Results:
How old are you: 24
Your age is 24
You were born on 1998
```

A.5 PART 2: IF FUNCTION

The 'if' function executes a group of functions only if the condition is true, as follows:

```
a = input('Input the GPA value: ') #Ask for input
if float(a) >3.6: #input in Python is always regarded as string,
thus must be identified as number
  print("You have excellent academic standing") #solution for if
  function
#Results
Input the GPA value: 3.67
You have excellent academic standing
```

The if function can be extended and other conditions can be added with 'elif' function, which stands for 'else if', and if the input doesn't follow any condition, a 'else' function is added. The previous example can be extended as follows:

```
a = input('Input the GPA value: ') #Ask for input
if float(a) >= 3.6: #input in Python is always regarded as
string, thus must be identified as number
  print("You have excellent academic standing") #solution for if
function
elif float(a) >= 3.0:
  print("You have good academic standing") #pay attention to tabs
elif float(a) >= 2.5:
  print("You have acceptable academic standing")
else:
print("You have bad academic standing")
#Answer 1
Input the GPA value: 3.8
You have excellent academic standing
#Answer 2
Input the GPA value: 3.5
You have good academic standing
#Answer 3
Input the GPA value: 2.8
You have acceptable academic standing
#Answer 4
Input the GPA value: 2
You have bad academic standing
```

Another example utilizing if, elif, and else functions is shown:

```
married = input('Are you married (Y/N):') #Question
if married == 'Y': #== indicate 'is it equal'
  print("Congratulations") #shows results
elif married == 'N': #elif stands for else if
  print("Good luck") #shows results
else: print("False input") #ways to fail a program is always
sought, thus must be considered
#Answering with 'Y'
Are you married (Y/N): Y
Congratulations
#Answering with 'N'
Are you married (Y/N): N
Good luck
#Answering with other
Are you married (Y/N): g
False input
```

A.5.1 For Loop

In terms of loops, the 'for' repeats a set of statements over a group of values specified. A variable name gives a value to each repetition to keep track of progress. The group of values can be a range of integers using the 'range' function.

```
for x in range(6):
 print(x)
#Results:
0
1
2
3
4
5 #The values start from 0 and don't end at 6 but at 5, but the
total number of results are 6
```

The 'for' loop statements can be saved with every repetition, as follows:

```
a = 0
for x in range(6):
 a = a+x
 print(x)
print(a)
#Results:
0
1
2
3
4
5
15 #The sum of all x values
```

Another loop used in Python is the 'while' loop, which executes a group of statements while the condition remains true and only stops when the condition is false. An example is as follows:

```
x = 1
while x < 100: #condition
  x = x*2
  print(x)
#Results:
2
4
8
16
32
64
128 #statements are run until condition is false
```

There are several logical expressions to be used with 'if' statements and loops and are shown in the following table:

Logical Expression	Meaning
==	Is it equal
!=	Is not equal
<	Less than
<=	Less than or equal
>	More than
>=	More than or equal

Combining the logical expressions with other operators can be very useful. Examples of operators are shown in the following table:

Operator	Meaning
and	Combines conditions and all have to be true
or	Anyone of the conditions can be true
not	Outputs 'True' if the expression is false

Loops can go forever, therefore loop controls are needed. Loop controls are shown in the following table:

Function	Meaning
break	Terminates the loop
continue	Ends the current iteration and moves forward
pass	Used as placeholder for future code

A.6 PART 3: DATA TYPES

Data type is a property of the object, not the variable. When inputting a value into a variable, Python always treats it as a string (text). Therefore, clarification is required to identify each variable. To do this, there are several data types and functions that can be used. The following table shows several functions, operators, and symbols used in Python.

Operator	Meaning
int(x)	Converts x into an integer
float(x)	Converts x into a floating number
xEy	E is the mathematical $\times 10^y$
x ** y	** is the exponential
J	Complex number

Complex number can be used in Python using 'j' to identify the complex number and value of $\sqrt{-1}$. An example of complex number as follows:

```
x = 1+2j
z = 2j
sum = x + z
print(sum)
#Results
(1+4j) #complex numbers are added
```

A.6.1 STRINGS

When defining a variable as a string, it can still use '+' operator, as follows:

```
x = 'Hello' #defining as a string
y = 'How are you'
z = x + ',' + y #adding variables
print(z) #print function
#Results:
Hello, How are you
```

A print function could be across several lines by using three quotes as follows:

```
print("'Hello
My name is Nayef
It is nice to meet you"')
#Results
Hello
My name is Nayef
It is nice to meet you
```

In a string, each character is numbered starting from zero. This is shown in the following examples.

Example 1:

```
x = "Apple"
for y in range(7):
  print(x[y])
#Results
A
P
P
L
e
```

Example 2:

```
x = "0123456789"
print(x[0]) #starts from 0
print(x[1]) #1 is the second character in the string
print(x[1:4]) #1, 2, and 3 are included
print(x[6:10])
#Results
0
1
123
6789
```

There are function used with strings in Python as shown in the following table:

Operator	Meaning
len(string)	Shows the number of characters in a string
str(string)	Identifies the variable as a string

The functions are shown in an example as follows:

```
x = "0123456789"
y = "CRC press"
z = 5
print(len(x)) #len shows number of characters
print(len(y))
c = z + 4
print(c) #to prove c is treated as a number
p = str(z) + ' ' + x #str treats a variable as a string
print(p)
#Results
10
9
9
5 0123456789
```

When writing a print function, the '%' operator can be used to include either a number or a specific string. The following example illustrates the point.

```
x = 1
y = 'One'
print("One is %d" %x)
print("1 is %s" %y)
#However writing the code as follows is wrong
#print("1 is %d" %y) #output an error
#Results
One is 1
1 is One
```

Types for data collection

There are three types of data collection.

First, a 'list' with an ordered set of data

```
list = ["Name1", "Name2", "Name3"] #[] brackets are used
print(list[0])
print(list[1])
print(list[2])
#Results
Name1
Name2
Name3
```

Second, a 'set' with unordered data

```
set = {"Name1", "Name2", "Name3", "Name4", "Name5"} #More names
are added to show randomness
print(set)
#Results
{'Name3', 'Name1', 'Name2', 'Name5', 'Name4'}
```

Third, a 'dictionary' with a pair of values stored

```
dictionary = {
    "Company": "Toyota",
    "Model": "Camry",
    "Year": "2022",
    "Color": "Black"
}
print(dictionary)
print(dictionary["Company"])
print(dictionary["Model"])
print(dictionary["Year"])
print(dictionary["Color"])
#Results
{'Company': 'Toyota', 'Model': 'Camry', 'Year': '2022',
'Color': 'Black'}
Toyota
Camry
2022
Black
```

There are several functions in a list as shown in the following table:

List Function	Meaning
list.append(x)	Add x item at the end of the list
list.insert(i, x)	Inserts an x item at the i position
list.remove(x)	Removes x value item from the list
list.pop(i)	Removes the item at position i and returns it
list.index(x)	Shows the index in first item with value x
list.count(x)	Shows the number of x appears in the list
list.sort()	Sorts items in ascending order
list.reverse()	Reverses items in the list

In dictionaries, addition can be done in the code as shown in the following example.

```
dic = {
"One": "1",
"Two": "2",
"Three": "3",
"Four": "90", #written wrong intentially
"Five": "5",
"Six": "6",
"Seven": "7",
"Eight": "8",
"Nine": "9",
}
print(dic["Four"])
dic["Four"] = 4 #values can be set
print(dic["Four"])
dic["Ten"] = 10 #values can be added
print(dic)
del(dic["One"]) #del deletes an element
print(dic)
#Results
90
4
{'One':'1', 'Two':'2', 'Three':'3', 'Four':4, 'Five':'5',
'Six':'6', 'Seven':'7', 'Eight':'8', 'Nine':'9', 'Ten':10}
{'Two': '2', 'Three': '3', 'Four': 4,'Five': '5','Six': '6',
'Seven': '7','Eight': '8', 'Nine': '9', 'Ten': 10}
```

Loops can be used in dictionaries as follows:

```
CarColors = {
"Togg": "Red",
"Camry": "Black",
"Tesla": "White",
"Hyundai": "Silver",
"Mercedes": "Gold",
"BMW": "Blue",
}
```

```
for k in CarColors.keys():
  print(k, ": ", CarColors[k])
#Results
Togg: Red
Camry: Black
Tesla: White
Hyundai: Silver
Mercedes: Gold
BMW: Blue
```

Lists and dictionaries can be copied. For a list, the function 'list' is used to copy e.i. I2=list(I1). For copying the dictionary, the function 'list.copy()' is used.

A.7 PART 4: FUNCTIONS

To create a function in Python, it is defined as 'def'. First, the function is defined then can be called whenever, as follows:

```
def test(): #defining a function
  print("This is a function test") #part of the function

print("To show space")
test() #calling a function
#Results
To show space
This is a function test
```

Functions are objects that can be assigned to a variable, as a parameter, and returned from a function. Functions require the () after them, and this is what distinguishes them from variables.

```
def names(name):
  print("Hello " + name)

names("Name1")
names("Name2")
names("Name3")
#Results
Hello Name1
Hello Name2
Hello Name3
```

Difference between parameters and arguments. A parameter is a variable listed inside the parentheses in the function definition. While an argument is the value that is sent to the function when it is called. A function can have two arguments, as follows

```
def APA(author,year):
  print("(" + author + "," + year +") ")
APA("Name1", "2022")
APA("Name2", "2020")
#Results
(Name1, 2022)
(Name2, 2020)
```

If the number of arguments is unknown, add '*' before the parameter name. Example as follows:

```
def kids(*names):
  print("The youngest child is " + names[2])
kids("OLd", "Middle", "Youngest")
#Results
The youngest child is Youngest
```

Assigning values to the function is also possible as follows:

```
def kids(name1, name2, name3):
  print("The youngest child is " + name3)
kids(name1 = "Name1", name2 = "Name2", name3 = "Name3")
#Results
The youngest child is Name3
```

A.7.1 Modules

A module is a Python program containing statements and definitions. In Python, one can call the module by different ways, such as.

Function	Meaning
import module	imports all statements in module
from module **import** part	imports apart from the module
import module **as** name	imports a module with a specified name

A.8 PART 5: TEXTS AND FILE PROCESSING

Strings start and end with quotations "" or apostrophe '' characters. The following is examples or writing a string:

```
"This is a string"
'This is a string'
"This outputs error'
"This output "error""
"This is 'a' string" #" and 'are treated as different
'This is "a" string'
```

The backlash '\' in a string has different properties depending on the proceeding character, as shown in the following example.

```
#Comments in print are added seperately
print("'
  Hello \\ #to print \
  Hello \t Hey #creates tap space
  Hello \n N #creates new line
  Hello \" #prints quotation
"')
```

```
#Results
  Hello\
  Hello     Hey
  Hello
N
  Hello "
```

Characters in a string are numbered with indexes starting from 0, as follows:
Name = "Mohamed"

Index	0	1	2	3	4	5	6	7
Character	M	o	h	a	m	e	d	"Space"

As can be seen in the following example.

```
name = "Mohamed"
print(name[0])
print(name[1])
print(name[2])
print(name[3])
print(name[4])
print(name[5])
print(name[6])
print(name[7])
#Results
M
o
h
a
m
e
d
```

When outputting a string, it can be manipulated with different functions. Several functions are shown in the following table.

Function	Meaning
capitalize	Capitalizes the first word
center	Centers a string
casefold	Makes all lowercase
count('x')	Counts how many "x"
endswith("x")	Checks if ends with x
find("x")	True or false if string ends with x
upper	Makes all characters upper case
lower	Makes all characters lower case

The following uses each function in an example.

```
text = "python is a great language to learn" #all lowercase
name = "my name is Nayef ghasem" #all lowercase
```

```
mssg = "HOW ARE YOU" #all caps

print(text.capitalize()) #capitalizes the first word
print(mssg.center(36, '*')) #centers a string
print(mssg.casefold()) #makes lowercase
print(mssg.count("O"))
print(mssg.count("o")) #upper and lower case sensitive
print(mssg.endswith("?")) #checks if ends with specified string
print(name.find("Nayef")) #outputs first index value of the
searched string
print(name.upper()) #all uppercase
print(name.lower()) #all lowercase
#Results
Python is a great language to learn
************HOW ARE YOU*************
how are you
2
0
False
11
MY NAME IS NAYEF GHASEM
my name is nayef ghasem
```

For loops can be used in strings as follows:

```
for x in "Tesla":
  print(x)
#Results
T
e
s
l
a
```

In Python, characters map to numbers using standardized mappings such as ASCII and Unicode. Therefore, every string character has a number as in the following example:

```
alp = "abcdxyz"
for x in alp:
  print(x)
  print(ord(x))
#Results
a
97
b
98
c
99
d
100
x
120
y
121
z
122
```

This method is reversable, as a number can be converted into a string, as follows:

```
a = 97
b = 98
c = 99
print(chr(a))
print(chr(b))
print(chr(c))
#Results
a
b
c
```

A.9 PART 6: OBJECTS AND CLASSES

Python is an object-oriented programming language, and almost everything in Python is an object, with its own properties and methods. A class in Python is like an object constructor, or a "blueprint" for creating objects.

To create a class the code is written as follows:

```
#class Name:
  #statements
class myclass:
  a = 1
  b = 2
  c = 3

T = myclass() #variable linked to class
print(T.a) #anything can be called from class
print(T.b)
print(T.c)
#Results
1
2
3
```

Using the link and call method, people can create classes and share them with different people. A class might sound similar to a module, but the difference is that a class is used to define a blueprint for a given object, whereas a module is used to reuse a given piece of code inside another program.

The __init__() function:

All classes have a function called __init__(), which is always executed when the class is being initiated. An example of the init function is as follows:

```
class information:
  def __init__(self, name, age): #the self reference is needed
    self.name = name
    self.age = age

T = information("Name1",24)
print(T.name)
print(T.age)
#Results
Name1
24
```

A.10 OBJECT METHODS

Objects can also contain methods (functions that belong to the object), as shown in the following example:

```
class information:
  def __init__(self, name, age):
    self.name = name
    self.age = age

  def myfunc(self):
  print("I am "+ self.name + "\n Nice to meet you.") #works with
  strings

T = information("Mohamed", 24)
T.myfunc()
#Results
I am Mohamed
Nice to meet you.
```

The "self" is used in the examples, but it can be anything specified by the user. The objective of the first variable is to reference the current instance of the class. As can be seen in the following example:

```
class information:
  def __init__(anything, name, age):
    anything.name = name
    anything.age = age
  def myfunc(anything):
    print("Hello "+ anything.name + "\n Nice to meet you.")
#works with strings

T = information("John",63)
T.myfunc()
#Results
Hello John
Nice to meet you.
```

Modifications can be done to the class, and there are three main operators as shown in the following example.

```
class information:
  def __init__(anything, name1, name2, name3):
    anything.name1 = name1
    anything.name2 = name2
    anything.name3 = name3

  def first(anything):
    print("I am "+ anything.name1)
  def second(anything):
    print("I am "+ anything.name2)
  def third(anything):
    print("I am "+ anything.name3)

list = ["Name1", "Name2", "Name3"]
T = information('name1', 'name2', 'name3')
```

```
name1 = list[0]
name2 = list[1]
name3 = list[2]
T.first()
T.second()
T.third()
T.name2 = "NMG" #things can be assigned
T.second()
#del T.name1 can be used to delete
#Results
I am Name1
I am Name2
I am Name3
I am NMG
```

In classes, the user can define any function so that Python's built-in operators can also be utilized.

Operator	Class Method
–	__neg__(self, cont), or __neg__(self)
+	__pos__(self, cont), or __pos__(self)
*	__mul__(self, cont)
/	__truediv__(self, cont)
==	__eq__(self, cont)
!=	__ne__(self, cont)
<	__lt__(self, cont)
>	__gt__(self, cont)
<=	__le__(self, cont)
>=	__ge__(self, cont)

A.11 CONCLUSION

Appendix A introduced Python, starting from setting up the program to explaining all the functions used in Python. Each function method of working is explained, and examples are provided to show actual results. The objective is to explain the main functions possible, but this information can be used and extended to no boundaries. This provides foundations, but there is no limit set for the users and their creativity and applications.

Appendix B
Introduction to Simulink

Simulink is a graphical extension to MATLAB for the modeling and simulation of systems. The system is drawn as a block diagram. Simulink is a software package that models and simulates a dynamic system whose output changes over time. Simulink is integrated with MATLAB, and the data can be easily transferred between the programs.

This appendix is written to introduce a tutorial for learners without prior experience with Simulink. In this appendix, we will apply Simulink to solve system model equations. Lines transmit signals in the direction indicated by the arrow, and lines must be combined using a block such as a summing junction.

There are several general classes of blocks:

1. Sources, such as Ramp, Random Generator, Step
2. Sink: Used to output or display signals
3. Discrete: Transfer functions, state space
4. Linear: summing, gains
5. Nonlinear: arbitrary functions, delay
6. Connections: Multiplexer, Mux.
7. Continuous and discrete dynamics blocks include Integrator, Transfer, and Transport Delay functions.
8. Math blocks: Sum and Product, and Add block.

First, start MATLAB (Figure B.1).
Individuals can start Simulink in two ways:

1. Click the Simulink icon on the MATLAB toolbar.
2. Enter the '>>Simulink' Command at the MATLAB prompt.

Starting Simulink displays the Simulink Library Browser (Figure B.2).

Click on the 'Blank Model' displays the Simulink Library Browser (Figure B.3).

Click on 'Library Browser' displays the Simulink Library Browser. The library Browser displays a tree-structured view of the Simulink block libraries installed on your system. One can build models by copying blocks from the library browser into a model window. The Simulink commonly used blocks (Figure B.3) contain the most frequently used Simulink block functions (Figure B.4).

The library lists the available 'continuous' function blocks (Figure B.5). The blocks are the most used in numerical methods and most often used are the transfer function (Transfer Fcn) and the 'State-Space'.

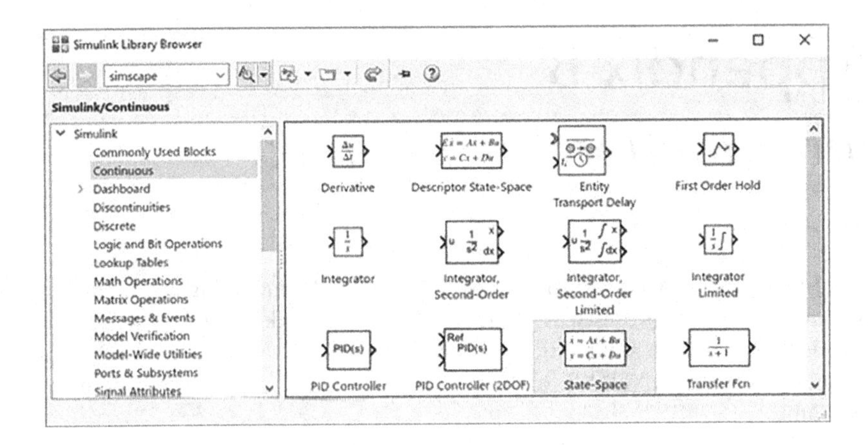

FIGURE B.1 MATLAB start window.

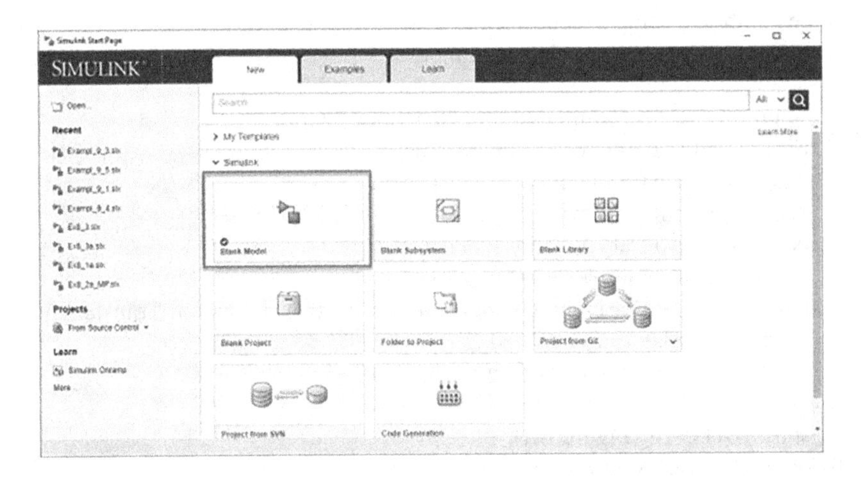

FIGURE B.2 Simulink start page.

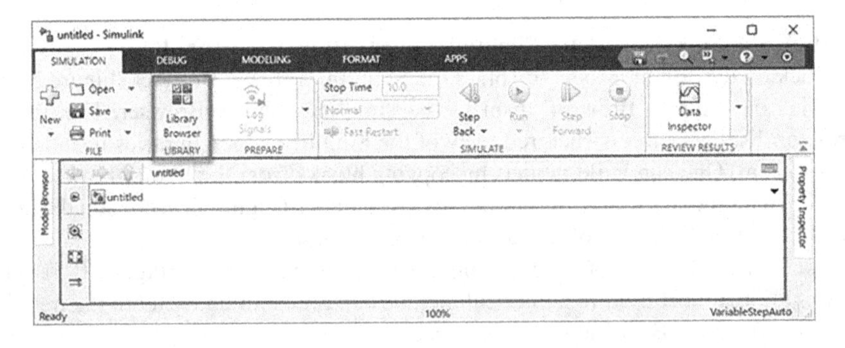

FIGURE B.3 Simulink untitled new file.

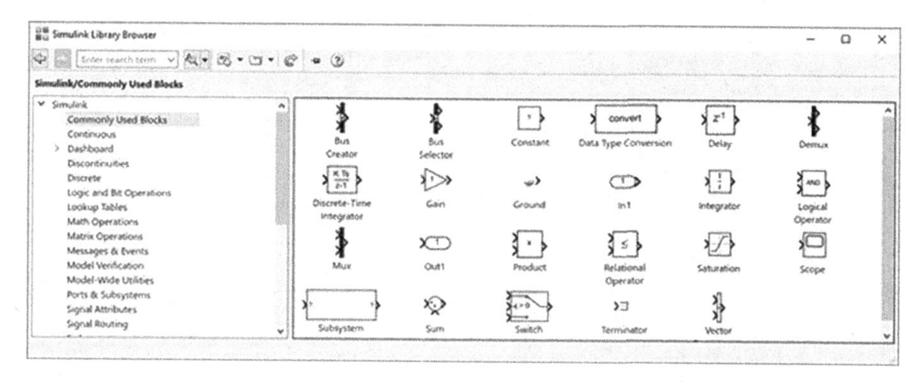

FIGURE B.4 Simulink commonly used blocks.

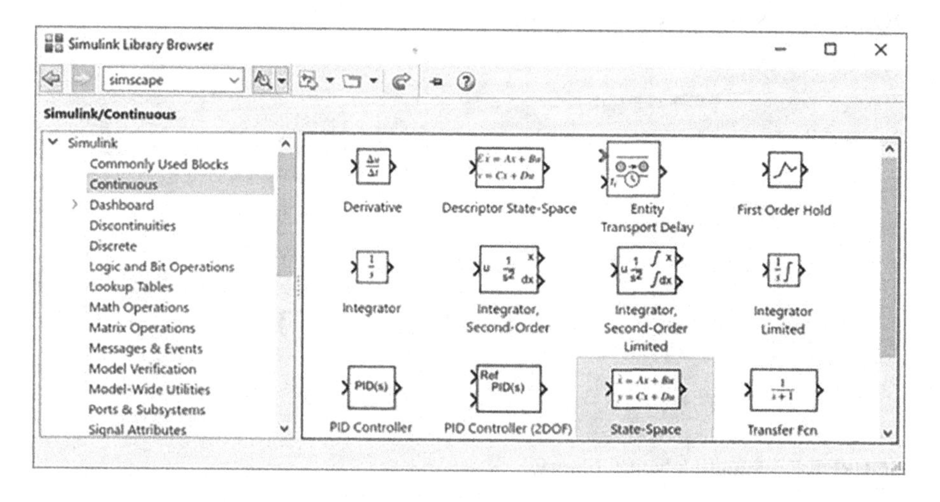

FIGURE B.5 The available Simulink 'continuous' function blocks.

The Sources icon yields the library shown in Figure B.6. The most used source is 'Clock', which is used to generate a time vector.

The 'Sinks' icons are shown in Figure B.7. The 'Scope' is the most common block used for plotting signals from a specific variable in the Simulink block diagram.

The Math library contains math functions block (Figure B.8).

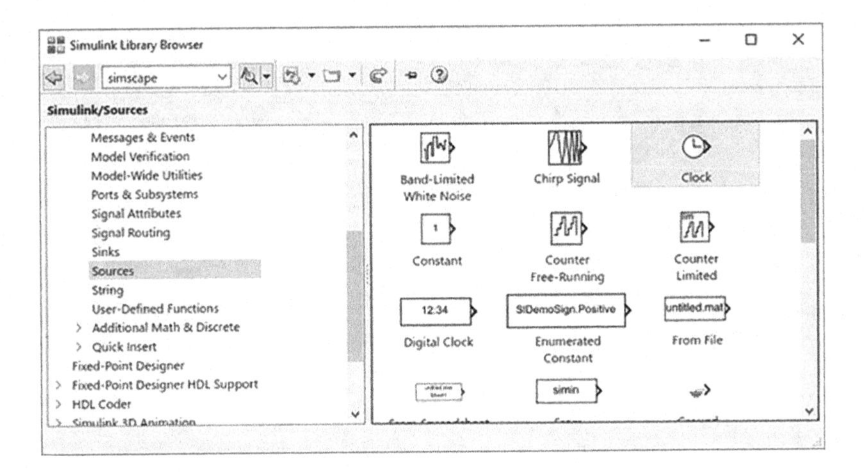

FIGURE B.6 Simulink 'sources' library.

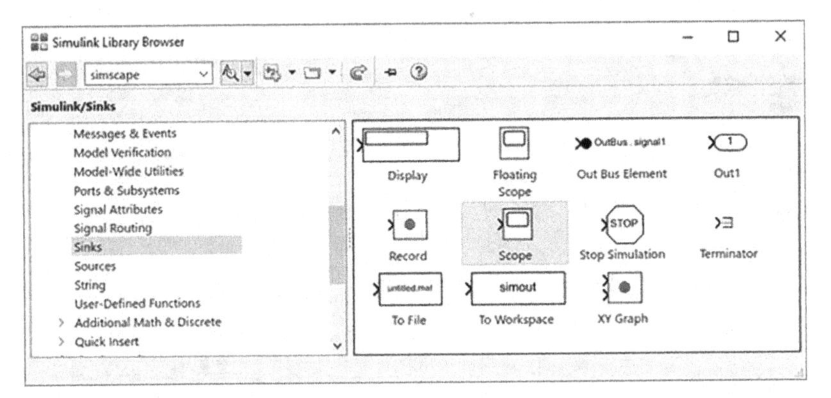

FIGURE B.7 Simulink 'sinks' library.

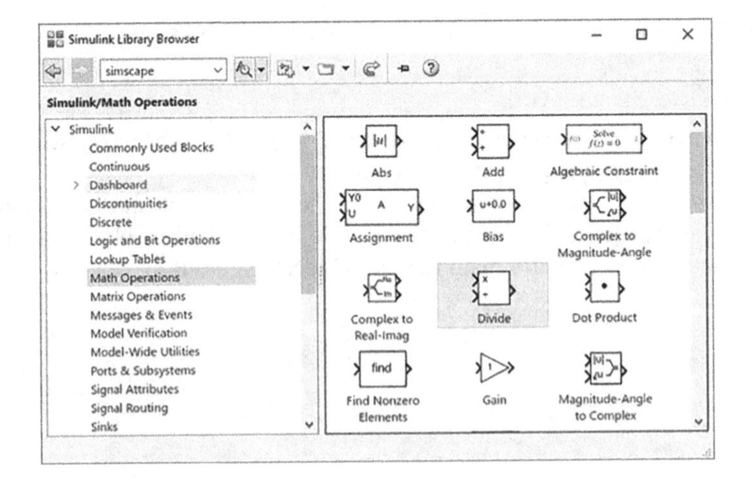

FIGURE B.8 Math operations library.

Example B.1 Convert Celsius (°C) to Fahrenheit (°F)

Model the following algebraic equation using Simulink:

$$T_C = (T_F - 32)/1.8$$

Solution

1. To create a new model, start MATLAB, click on **Simulink,** then **Blank Model**.
2. Click the **New** button in the library browser toolbar. Simulink opens a new model window (Figures B.9 and B.10).
3. Click on **the Library Browser** button.

To create this model, you need to drag and drop the blocks into the model from the following Simulink block libraries:

1. Constant block to define a constant of thirty-two (32), copy from the **Sources Library**.
2. Gain block to multiply the input spinal by 1.8 from the **Math** library.
3. Sum block to add the two quantities, also from the Math library.
4. Scope block to display the output from the Sinks library.

Next, gather the blocks into the model window (Figure B.11)

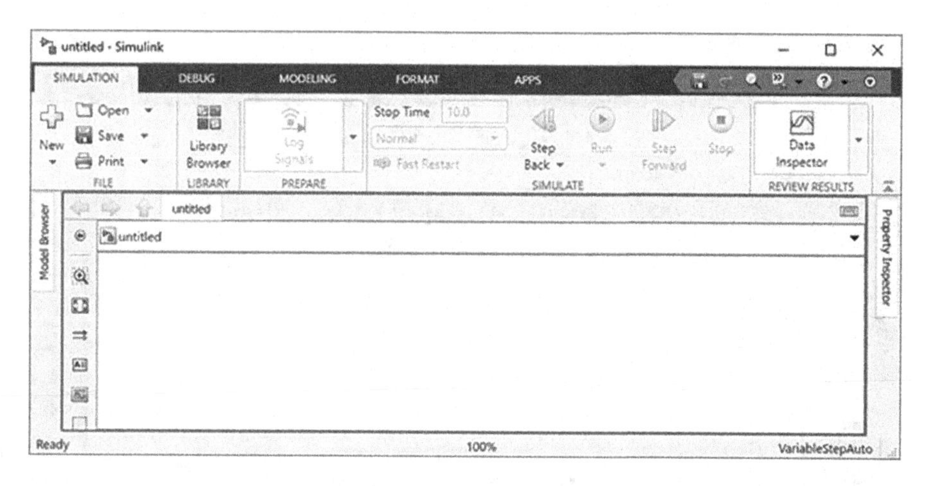

FIGURE B.9 New untitled blank page of Simulink file.

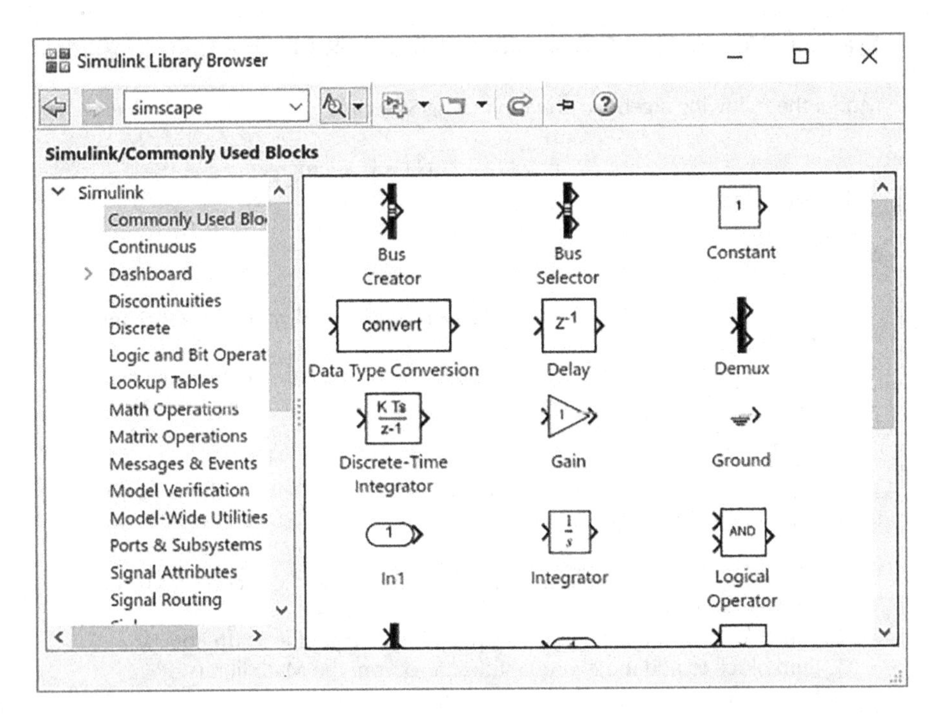

FIGURE B.10 Commonly used blocks.

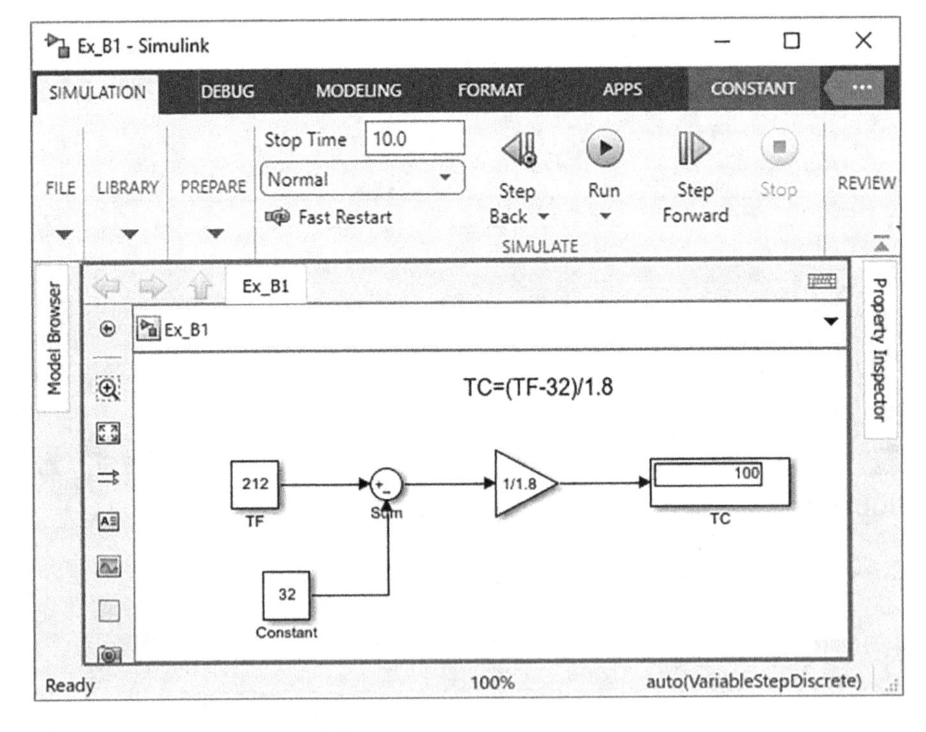

FIGURE B.11 Simulink solution of Example B1.

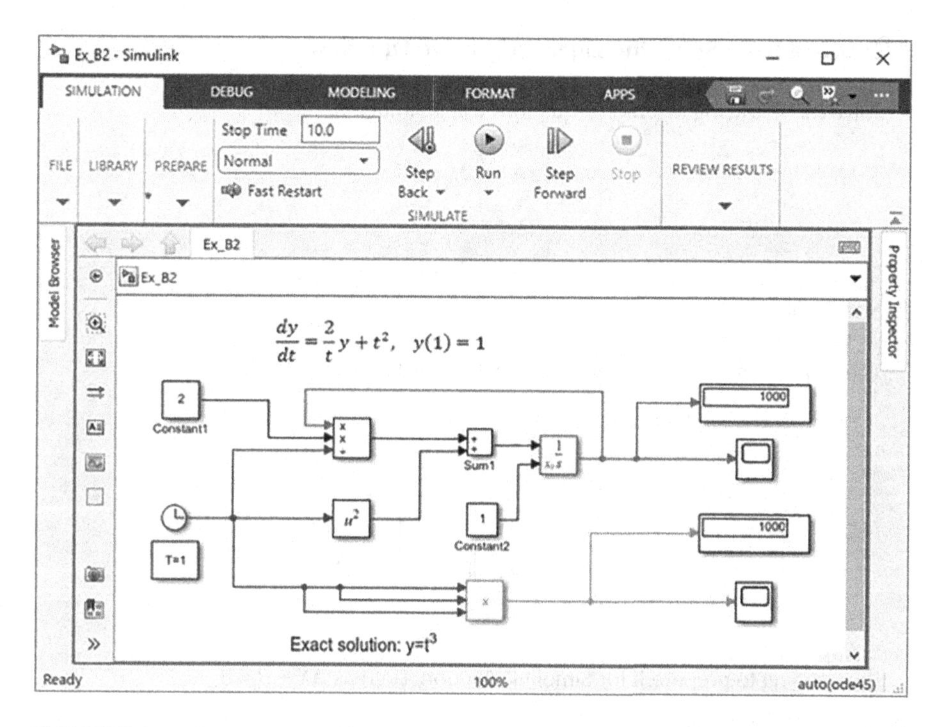

FIGURE B.12 Simulink solution of linear order differential equation model.

Example B.2 Solving Single ODE

Solve the linear first-order differential equation:

$$\frac{dy}{dt} = \frac{2}{t}y + t^2,\ y(1) = 1$$

Solution

The equation is solved using Simulink as follows:

Figure B.12 shows the model of this Example. By running the simulation, we obtained the numerical solution as well as the exact solution ($y = t^3$). The model time interval should start from 1, as the initial condition is $y(1) = 1$. To do that, while on the server, select the start time to 1, not zero. To perform better clipper.

Modeling/Model setting

Set the value of the Start Time: 1

Example B.3 Solve the Linear Algebraic Equation

Solve the following algebraic equations using Simulink:

$$2x + y - 2z = 3$$

$$x - y - z = 0$$

$$x + y + 3z = 12$$

Solution

The problem can be arranged in matrix form as follows:

$$AX = B$$

$$\begin{pmatrix} 2 & 1 & -2 \\ 1 & -1 & -1 \\ 1 & 1 & 3 \end{pmatrix}\begin{pmatrix} x \\ y \\ z \end{pmatrix} = \begin{pmatrix} 3 \\ 0 \\ 12 \end{pmatrix}$$

Rearranging to prepare it for Simulink solution, such as $AX - B = 0$

$$\begin{pmatrix} 2 & 1 & -2 \\ 1 & -1 & -1 \\ 1 & 1 & 3 \end{pmatrix}\begin{pmatrix} x \\ y \\ z \end{pmatrix} - \begin{pmatrix} 3 \\ 0 \\ 12 \end{pmatrix} = 0$$

Simulink Solution

Open the new Simulink page and select the blocks shown in Figure B.13.

Select the gain, double click, and input the matrix in the Gain windows (Figure B.14).

[2 1-2; 1-1-1; 1 1 3]

Multiplication: Matrix(*K**u*) (*u* vector)

Select constant, double click, and enter the constant as shown in Figure B.15.

Example B.4 Solving First-Order Differential Equation

Solve the first-order differential equation in MATLAB Simulink:

$$\frac{dy}{dt} = \frac{1}{t^2}y + \frac{3}{t},\ y(1) = -2$$

Build the Simulink block diagram and run it for the range of 1–4 seconds (Figure B.16).

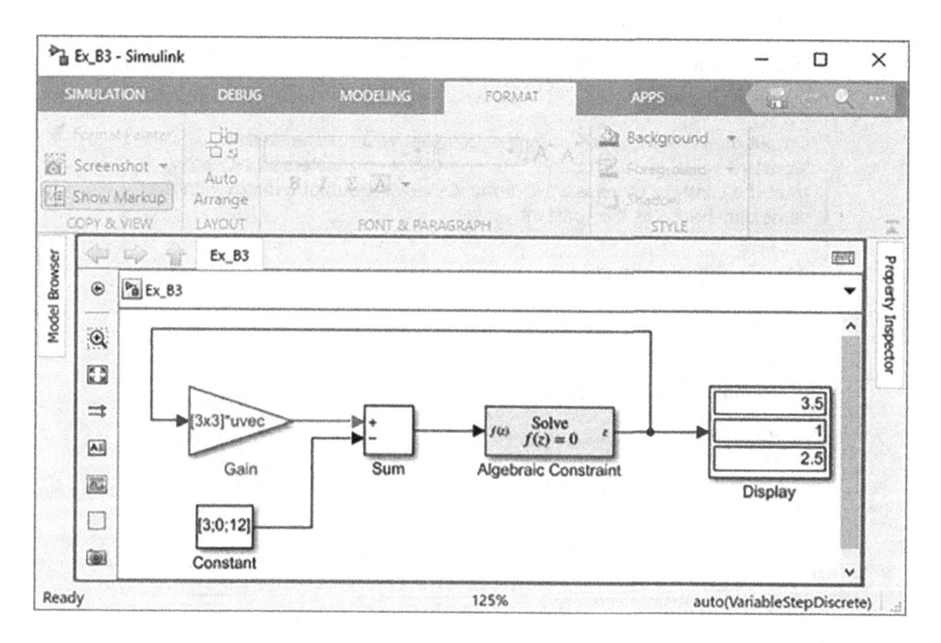

FIGURE B.13 Solution of the matrix of Example B.3.

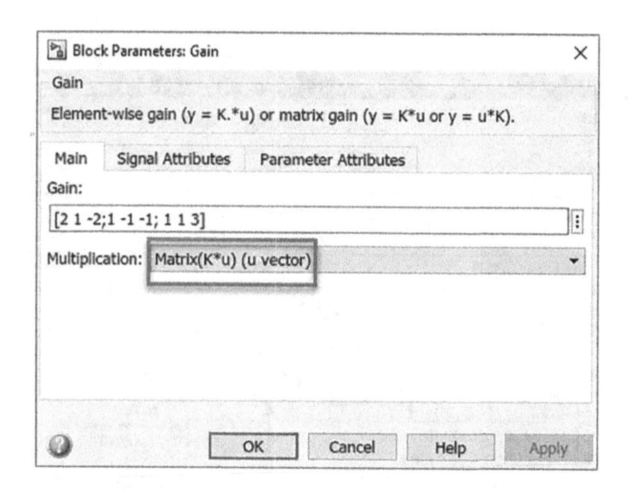

FIGURE B.14 Matrix coefficient and multiplication (u vector).

Solution

Set the modeling simulation period between 0 and 4. The setting of the configuration parameters can be modified through MODELING/Model setting (Figures B.17 and B.18).

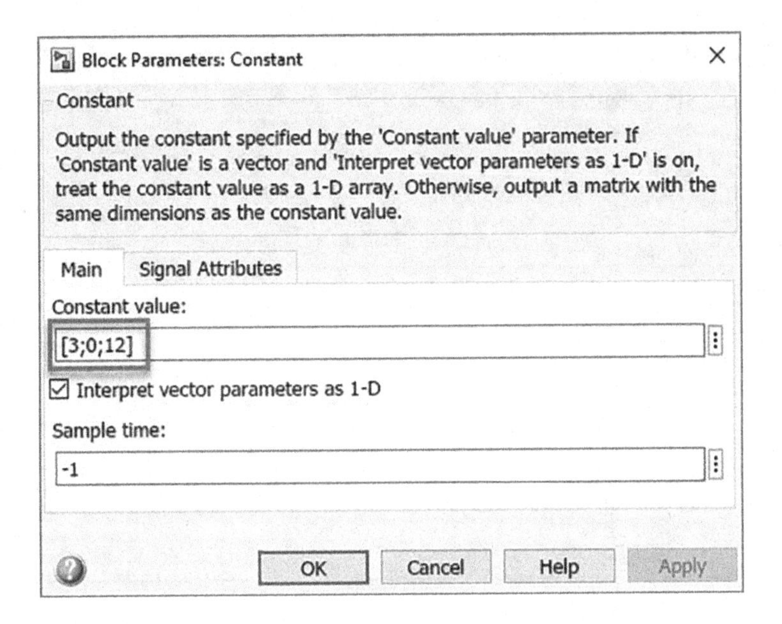

FIGURE B.15 Matrix constant coefficient.

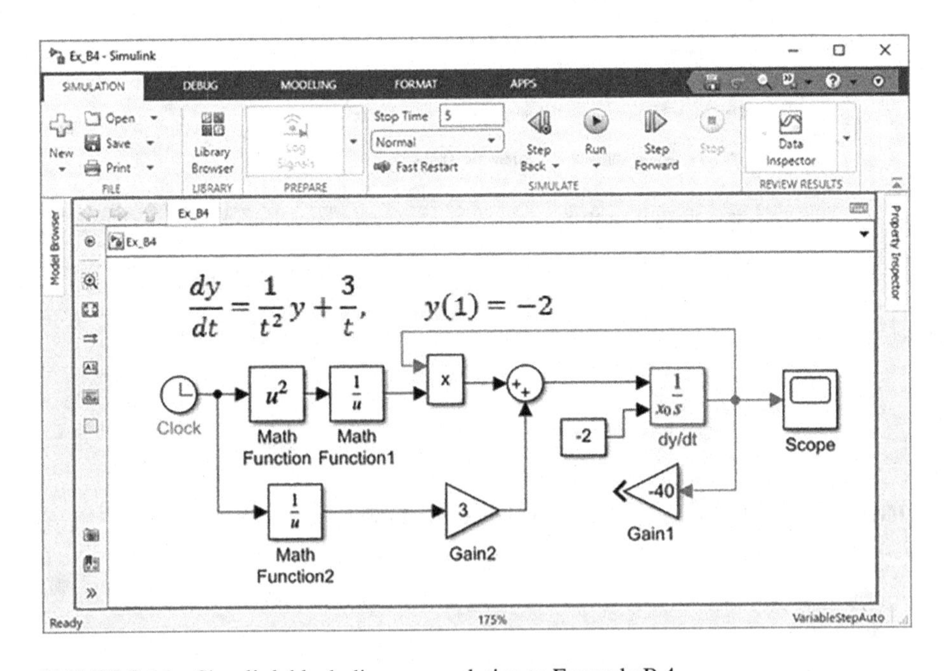

FIGURE B.16 Simulink block diagram a solution to Example B.4.

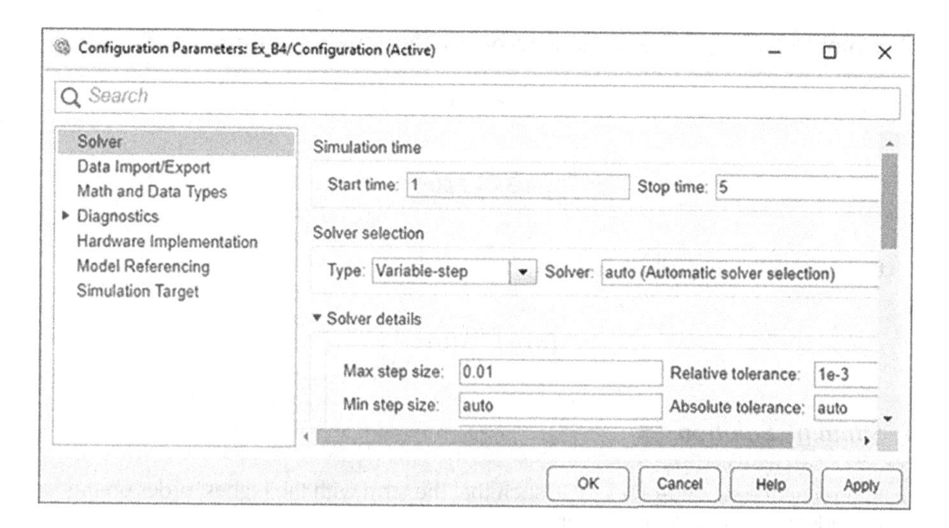

FIGURE B.17 Changing the start of the simulation time.

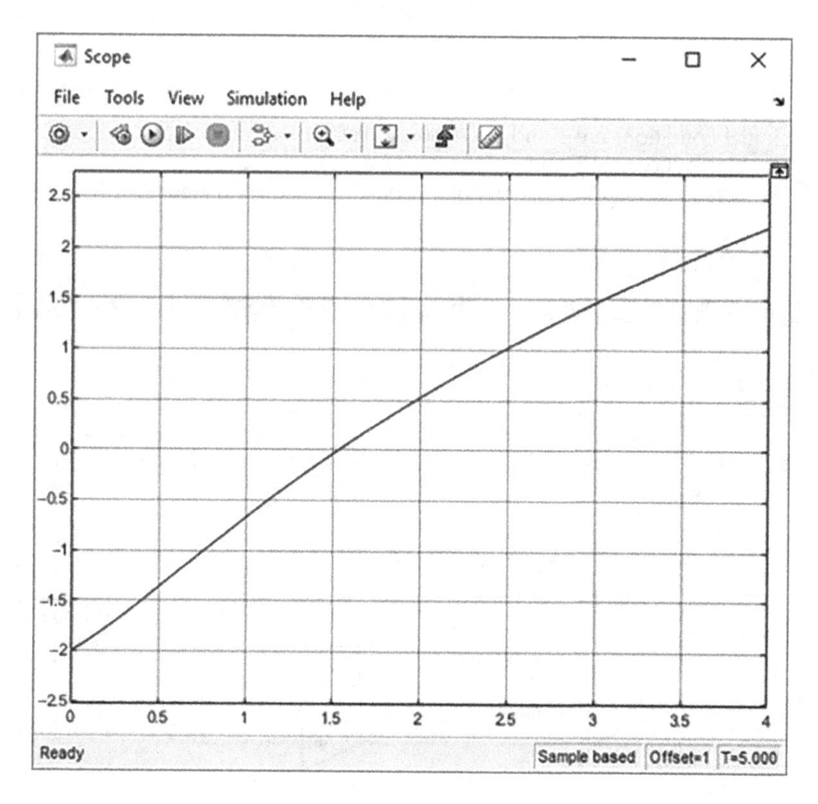

FIGURE B.18 Plot of y versus time.

Example B.5 Solving Second-Order Differential Equation

Solve the following second order differential equation using Simulink:

$$\frac{d^2y}{dt^2}+3\frac{dy}{dt}+40y=0$$

Using the following initial conditions,

$$y(0)=1, \frac{dy}{dt}(0)=\frac{1}{3}$$

Simulink Solution

Rearrange the second-order ODE such that the term with the highest order on the left side equals the rest of the terms.

$$\frac{d^2y}{dt^2}=-3\frac{dy}{dt}-40y$$

The initial conditions are 1 and 1/3 (Figure B.19).

The plot is shown in Figure B.20.

The default scope background color is black, as shown in Figure B.21.

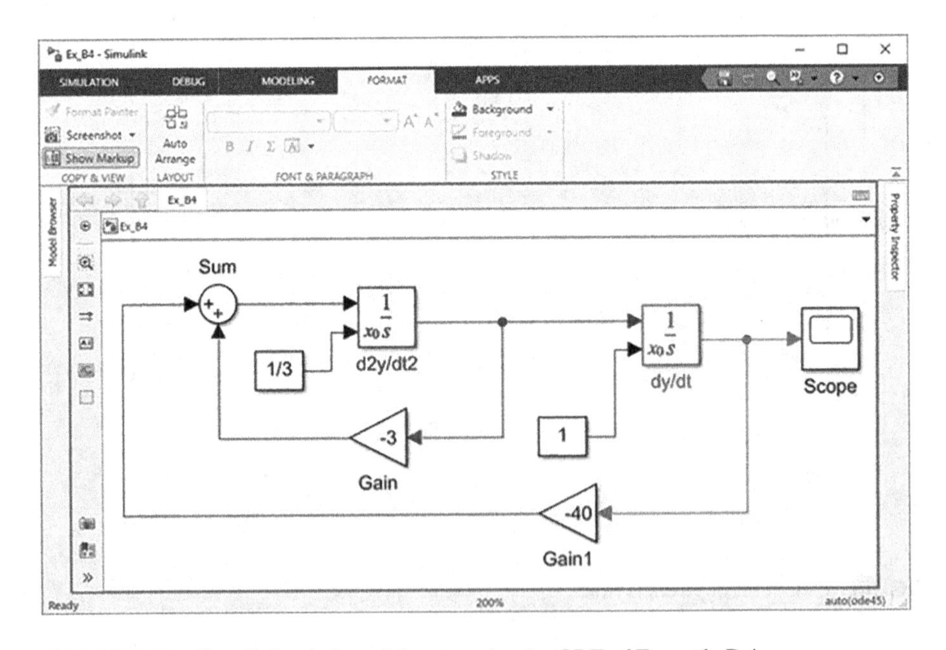

FIGURE B.19 Simulink solution of the second order ODE of Example B.4.

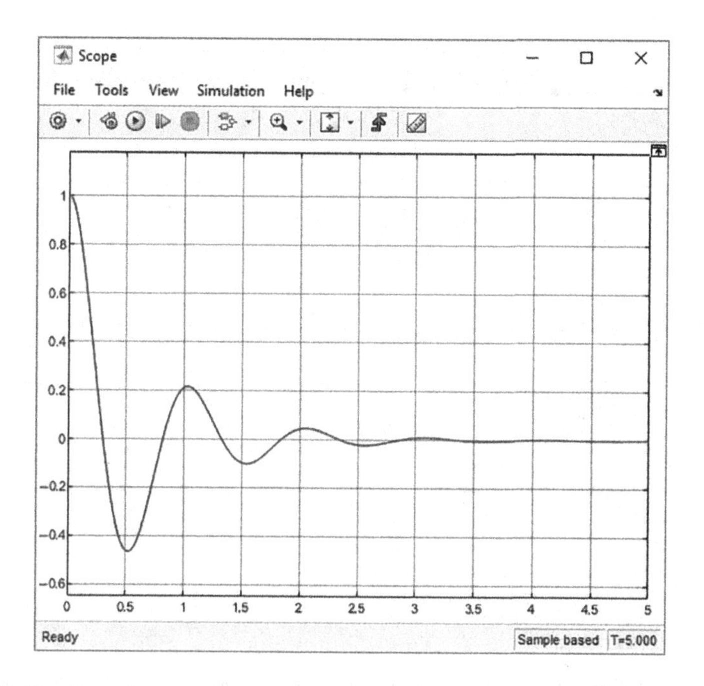

FIGURE B.20 Plot of y versus time.

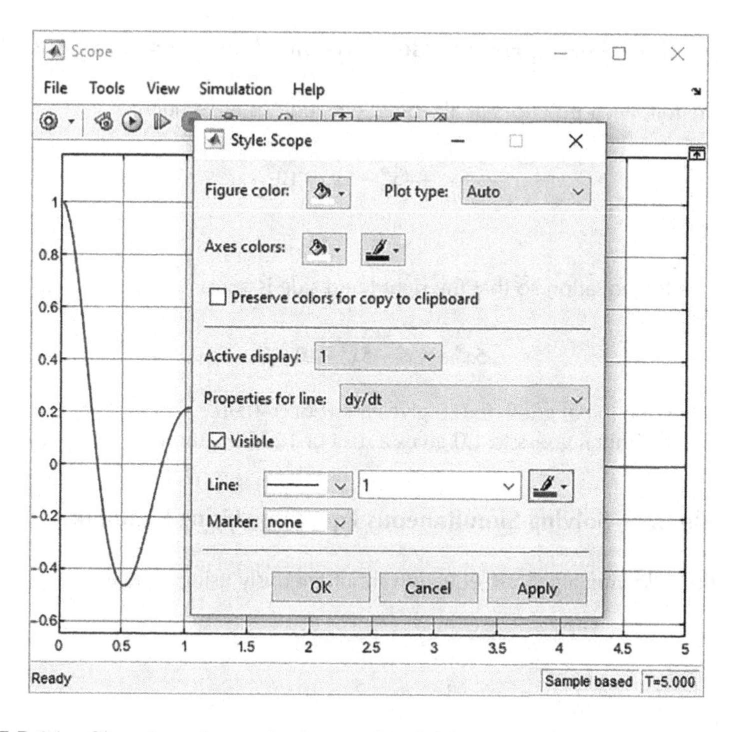

FIGURE B.21 Changing of scope background and color using view/style.

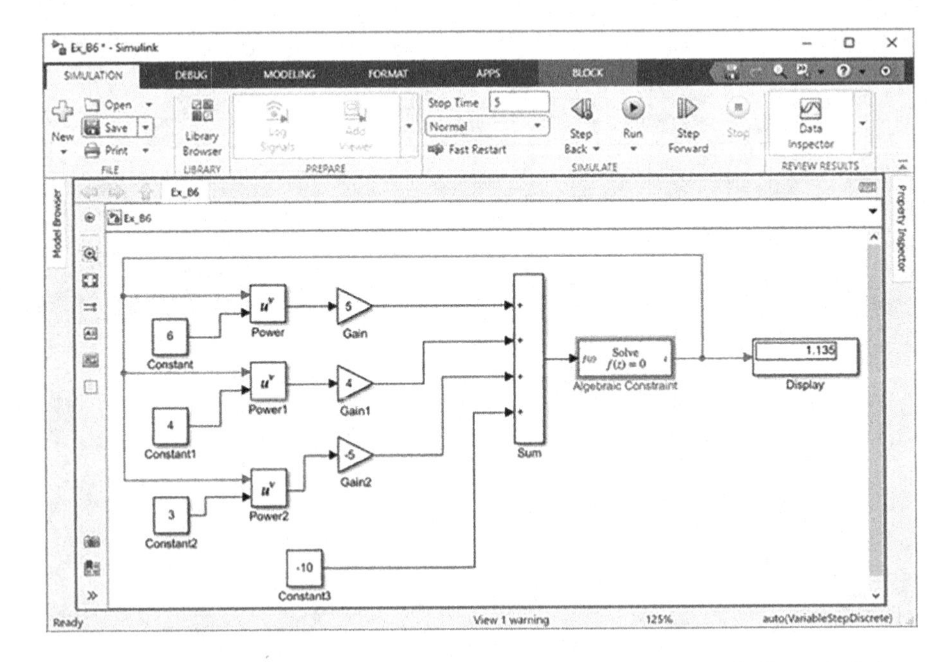

FIGURE B.22 Simulink solution of the polynomial equation presented in Example B.6.

Example B.6 Solving High-Order Polynomial Equation Using Simulink

Solve the following polynomial algebraic equation using Simulink:

$$5x^6 + 4x^4 - 5x^3 = 10$$

Solution

Rearrange the equation so that the right-hand side is zero.

$$5x^6 + 4x^4 - 5x^3 - 10 = 0$$

Substituting the initial guess to 0.0 gives the root 0–0.96.

Setting the initial guess to 1.0 gives a root of 1.135 (Figure B.22).

Example B.7 Solving Simultaneous Equations Using Simulink

Solve the following algebraic equation simultaneously using Simulink:

$$5x + 4y - 5z = -5$$

$$3x - 8y + 2z = 10$$

$$7x - 4y - z = -14$$

Solution

Rearrange the equation so that the right-hand side is zero.

$$5x + 4y - 5z + 5 = 0$$
$$3x - 8y + 2z - 10 = 0$$
$$7x - 4y - z + 14 = 0$$

The solution of the set of algabriac equations are solved in Figure B.23.
An alternative way is to put it in Matrix form is as follows:

$$\boldsymbol{AX} - \boldsymbol{B} = 0$$
$$5x + 4y - 5z + 5 = 0$$
$$3x - 8y + 2z - 10 = 0$$
$$7x - 4y - z + 14 = 0$$

$$\begin{pmatrix} 5 & 4 & -5 \\ 3 & -8 & 2 \\ 7 & -4 & -1 \end{pmatrix}\begin{pmatrix} x \\ y \\ z \end{pmatrix} - \begin{pmatrix} -5 \\ 10 \\ -14 \end{pmatrix} = 0$$

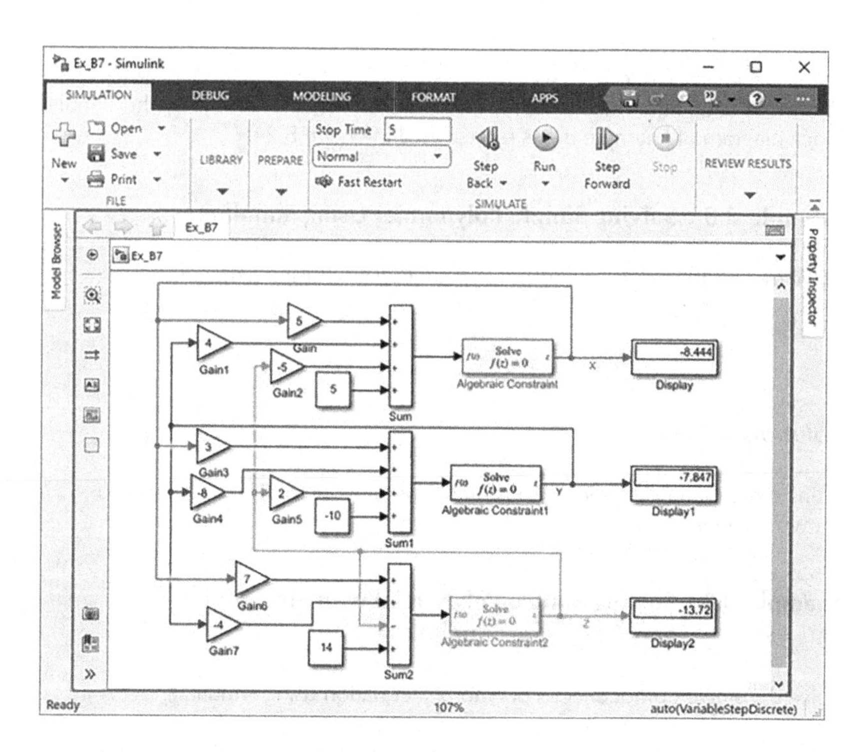

FIGURE B.23 Simulink solution of Example B.7.

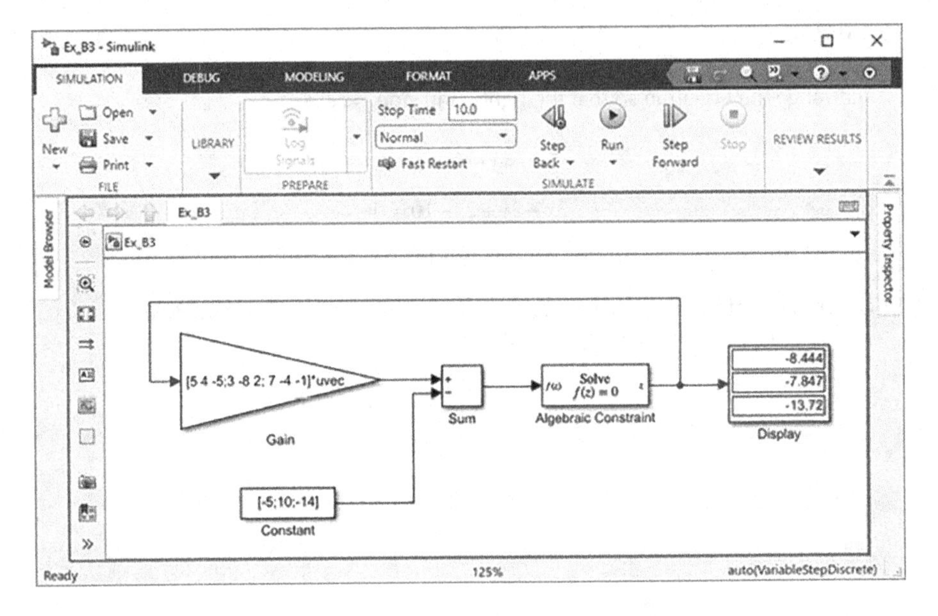

FIGURE B.24 Simulink alternative solution of the algebraic equations defined in Example B.7.

The more straightforward solution and using the matrix structure, the Simulink block diagram of Example B.7 is represented in Figure B.24.

Example B.8 Solving Simple Polynomial Using Simulink

Solve the single homogeneous polynomial equation using Simulink:

$$x^2 - x - 2 = 0$$

Solution

Using the Simulink Algebraic Constraint function, one of the roots is found (-1), as shown in Figure B.25.

Example B.9 Solving Simple Polynomial with Sin and Cos Using Simulink

Solve the single homogeneous polynomial equation using Simulink:

$$3x^3 + 5\sin(x) - 7\cos(2x) = 0$$

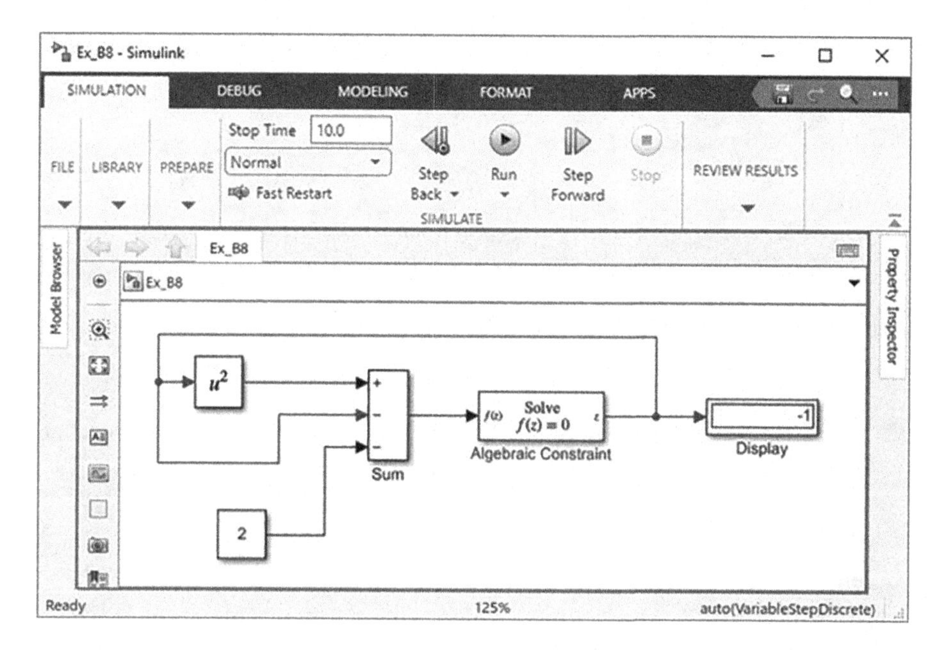

FIGURE B.25 Simulink solution of the single homogeneous polynomial equation of Example B.8.

Solution

Figure B.26 shows the Simulink Algebraic Constraint function for solving a polynomial equation with trig functions (i.e., sin(x), cos($2x$)).

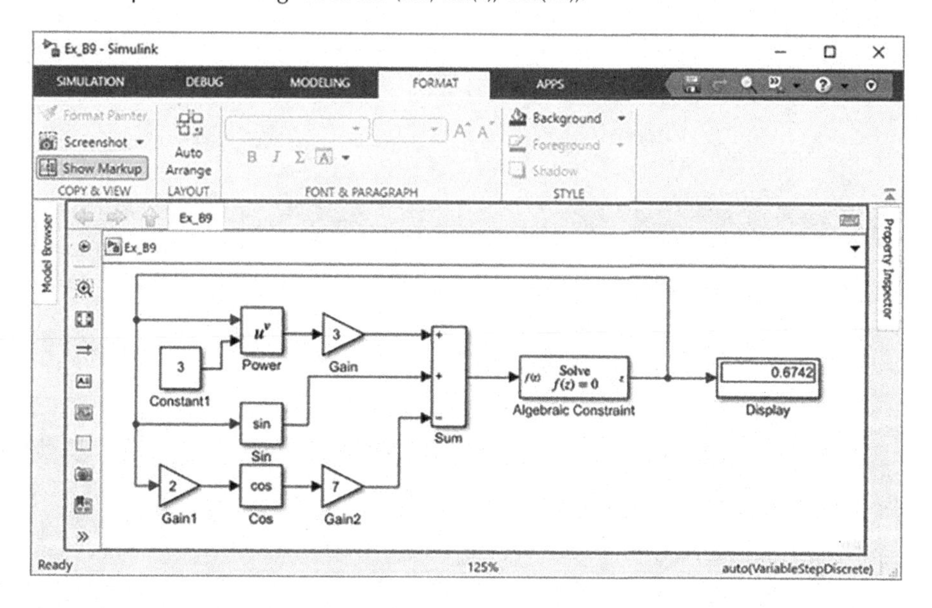

FIGURE B.26 Simulink solution of the cos and sin function of Example B.9.

Index

Note: **Bold** page numbers refer to tables and *italic* page numbers refer to figures.

Printed in the United States
by Baker & Taylor Publisher Services